Rainer Pöttgen, Thomas Jüstel, Cristian A. Strassert (Eds.)
Applied Inorganic Chemistry

Also of interest

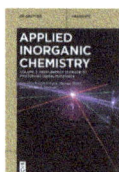

Applied Inorganic Chemistry
Volume 2: From Energy Storage to Photofunctional Materials
Rainer Pöttgen, Thomas Jüstel, Cristian A. Strassert (Eds.), 2023
ISBN 978-3-11-079878-4, e-ISBN 978-3-11-079889-0

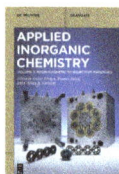

Applied Inorganic Chemistry
Volume 3: From Magnetic to Bioactive Materials
Rainer Pöttgen, Thomas Jüstel, Cristian A. Strassert (Eds.), 2023
ISBN 978-3-11-073837-7, e-ISBN 978-3-11-073347-1

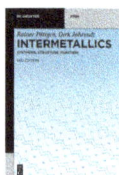

Intermetallics
Synthesis, Structure, Function
Rainer Pöttgen, Dirk Johrendt, 2019
ISBN 978-3-11-063580-5, e-ISBN 978-3-11-063672-7

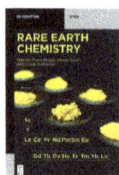

Rare Earth Chemistry
Rainer Pöttgen, Thomas Jüstel, Cristian A. Strassert (Eds.), 2020
ISBN 978-3-11-065360-1, e-ISBN 978-3-11-065492-9

Zeitschrift für Kristallographie - Crystalline Materials
Rainer Pöttgen (Editor-in-Chief)
ISSN 2194-4946, e-ISSN 2196-7105

Applied Inorganic Chemistry

Volume 1: From Construction Materials
to Technical Gases

Edited by
Rainer Pöttgen, Thomas Jüstel, Cristian A. Strassert

DE GRUYTER

Editors

Prof. Dr. Rainer Pöttgen
Institut für Anorganische und Analytische Chemie
Westfälische Wilhelms-Universität Münster
Corrensstraße 30
48149 Münster
Germany
E-mail: pottgen@uni-muenster.de

Prof. Dr. Thomas Jüstel
Fachbereich Chemieingenieurwesen
Fachhochschule Münster
Stegerwaldstraße 39
48565 Steinfurt
Germany
E-mail: tj@fh-muenster.de

Prof. Dr. Cristian A. Strassert
Institut für Anorganische und Analytische
Chemie
CiMIC – CeNTech – SoN
Westfälische Wilhelms-Universität Münster
Corrensstraße 28/30
48149 Münster
Germany
E-mail: ca.s@wwu.de

ISBN 978-3-11-073814-8
e-ISBN (PDF) 978-3-11-073314-3
e-ISBN (EPUB) 978-3-11-073332-7

Library of Congress Control Number: 2022935001

Bibliographic information published by the Deutsche Nationalbibliothek
The Deutsche Nationalbibliothek lists this publication in the Deutsche Nationalbibliografie; detailed bibliographic data are available on the Internet at http://dnb.dnb.de.

© 2023 Walter de Gruyter GmbH, Berlin/Boston
Cover image: Zementwerk Geseke, HeidelbergCement AG/Steffen Fuchs
Typesetting: Integra Software Services Pvt. Ltd.
Printing and binding: CPI books GmbH, Leck

www.degruyter.com

Preface

The Periodic Table meanwhile lists 118 chemical elements, which leads to a vast number of inorganic compounds. Many of them have well-defined physicochemical properties, which are exploited for the realization of functional materials we all comfortably use in daily life without even thinking about it, including magnetic and optical materials, construction materials, materials for energy storage and conversion – just to name a few remarkable examples. The impact of inorganic chemistry in human evolution cannot be overstated, and is proven by the designation of historical ages, such as stone, copper, bronze or iron age (even golden ages and gold rush), or by geographical locations (such as the Silicon Valley and Argentina). While carbon-based organic chemistry has provided incredible breakthroughs in medicinal chemistry and plastic materials, there is no doubt that the solution of the most urgent problems currently faced by humanity will stem from inorganic chemistry providing high-density/high-stability materials for construction, information technologies, energy storage and conversion.

Chemical sciences and industries are often demonized, but the many indispensable materials we use in daily life impressively show how significantly they influence our society. Ecosystems, metabolic and pathophysiological processes, food production, construction in its broadest sense, mobility and energy conversion are determined by chemistry – these facts cannot simply be ignored! The present book summarizes the many basic examples of inorganic materials we use on a large scale in everyday life, but also niche products with thoroughly optimized properties. Many subchapters are written by experts from academia and industry. We tried to ensure a proper balance of topics, even though it is simply impossible to cover all aspects of applied inorganic chemistry. Nonetheless, we hope that we made a good compromise – if any topic is missing, this was unintentional. The final chapter focusses on energy flows and resources, which constitutes one of the most urgent topics. As a kind of appetizer for the following 16 chapters, we briefly summarize some applications for the elements of the first four rows of the Periodic Table. Several of these topics are picked up again in the following chapters:

Hydrogen: energy source; **helium:** low-temperature refrigerant; ballon gas, **lithium:** anode materials for lithium-ion batteries; **beryllium:** hardening component for light-weight alloys, non-spark alloys, X-ray windows; **boron:** hardening component for intermetallics; **carbon:** electrode materials, black pigment; **nitrogen:** source for ammonia and nitrate fertilizers, protective gas, low-temperature cooling; **oxygen:** medical gas, liquid oxygen for the Linz-Donauwitzer process in steel refinement; **fluorine:** uranium hexafluoride production; **neon:** helium-neon lasers; **sodium:** reducing agent; **magnesium:** alloying component and sacrificial anodes; **aluminum:** light-weight alloys, construction material; **silicon:** semiconductors; **phosphorus:** synthesis of phosphoric acid; matches; **sulfur:** vulcanization of rubber; **chlorine:** disinfection of water; **argon:** protective gas in chemical synthesis

https://doi.org/10.1515/9783110733143-202

and arc-welding; **potassium:** liquid sodium-potassium alloys as coolants in nuclear reactors; **calcium:** reducing agent in metallurgy; **scandium:** additive for aluminum-based alloys, component of electron emitters; **titanium:** steel additive, corrosion resistant alloys; **vanadium:** high-speed tool steels; **chromium:** stainless steel and chromium plating; **manganese:** ferromanganese, activator in LED phosphors; **iron:** steel and cast iron; **cobalt:** superalloys and samarium-cobalt magnets; **nickel:** catalysis and anti-corrosion coatings; **copper:** cables and water tubes; **zinc:** facade cladding, corrosion protection; **gallium:** gallium nitride, phosphide or arsenide semiconductors; **germanium:** semiconductors and detection technology; **arsenic:** doping of semiconductors; **selenium:** II-VI semiconductors and alloy additive for free cutting steel; **bromine:** special disinfection products and synthesis of flame retardants; **krypton:** excimer lasers, KrCl excimer discharge lamps. The reader might notice that transition metals and lanthanides are not even mentioned here; there would not be sufficient space in a preface to list their impact!

Such a book project is not realizable without the help of numerous colleagues and co-workers. We thank Gudrun Lübbering for continuous help with literature search and text processing and Thomas Fickenscher for providing with many photos of materials and devices. We are especially grateful to our colleagues for their immediate agreements to write up a subchapter. It is always challenging to compile a concise Table of Contents and find the right co-authors. We are indebted to the editorial and production staff of De Gruyter. Our particular thanks go to Kristin Berber-Nerlinger, Dr. Vivien Schubert and Melanie Götz for their continuous support during conception, writing and producing the present book.

Münster, Steinfurt, June 2022
Thomas Jüstel, Rainer Pöttgen, Cristian A. Strassert

This book contains two different tokens, pointing to:

list of references

recommended literature for further reading; i.e. relevant text books, review articles or important original articles

Contents

Volume 2 (From Energy Storage to Photofunctional Materials)

Volume 3 (From Magnetic to Bioactive Materials)

List of contributors

Ackermann, Dr. habil. Lothar
Deutsche Stiftung Edelsteinforschung (DSEF)
Professor-Schlossmacher-Straße 1
55743 Idar-Oberstein
Germany
E-mail: lackermann@outlook.de

Agne, Dr. Matthias
Forschungszentrum Jülich GmbH
Helmholtz-Institut Münster (HI MS, IEK-12)
Corrensstraße 46
48149 Münster
Germany
E-mail: m.agne@fz-juelich.de

Apaydin, Dr. Dogukan H.
Institut für Materialchemie
TU Wien
Getreidemarkt 9/165
1060 Wien
Austria
E-mail: dogukan.apaydin@tuwien.ac.at

Arnault, Dr. Jean-Charles
Laboratoire des Edifices Nanométriques
Université Paris-Saclay, CEA, CNRS, NIMBE
91191 Gif sur Yvette
France
E-mail: jean-charles.arnault@cea.fr

Banik, Dr. Ananya
Institut für Anorganische und Analytische Chemie
Westfälische Wilhelms-Universität Münster
Corrensstraße 30
48149 Münster
Germany
E-mail: banik@uni-muenster.de

Bauer, Dr. Thomas
Deutsches Zentrum für Luft- und Raumfahrt (DLR)
Institut für Technische Thermodynamik
Thermische Prozesstechnik
Linder Höhe, Gebäude 26
51147 Köln
Germany
E-mail: thomas.bauer@dlr.de

Baur, Dr. Florian
Fachbereich Chemieingenieurwesen
Fachhochschule Münster
Stegerwaldstraße 39
48565 Steinfurt
Germany
E-mail: florian.baur@fh-muenster.de

Bayer, Dr. Bernhard
Institut für Materialchemie
TU Wien
Getreidemarkt 9/165
1060 Wien
Austria
E-mail: bernhard.bayer-skoff@tuwien.ac.at

Behrend, Prof. Dr.-Ing. habil. Detlef
Lehrstuhl Werkstoffe für die Medizintechnik
Fachbereich Maschinenbau und Schiffstechnik
Universität Rostock
Friedrich-Barnewitz-Straße 4
18119 Rostock
Germany
E-mail: detlef.behrend@uni-rostock.de

Behrens, Dr. Rainer
VDM Metals International GmbH
Kleffstraße 23
58762 Altena
Germany
E-mail: rainer.behrens@vdm-metals.com

Bertau, Prof. Dr. rer. nat. habil. Martin
Institut für Technische Chemie
TU Bergakademie Freiberg
Leipziger Straße 29
09599 Freiberg
Germany
and
Fraunhofer Technology Center for High-Performance Materials THM
Fraunhofer Institut for Ceramic Technologies and Systems IKTS
Am St.-Niclas-Schacht 13
09599 Freiberg
Germany
E-mail: Martin.Bertau@chemie.tu-freiberg.de

https://doi.org/10.1515/9783110733143-204

Binnewies, Prof. Dr. Michael
Institut für Anorganische Chemie
Naturwissenschaftliche Fakultät
Leibnitz Universität Hannover
Callinstraße 3-9
30167 Hannover
Germany
E-mail: michael.binnewies@aca.uni-hannover.de

Boos, Dr. Markus
Remmers GmbH
Bernhard-Remmers-Straße 13
49624 Löningen
Germany
E-mail: mboos@remmers.de

Boos, Dr. Peter
HeidelbergCement AG
Zur Anneliese 11
59320 Ennigerloh
Germany
E-mail: Peter.Boos@heidelbergcement.com

Bredol, Prof. Dr. Michael
Fachbereich Chemieingenieurwesen
Fachhochschule Münster
Stegerwaldstraße 39
48565 Steinfurt
Germany
E-mail: bredol@fh-muenster.de

Broll, Dr. Sascha
Broll-Buntpigmente GmbH & Co. KG
Karl-Winnacker-Straße 2-4
36396 Steinau
Germany
E-mail: drsascha@broll-buntpigmente.de

Buchner, Dr. Magnus R.
Anorganische Chemie, Fluorchemie
Philipps-Universität Marburg
Hans-Meerwein-Straße 4
35032 Marburg
Germany
E-mail: magnus.buchner@chemie.uni-marburg.de

Busch, Dr. Frank
Materialprüfungsamt Nordrhein-Westfalen
Marsbruchstraße 186
44287 Dortmund
Germany
E-mail: busch@mpanrw.de

Buttler, Dr.-Ing. Torben Alexander
-ISAF- Institut für Schweißtechnik und
Trennende Fertigungsverfahren
Technische Universität Clausthal
Agricolastraße 2
38678 Clausthal-Zellerfeld
Germany
E-mail: buttler@isaf.tu-clausthal.de

Dewalsky, Dr. Martin V.
Am Gemeindeholz 6
82205 Gilching
Germany
E-mail: martinvdew@gmail.com

Dorsch, Leonhard Yuuta
Institut für Anorganische Chemie
Universität Leipzig
Johannisallee 29
04103 Leipzig
Germany
E-mail: leonhard.dorsch@uni-leipzig.de

Dramicanin, Prof. Dr. Miroslav
Vinca Institute of Nuclear Sciences
University of Belgrade,
PO Box 522
11001 Belgrade
Serbia
E-mail: dramican@vinca.rs

Eckert, Prof. Dr. Hellmut
Institut für Physikalische Chemie
Westfälische Wilhelms-Universität Münster
Corrensstraße 30
48149 Münster
Germany
and
Instituto de Física de Sao Carlos
Universidade de Sao Paulo
Avenida Trabalhador Saocarlense 400
Sao Carlos, SP 13566-590
Brasil
E-mail: eckerth@uni-muenster.de

Eder, Prof. Dr. Dominik
Institut für Materialchemie
TU Wien
Getreidemarkt 9/165
1060 Wien
Austria
E-mail: dominik.eder@tuwien.ac.at

Engel, Stefan
Universität des Saarlandes
Anorganische Festkörperchemie
Campus C4 1
66123 Saarbrücken
Germany
E-mail: stefan.engel@uni-saarland.de

Engels, Ir. Marcel
Forschungsinstitut für Glas | Keramik (FGK)
Heinrich-Meister-Straße 2
56203 Höhr-Grenzhausen
Germany
E-mail: marcel.engels@fgk-keramik.de

Epple, Prof. Dr. Matthias
Anorganische Chemie
Fakultät für Chemie
Universität Duisburg-Essen
Universitätsstraße 7
45141 Essen
Germany
E-mail: matthias.epple@uni-due.de

Faust, PD Dr. Andreas
European Institute for Molecular Imaging (EIMI)
Waldeyerstraße 15
48149 Münster
Germany
E-Mail: faustan@uni-muenster.de

Feser, Prof. Dr.-Ing. Ralf
Fachbereich Informatik und
Naturwissenschaften
Fachhochschule Südwestfalen
Frauenstuhlweg 31
58644 Iserlohn
Germany
E-Mail: feser.ralf@fh-swf.de

Fickenscher, Thomas
Institut für Anorganische und Analytische
Chemie
Westfälische Wilhelms-Universität Münster
Corrensstraße 30
48149 Münster
Germany
E-mail: thomasfi@uni-muenster.de

Fröhlich, Dr. Peter
Institut für Technische Chemie
TU Bergakademie Freiberg
Leipziger Straße 29
09599 Freiberg
Germany
E-Mail: peter.froehlich@chemie.tu-freiberg.de

Ghidiu, Dr. Michael
Institut für Anorganische und Analytische
Chemie
Westfälische Wilhelms-Universität Münster
Corrensstraße 30
48149 Münster
Germany
E-mail: ghidiu@uni-muenster.de

Glaum, Prof. Dr. Robert
Institut für Anorganische Chemie
Rheinische Friedrich-Wilhelms-Universität
Gerhard-Domagk-Straße 1
53121 Bonn
Germany
E-mail: rglaum@uni-bonn.de

Grönefeld, Dr. Martin
Magnetfabrik Bonn GmbH
Dorotheenstraße 215
53119 Bonn
Germany
E-Mail: Martin.Groenefeld@Magnetfabrik.de

Haberkamp, Prof. Dr.-Ing. Jens
Fachbereich Bauingenieurwesen
Fachhochschule Münster
Corrensstraße 25
48149 Münster
Germany
E-mail: haberkamp@fh-muenster.de

Haneklaus, Dr. Nils
Td Lab Sustainable Mineral Resources
Universität für Weiterbildung Krems
Dr.-Karl-Dorrek-Straße 30
3500 Krems an der Donau
Austria
E-mail: nils.haneklaus@donau-uni.ac.at
and
Institut für Technische Chemie
TU Bergakademie Freiberg
Leipziger Straße 29
09599 Freiberg
Germany

Hayen, Prof. Dr. Heiko
Institut für Anorganische und Analytische
Chemie
Westfälische Wilhelms-Universität Münster
Corrensstraße 48
48149 Münster
Germany
E-mail: heiko.hayen@uni-muenster.de

Hendriks, Dr. Theodoor
Forschungszentrum Jülich GmbH
Helmholtz-Institut Münster (HI MS, IEK-12)
Corrensstraße 46
48149 Münster
Germany
E-mail: t.hendriks@fz-juelich.de

Hermes, Dr. Wilfried
trinamiX GmbH
Industriestraße 35
67063 Ludwigshafen
Germany
E-mail: wilfried.hermes@trinamix.de

Herrmann, Dr. Fabian
Institut für Pharmazeutische Biologie und
Phytochemie
Westfälische Wilhelms-Universität Münster
Corrensstraße 48
48149 Münster
Germany
E-mail: fabian.herrmann@wwu.de

Hosono, Prof. Dr. Hideo
Materials Research Center for Element
Strategy (MCES)
Tokyo Institute of Technology
SE-1, 4259 Nagatsuta-cho
Midori-ku, Yokohama, Kanagawa, 226-8503
Japan
E-mail: hosono@mces.titech.ac.jp

Huppertz, Prof. Dr. Hubert
Institut für Allgemeine, Anorganische und
Theoretische Chemie
Universität Innsbruck
Innrain 80–82
6020 Innsbruck
Austria
E-mail: Hubert.Huppertz@uibk.ac.at

Janiak, Prof. Dr. Christoph
Institut für Anorganische Chemie und
Strukturchemie
Lehrstuhl für nanoporöse und nanoskalierte
Materialien
Heinrich-Heine-Universität Düsseldorf
Universitätsstraße 1
40225 Düsseldorf
Germany
E-mail: janiak@hhu.de

Janka, PD Dr. Oliver
Universität des Saarlandes
Anorganische Festkörperchemie
Campus C4 1
66123 Saarbrücken
Germany
E-mail: oliver.janka@uni-saarland.de

Jin, Wenqi
Xinjiang Technical Institute of Physics &
Chemistry
Chinese Academy of Sciences
40-1 South Beijing Road
830011 Urumqi
China
E-Mail: jwqineni@qq.com

Johrendt, Prof. Dr. Dirk
Department Chemie
Ludwig-Maximilians-Universität München
Butenandtstraße 5-13 (Haus D)
81377 München
Germany
E-mail: johrendt@lmu.de

Jüstel, Prof. Dr. Thomas
Fachbereich Chemieingenieurwesen
Fachhochschule Münster
Stegerwaldstraße 39
48565 Steinfurt
Germany
E-mail: tj@fh-muenster.de

Klapötke, Prof. Dr. Thomas M.
Department Chemie
Ludwig-Maximilians-Universität München
Butenandtstraße 5-13 (Haus D)
81377 München
Germany
E-mail: tmk@cup.uni-muenchen.de

Kohlmann, Prof. Dr. Holger
Institut für Anorganische Chemie
Universität Leipzig
Johannisallee 29
04103 Leipzig
Germany
E-mail: holger.kohlmann@uni-leipzig.de

Koller, PD Dr. Hubert
Institut für Physikalische Chemie
Westfälische Wilhelms-Universität Münster
Corrensstraße 30
48149 Münster
Germany
E-mail: hkoller@uni-muenster.de

Kratz, Dr. Nadja
Forschungsinstitut für Glas | Keramik (FGK)
Heinrich-Meister-Straße 2
56203 Höhr-Grenzhausen
Germany
E-mail: nadja.kratz@fgk-keramik.de

Kränkel, PD Dr. Christian
Leibniz-Institut für Kristallzüchtung (IKZ)
Max-Born-Straße 2
12489 Berlin
Germany
E-mail: christian.kraenkel@ikz-berlin.de

Krzywinski, Jacek
Magnetfabrik Bonn GmbH
Dorotheenstraße 215
53119 Bonn
Germany
E-Mail: Jacek.Krzywinski@Magnetfabrik.de

Langner, Dr. Bernd E.
Glockenheide 11
21423 Winsen
Germany
E-mail: langner@understanding-copper.com

Letz, Dr. Martin
SCHOTT AG
Research and Technology Development
Hattenbergstraße 10
55122 Mainz
Germany
E-mail: martin.letz@schott.com

Lider, Konstantin
Diener & Rapp GmbH & Co. KG Eloxalbetrieb
Junkersstraße 39
78056 Villingen-Schwenningen
Germany
E-mail: konstantin.lider@dienerrapp.de

Lox, Prof. Dr. Ir. Egbert S. J.
Am Laerchentor 8
36355 Grebenhain-Hochwaldhausen
Germany
E-mail: Egbert.Lox@gmail.com

Lovrincic, Dr. Robert
trinamiX GmbH
Industriestraße 35
67063 Ludwigshafen
Germany
E-mail: robert.lovrincic@trinamix.de

Maletz, Prof. Dr. Reinhard
VOCO GmbH
Anton-Flettner-Straße 1-3
27472 Cuxhaven
Germany
E-mail: r.maletz@voco.de

Matschke, Dr. Christian
BERLIN-CHEMIE AG
Glienicker Weg 125
12489 Berlin
Germany
E-mail: cmatschke@berlin-chemie.de

Mertens, Prof. Dr. Konrad
Fachbereich Elektrotechnik und Informatik
Fachhochschule Münster
Stegerwaldstraße 39
48565 Steinfurt
Germany
E-mail: mertens@fh-muenster.de

Mudryk, Dr. Yaroslav
Ames Laboratory, U.S. Department of Energy
Iowa State University
254 Spedding
Ames, IA 50011-2416
USA
E-mail: slavkomk@ameslab.gov

Niehaus, Dr. Oliver
Umicore AG & Co. KG
Rodenbacher Chaussee 4
63457 Hanau
Germany
E-mail: Oliver.Niehaus@eu.umicore.com

Pan, Prof. Dr. Shilie
Xinjiang Technical Insitute of Physics &
Chemistry
Chinese Academy of Sciences
40-1 South Beijing Road
830011 Urumqi
China
E-mail: slpan@ms.xjb.ac.cn

Pavón Regaña, Dr. Ing. Sandra
Fraunhofer-Institut für Keramische
Technologien und Systeme IKTS
Fraunhofer-Technologiezentrum
Hochleistungsmaterialien THM
Am St.-Niclas-Schacht 13
09599 Freiberg
Germany
and
Institut für Technische Chemie
TU Bergakademie Freiberg
Leipziger Straße 29
09599 Freiberg
Germany
E-mail: sandra.pavon.regana@ikts.
fraunhofer.de

Pecharsky, Prof. Dr. Vitalij K.
Ames Laboratory
Iowa State University
Ames, IA 50011-2416
USA
E-mail: vitkp@ameslab.gov

Piribauer, Dipl.-Ing. Dr. rer. nat. Christoph
Forschungsinstitut für Glas | Keramik (FGK)
Heinrich-Meister-Straße 2
56203 Höhr-Grenzhausen
Germany
E-mail: christoph.piribauer@fgk-keramik.de

Pöttgen, Prof. Dr. Rainer
Institut für Anorganische und Analytische
Chemie
Westfälische Wilhelms-Universität Münster
Corrensstraße 30
48149 Münster
Germany
E-mail: pottgen@uni-muenster.de

Quirmbach, Prof. Dr. rer. nat. Dr. h.c. Peter
Technische Chemie und
Korrosionswissenschaften
Universität Koblenz-Landau
Universitätsstraße 1
56070 Koblenz
Germany
E-mail: pquirmbach@uni-koblenz.de

Reiss, Prof. Dr. Günter
Physics Department
Center for Spinelectronic Materials and
Devices
Universitätsstraße 25
33615 Bielefeld
Germany
E-mail: guenter.reiss@uni-bielefeld.de

Riedel, Prof. Dr. Sebastian
Institut für Chemie und Biochemie
Anorganische Chemie
Freie Universität Berlin
Fabeckstraße 34/36
14195 Berlin
Germany
E-mail: s.riedel@fu-berlin.de

Rieger, Dr. Thorsten
VDM Metals International GmbH
Kleffstraße 23
58762 Altena
Germany
E-mail: Torsten.Rieger@vdm-metals.com

Salvermoser, Dr. Manfred
Riedel Filtertechnik GmbH
Westring 83
33818 Leopoldshöhe
Germany
E-mail: manfred.salvermoser@riedel-
filtertechnik.com

Sax, Dr.-Ing. Almuth
Technische Chemie und
Korrosionswissenschaften
Universität Koblenz-Landau
Universitätsstraße 1
56070 Koblenz
Germany
E-mail: asax@uni-koblenz.de

Schäferling, Prof. Dr. Michael
Fachbereich Chemieingenieurwesen
Fachhochschule Münster
Stegerwaldstraße 39
48565 Steinfurt
Germany
E-mail: michael.schaeferling@fh-muenster.de

Schmid, Jonas R.
Institut für Chemie und Biochemie
Anorganische Chemie
Freie Universität Berlin
Fabeckstraße 34/36
14195 Berlin
Germany
E-mail: jonas.schmid@fu-berlin.de

Schramm, Dr. Stefan
Merck KGaA
Frankfurter Straße 250
64293 Darmstadt
Germany
E-mail: stefan.schramm@merckgroup.com

Schupp, Prof. Dr. Thomas
Fachbereich Chemieingenieurwesen
Fachhochschule Münster
Stegerwaldstraße 39
48565 Steinfurt
Germany
E-mail: thomas.schupp@fh-muenster.de

Seifert, Dr. Markus
TU Dresden
Walther-Hempel-Bau
Mommsenstraße 4
01069 Dresden
Germany
E-mail: markus.seifert1@tu-dresden.de

Slabon, Prof. Dr. Adam
Chair of Inorganic Chemistry
University of Wuppertal
Gaußstraße 20
42119 Wuppertal
Germany
E-mail: slabon@uni-wuppertal.de

Staffel, Prof. Dr. Thomas
Research & Development, Phosphate
solutions
BK Giulini GmbH, ICL Group Ltd.
Dr.-Albert-Reimann-Straße 2
68526 Ladenburg
Germany
E-mail: thomas.staffel@icl-group.com

Stephan, Dr. Tom
Deutsche Gemmologische Gesellschaft e.V.
(DGemG)
Prof.-Schlossmacher-Straße 1
55743 Idar-Oberstein
Germany
E-mail: t.stephan@dgemg.com

Stengel, Dr. Ilona
Merck KGaA
Frankfurter Straße 250
64293 Darmstadt
Germany
E-mail: ilona.stengel@merckgroup.com

Stöwe, Prof. Dr. Klaus
Faculty of Natural Sciences
Institute of Chemistry, Chemical Technology
Technische Universität Chemnitz
09107 Chemnitz
Germany
E-mail: klaus.stoewe@chemie.tu-chemnitz.de

Strassert, Prof. Dr. Cristian A.
Institut für Anorganische und Analytische
Chemie
CiMIC – CeNTech – SoN
Westfälische Wilhelms-Universität Münster
Corrensstraße 28/30
48149 Münster
Germany
E-mail: ca.s@wwu.de

Teliban, Dr. Iulian
Magnetfabrik Bonn GmbH
Dorotheenstraße 215
53119 Bonn
Germany
E-Mail: Iulian.Teliban@Magnetfabrik.de

Termath, Dr. Andreas
Clariant Plastics & Coatings (Deutschland) GmbH
Chemiepark Knapsack
Industriestraße 149
Gebäude 2703, R. 128
50354 Hürth
Germany
E-mail: andreas.termath@clariant.com

Trodler, Dr. Jörg
Trodler-EAVT
Technische Beratung für die Aufbau und
Verbindungstechnik in der Elektronik
Grüner Weg 18/19
15712 Königs Wusterhausen
Germany
E-mail: joerg.trodler@trodler-eavt.de

Voigt, Dominik
Fachbereich Chemieingenieurwesen
Fachhochschule Münster
Stegerwaldstraße 39
48565 Steinfurt
Germany
E-mail: dv009200@fh-muenster.de

Voigt, Prof. Dr. Ingolf
Fraunhofer-Institut für Keramische
Technologien und Systeme IKTS
Michael-Faraday-Straße 1
07629 Hermsdorf
Germany
E-mail: ingolf.voigt@ikts.fraunhofer.de

Warkentin, apl. Prof. Dr.-Ing. habil. Dr. rer. nat. Mareike
Fakultät für Maschinenbau und
Schiffstechnik
Universität Rostock
Friedrich-Barnewitz-Straße 4
18119 Rostock
Germany
E-mail: mareike.warkentin@uni-rostock.de

Weigand, Prof. Dr. Jan J.
TU Dresden
Walther-Hempel-Bau
Mommsenstraße 4
01069 Dresden
Germany
E-mail: jan.weigand@tu-dresden.de

Werner, Prof. Dr. Jan
Forschungsinstitut für Glas | Keramik (FGK)
Heinrich-Meister-Straße 2
56203 Höhr-Grenzhausen
Germany
E-mail: jan.werner@fgk-keramik.de

Wendel, Dr. Jörg
Wendel GmbH Email- und Glasurenfabrik
Am Güterbahnhof 30
35683 Dillenburg
Germany
E-mail: joerg.wendel@wendel-email.de

Wilhelm, Dr. Dominik
TYROLIT - Schleifmittelwerke Swarovski K.G.
Swarovskistraße 33
6130 Schwaz
Austria
E-mail: Dominik.Wilhelm@Tyrolit.com

Winter, Dr. Florian
Culimeta Textilglas-Technologie GmbH & Co. KG
Werner-von-Siemens-Straße 9
49593 Bersenbrück
Germany
E-mail: fwinter@culimeta.de

Yang, Prof. Dr. Zhihua
Xinjiang Technical Institute of Physics &
Chemistry
Chinese Academy of Sciences
40-1 South Beijing Road
Urumqi 830011
China
E-Mail: zhyang@ms.xjb.ac.cn

Zeier, Prof. Dr. Wolfgang
Institut für Anorganische und Analytische
Chemie
Westfälische Wilhelms-Universität Münster
Corrensstraße 30
48149 Münster
Germany
E-mail: wzeier@uni-muenster.de

Ziegler, Raimund
Institut für Allgemeine, Anorganische und
Theoretische Chemie
Universität Innsbruck
Innrain 80–82
6020 Innsbruck
Austria
E-mail: Raimund.Ziegler@uibk.ac.at

Zumdick, Dr. Markus
H.C. Starck Tungsten GmbH
Im Schleeke 78-91
38642 Goslar
Germany
E-mail: markus.zumdick@hcstarck.com

1 Construction materials and coatings

1.1 Basics of cement chemistry

Peter Boos

Cement is the (artificial) inorganic, non-metallic, finely ground hydraulic binder of concrete. A hydraulic binder is defined as a material that sets and hardens through a chemical reaction with water and remains hard under water as well as in the air [1]. During this reaction, called hydration, cement minerals form a stone-like mass.

The oldest cement-like bound product discovered so far dates back to around 7000 BC. It was found during road works at Yiftah El in Galilee, Israel [2]. Hydraulic hardening binder was often used as early as the Roman times. Therefore, the name concrete is derived from the Latin word "concretus" and means "grown together" or "composed". The Romans perfected their ways of using Pozzolan (a volcanic ash found near Pozzuoli). Therefore, they mixed this volcanic ash with water, lime, sand and gravel, and the resulting product was much stronger than what they had previously produced [2]. With this product, which was called "Opus Caementitium", the Romans created great structures such as the Pantheon in Rome, the Pont du Gard near Nimes (Figure 1.1.1), the Eifel water pipeline to Cologne and many more buildings, which still exist today [3].

Figure 1.1.1: Pont du Gard near Nimes, France, made with the Roman concrete.
Source: Vera Hesse, Telgte.

The crystals, which are responsible for the strength of the binders used by the Romans are (mineralogically and chemically) largely identical to those that are formed in the

https://doi.org/10.1515/9783110733143-001

hardening process of today's cements [4]. Today's modern cements are characterized by a high degree of uniformity, so that concretes produced with them are characterized by outstanding workability (flowability) and controlled setting, hardening and durability behavior. These controllable concrete properties and the fact that today's cements develop higher strengths have led to the global success of concrete construction.

Since there is a wide range of suitable and accepted cement constituents, today's modern cements are multi-component systems. Typical cement constituents are ground granulated blast furnace slag (GGBFS), limestone, pozzolan, fly ash from e.g. coal-fired power stations, etc. Even to the present day, Portland cement clinker is still the essential constituent in all cement formulas everywhere. Due to this, Portland cement clinker still has a significant impact on the technical properties of the cement in use and is the main topic of this chapter.

Please note that it is customary in cement chemistry to specify chemical compositions as oxides, and cement chemical abbreviations for the oxides (Table 1.1.1) have become established [5]. However, the usual abbreviations are largely avoided in this article.

Table 1.1.1: Cement chemical abbreviations [1, 4, 5, 7].

Oxide	Cement chemistry abbreviation	Example
CaO	C	$Ca_3SiO_5 \equiv 3\,CaO + SiO_2 \equiv C_3S$
SiO_2	S	
Al_2O_3	A	
Fe_2O_3	F	
MgO	M	
$CaSO_4$	Cs	
H_2O	H	
Na_2O	N	
K_2O	K	

1.1.1 Portland cement clinker

Portland cement clinker is an artificially produced, polycrystalline, heterogeneous stone. The clinker essentially consists of four minerals. Swedish geologist Alfred Elis Törnebohm described the technical clinker microscopically as early as 1897. For the sake of simplicity, he referred to the mineral that is most frequently seen in terms of quantity, using the first letter in the alphabet "A" and the ancient Greek term for stone "lithos" as "alite" and the second most common as "belite" [1, 7]. It was later discovered that the minerals called alite and belite are both calcium silicates (tricalcium silicate and dicalcium silicate) with some minor impurities in the structures by Al, Mg, Fe, Na, K, Cr, Ti, Mn, P, etc. [6].

Due to these different foreign ions incorporated into the structures, the names alite or belite are still used today in order to distinguish between the industrially produced (usually contaminated) clinker minerals and the pure calcium silicates.

Alite is the mineral to which modern Portland cement clinker owes its decisive properties. When mixed with water, alite reacts quickly and forms a solid structure of water-containing minerals called "hydrate phases" (calcium silicate hydrates, Figure 1.1.3) [7]. The ground belite crystals form with water the same hydrate minerals as the alite, but the reaction time of the belite is significantly lower [7].

In addition to the two calcium silicates, technical clinker also contains a variation of calcium aluminate and calcium aluminoferrite [1].

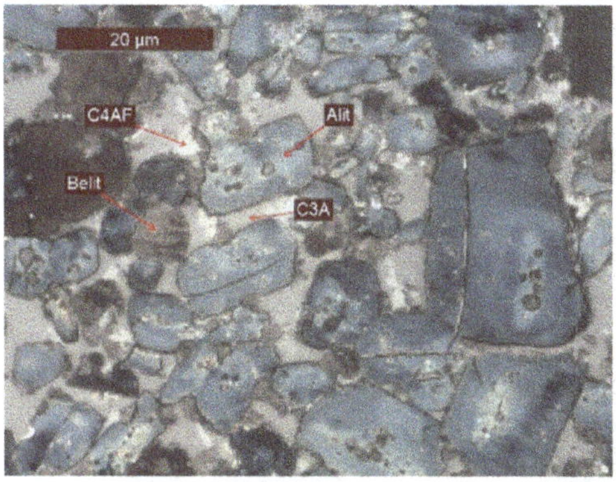

Figure 1.1.2: Microsection of a typical technical Portland cement clinker. The four main mineral phases alite, belite, C_3A (tricalcium aluminate) and C_4AF (tetracalcium aluminoferrite) are marked. Source: HeidelbergCement AG, Katja Steller.

The tricalcium aluminate reacts immediately with water and its reaction must therefore be controlled with calcium sulfate (see Section 1.1.7). In cement, tricalcium aluminate is essentially responsible for the setting and the onset of the hardening of the cement paste (mixture of cement with water). The hydrate phases (Figure 1.1.3) that form play a subordinate role in the later strength [5, 7].

Finely ground tetracalcium aluminate ferrite reacts very slowly with water. The reactivity also decreases with the iron content [7]. Except for the tetracalcium aluminate ferrite, all main mineral phases are white. Therefore, the amount and color of the tetracalcium aluminate ferrite determines the color of the cement. In particular, the color of the cement clinker (Table 1.1.2) is influenced by the burning conditions (oxidizing or reducing) and the cooling conditions (fast or slow) during production [17, 18].

Table 1.1.2: Essential data about the clinker phases [5].

Designation	Alite, tricalcium silicate	Belite, dicalcium silicate	Aluminate, tricalcium aluminate	Ferrite, calcium aluminoferrite
Chemical composition of the pure phase	$3CaO \times SiO_2$	$2CaO \times SiO_2$	$3CaO \times Al_2O_3$	$2CaO\,(Al_2O_3, Fe_2O_3)$
Possible foreign ions	Mg, Al, Fe	Alkalis, Al, Fe, fluoride	Alkalis, Fe, Mg	Si, Mg
Mineralogical modification	6	5	3	1
Modification occurring in technical clinker	Monoclinic (M II), trigonal (R)	ß, monoclinic	Cubic, orthorhombic, tetragonal	Orthorhombic
Color	White	White	White	Dark brown, due to MgO dark greyish-green
Extract of technical properties in cement	Rapid reaction, high early and final strength	Slow reaction, high final strength	Rapid reaction, high heat of hydration, promotes early strength	Very slow reaction

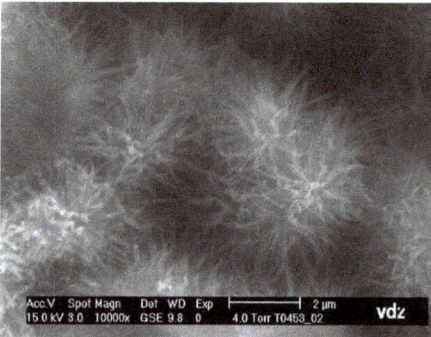

Figure 1.1.3: Scanning electron microscope image of calcium silicate hydrate.
Source: Forschungsinstitut der Zementindustrie/VDZ, Düsseldorf.

1.1.2 Clinker production: the chemical composition of the raw meal

By definition, Portland cement clinker consists of two thirds of calcium silicates [8]. To meet this requirement, the clinker must have a chemical composition as shown in Table 1.1.3. To achieve this specified chemical clinker composition, limestone contents of 75–79% in mass are necessary in the raw mixture. Due to that, cement clinker plants are usually located close to huge limestone, marl or chalk deposits. In addition to calcium carbonate, sources for silica and iron and aluminum are required; for example, clay and quartz. Beside these primary raw materials, today, alternative raw materials are also increasingly being used, such as various slag and fly ash types, mill scale, used foundry sands and by-products from drinking water production, etc. The ash from the fuel used for the production, also influences the chemical composition of the resulting clinker and must be considered in the raw meal composition. Table 1.1.3 shows the concentration ranges of the essential chemical components contained in clinker.

Table 1.1.3: Typical concentrations of the main chemical components of clinker [7].

	Minimum	Maximum	Average	
CaO	63	70	66.5	mass %
SiO_2	19	24	21.5	mass %
Al_2O_3	3	7	5.5	mass %
Fe_2O_3	1	5	2.5	mass %

To generate the correct proportions of the composition, chemical indicators are used in cement production, which place the chemical oxides in relation to each other. These relationships are called modules [1, 4, 5, 7, 9, 10, 13]. Since the modules represent the oxidic conditions and since the loss on ignition has the same effect as a factor on all oxides, they can be used for both raw meal and clinker. Undisturbed cement clinker production is achieved if the cement plants and raw material-specific modules are adhered to with little variation. It happens repeatedly that raw material fluctuations, in the modules of even a few tenths of a percent, have a massive impact on production (e.g. energy requirements), the product (clinker), the processability of the product, its performance and the production costs. Since the natural raw materials fluctuate more or less in their chemical composition, the cement industry works with corrective substances as well as with homogenization steps, such as the raw gravel blending beds (Figure 1.1.4).

Table 1.1.4 lists common modules, defines optimal areas for production and describes what influence the module has on the process or the product.

Figure 1.1.4: Blending bed of HeidelbergCement plant, Burglengenfeld: Raw material piled up in horizontal layers, degraded vertically and blended in front of the raw mill, so that a raw meal is produced that has the correct and uniform chemical composition for clinker production. **Source**: HeidelbergCement AG, S. Fuchs.

Table 1.1.4: Modules for raw meal [1, 4, 5, 7, 9, 10, 13].

Abbreviation	Moduli	Formula	Common range	Module explanation
LSt I	Lime standard	$\dfrac{100\ CaO}{2.8\ SiO_2 + 1.1\ Al_2O_3 + 0.35\ Fe_2O_3}$	92 . . . 102	Percentage CaO content in the raw material that can be bound by the clinker minerals under optimal production conditions
LSt II		$\dfrac{100\ CaO}{2.8\ SiO_2 + 1.18\ Al_2O_3 + 0.65\ Fe_2O_3}$		
LSt III		$\dfrac{100\ (CaO + 0.75\ MgO)}{2.8\ SiO_2 + 1.18\ Al_2O_3 + 0.65\ Fe_2O_3}$ max. considered MgO content 2%	90 . . . 102	
SR	Silica ratio	$\dfrac{SiO_2}{Al_2O_3 + Fe_2O_3}$	1.8 . . . 3.4	Reference value for the quantitative ratio of the silica bound predominantly in the solid silicates at the sintering temperature and the Al_2O_3 and Fe_2O_3 content in the melt.
AR	Alumina ratio	$\dfrac{Al_2O_3}{Fe_2O_3}$	1.8 . . . 2.8	Specification of a composition with which a melt is created at the lowest possible temperature

Table 1.1.4 (continued)

Abbreviation	Moduli	Formula	Common range	Module explanation
SG	Sulfatization degree	$\dfrac{SO_3}{K_2O + 1.52\ Na_2O}$	≥1	SR describes how many alkalis are theoretically bound to sulfate in the process

1.1.3 Raw material for the clinker production: extraction and processing

The material usually blasted in the quarry is transported by truck to a crusher (jaw crusher, hammer crusher, etc.), where it is crushed into about 120–200 mm large stones, maximum. The crushed material is either transported to the blending bed (Figure 1.1.4) or stored in silos. From there the material is conveyed to the raw mill and ground to a fine powder. As a rule, kiln exhaust gases or a hot gas generator are used during grinding to dry the raw material during this process (grinding drying). Drying is often necessary to prevent the grinding balls from being coated by the limestone powder (which significantly reduces the efficiency of grinding) and to prevent the mill from filling up during the grinding process (i.e. the ground material no longer flows out of the mill) and shutting down the grinding system. The chemical composition, fineness and homogeneity of the raw material determine the burnability of the material [4]. In the case of raw meal with a 90 μm sieve residue of 5–20%, even the coarsest proportions usually react completely in a technical process [9].

The ground raw material, which is called raw meal, is transported from the raw mill into silos. The silos are often designed in such a way that a further homogenization of the raw meal is achieved. The storage capacity of the silos is designed to be a buffer between the mill operation and kiln, raw mill needs maintenance weekly, while the kiln runs at a high availability thus requiring buffer volume to allow downtime on the mill.

1.1.4 Clinker burning process: the chemical transformation

Technical Portland cement clinker is now almost exclusively produced using the dry process in rotary kilns, which are preceded by a preheater. The most energy-efficient rotary kiln systems consist of cyclone preheaters with a calciner (combustion chamber for calcination purposes), a tertiary air duct and a grate cooler (Figure 1.1.5). In principle, the raw meal is fed into the upper (cold) end of the preheater tower and, due to gravity, moves downwards against the flow of hot gas (hence preheater) to the kiln inlet of the rotary kiln. The material is on its way down in the preheater, dispersed by the cyclones, then partly returned up with the gas, so that a very good heat transfer from the hot gas to the raw meal is achieved. In the lower area of the cyclone preheater, the material (now referred to as hot meal) enters the rotary kiln via the kiln inlet. Rotary kilns are steel tubes lined with refractory bricks (diameter: approx. 4 to 6 m) which are inclined approx. 2.5° to 3.5° and rotate around the longitudinal axis at 1.5 to 4.5 revolutions per minute. At the lower end of the kiln there is a burner flame which produces a gas temperature of 1800–2000 °C (resulting in a firing temperature of 1350–1500 °C).

Figure 1.1.5: Construction drawing of a cyclone preheater without a calciner and plant photo of the associated furnace system.
Source: HeidelbergCement AG, S. Fuchs.

Due to the inclination and rotation of the rotary kiln, the material moves in bulk in the direction of the primary burner to the lower end of the kiln. Clinker granules (or nodules) wind up as a result of the rolling motion and the beginning of the melting formation [1, 5, 7, 9, 13].

The clinker manufacturing process is a so-called sintering process, i.e. even in the hottest part of the kiln only part of the raw material is melted while the rest is solid. Due to the presence of the melt, the desired mineral phase alite (tricalcium silicate) is formed, at a sufficient rate, in the temperature range of 1250–1450 °C. Above this temperature range, it melts (incongruently) and below this it decomposes [6, 7, 22]. Above 1400 °C, Al_2O_3 and Fe_2O_3 are almost completely melted beside small parts of CaO and SiO_2 [1, 9, 5, 7].

The raw material is subjected to various chemical processes while passing through the cold part of the kiln to the hottest kiln area (the sintering zone). The free and adsorbed water is released between 100 and 400 °C. In the temperature range from 400 to 900 °C, the clay minerals give off interstitial water and the chemically bound water, while the clay structures are destroyed. From 550 °C on, the $CaCO_3$ also starts to dissociate into CaO (free lime) and gaseous CO_2. The fact that the dissociation of $CaCO_3$ already starts at this low temperature is related to the silica, aluminum oxide and iron oxide, which displace the CO_2 from the bond with calcium oxide [11]. The CaO content increases sharply around 800–900 °C, and more frequent calcium silicates, calcium aluminates and calcium aluminate ferrites are formed. From 1250 °C on, tricalcium silicate is formed from dicalcium silicate and CaO. At approx. 1280 °C, the first melt is formed, this significantly promotes the tricalcium silicate formation (solid state reaction), as CaO and dicalcium silicate only dissolve in the presence of the melt at this low temperature [11]. The CaO and dicalcium silicate contents considerably exceed the solubility for tricalcium silicate (supersaturation). In these areas, tricalcium silicate crystallizes. This process increases with increasing temperature [21].

In addition, to improve the formation of alite, the melt also has a procedural function. The melt inside the rotary kiln forms a solid coating on the refractories that protects both the refractory bricks and the furnace shell.

1.1.5 Clinker burning process: the importance of the melt

The reactions that take place during firing and cooling in the kiln are decisive for the properties of the clinker. These reactions are mainly influenced by the clinker melt, the importance of which is discussed in more detail in this section.

As already described, when the clinker is burned, the melt promotes the formation of tricalcium silicate. For energy reasons it is important that the melt is formed at the lowest possible temperature. The phase diagram in Figure 1.1.6 shows that the clinker melt is an eutectic mixture. The melt that forms at the lowest temperature has a composition of Al_2O_3/Fe_2O_3 in the ratio of 1.63/1, i.e. such a melt has an alumina ratio (AR) of 1.63 [12].

Technical clinker usually has an alumina ratio that is greater than 1.63. This means that at the lowest melting temperature, all the iron oxide has melted, but the aluminum oxide is still solid. To also melt the aluminum oxide, the burning temperature must be increased [7]. The maximum possible amount of melt at the lowest temperature is therefore present when the alumina ratio of the raw meal is exactly 1.63. As technical raw meals usually have higher alumina ratio, this means that a higher amount of melt can be produced at a lower temperature with the addition of iron oxide; iron oxide acts as a so-called fluxing agent. At the same time, it reduces the viscosity of the melt, which also promotes the formation of tricalcium silicate [11, 12], but may reduce the buildup that protects the furnace. Therefore, an optimum must be set for a furnace system, and this can deviate from the theoretical optimum.

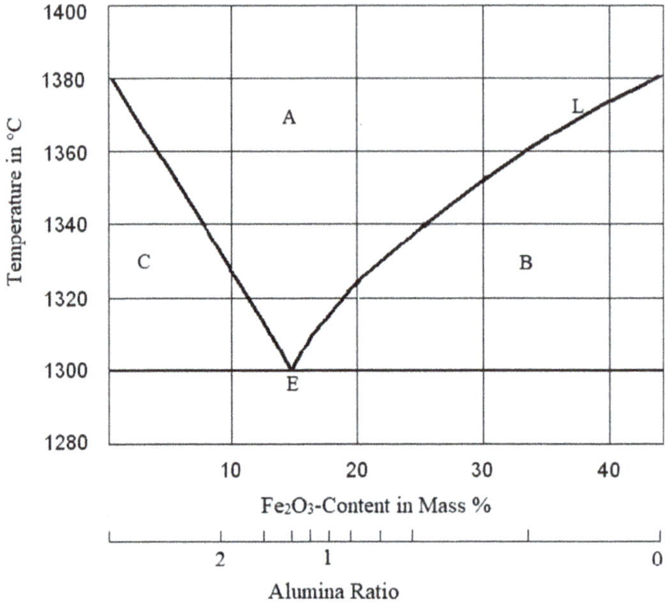

Figure 1.1.6: The quaternary system CaO-Al$_2$O$_3$-Fe$_2$O$_3$-SiO$_2$ with 5% magnesia in accordance with [12]. It is an eutectic mixture, i.e. the melting point of the pure component (e.g. Fe$_2$O$_3$) is lowered by the second component (Al$_2$O$_3$). A: Fe$_2$O$_3$ and Al$_2$O$_3$ melted, B: C$_3$S+C$_2$S+C$_4$AF solid materials, C: C$_3$S+C$_2$S+C$_3$A solid materials, E: Eutectic composition; AR = 1.63, L: Liquidus line.

The lowest temperature of formation of tricalcium silicate is 1250 °C [22]. As can be seen from the phase diagram (Figure 1.1.6), the lowest melting temperature is 1300 °C. However, this is reduced by around 20 K due to the presence of alkalis (Na, K) [23].

For the tricalcium silicate formation to take place sufficiently and quickly, burning is carried out in the temperature range between 1450 and 1500 °C.

Burning temperatures that are too high are not optimal, as they put a lot of strain on the furnace and the refractory lining in the rotary kiln. In addition, at higher temperatures it is generally more difficult to ensure the oxidic conditions necessary for clinker firing, which are necessary, for example, to prevent the formation of divalent iron in the clinker [11].

Part of the melt solidifies on the surface of the rotary kiln and forms the so-called coating that covers the refractory material as a protective layer. If too much build-up occurs, the furnace closes, which leads to operational disruptions.

1.1.6 Clinker cooling process

At the end of the rotary kiln there is a cooler that cools the clinker quickly (ideally). The cooling capacity depends on the design of the cooler [16]. Today, the efficient grate coolers are built almost exclusively; however, systems with planetary coolers and tube coolers are still operated, the cooling capacity of which is less efficient [14–16]. In addition, the two last-mentioned coolers have a longer pre-cooling zone in the furnace due to the burner position. Furthermore, to the type of construction, the cooling capacity also depends on the size of the clinker granules. Large clinker granules cool much more slowly than fine granules. In the case of dusty clinker, the cooling performance of the system also decreases due to the poorer heat transfer to the cooling air.

The clinker generally must be cooled after sintering in order to maintain the tricalcium silicate content, which is only stable up to a temperature of approx. 1250 °C. If the cooling is too slow, the melt would promote the decomposition of tricalcium silicate into dicalcium silicate and CaO [6, 12].

Furthermore, the cooling speed affects the crystallization of the calcium aluminate and calcium aluminate ferrite crystals from the melt. At a very high cooling rate, the melt solidifies in an almost glass-like manner. The slower the melt cools, the more differentiated the clinker structure appears and the larger the calcium aluminate and calcium aluminate ferrite crystals develop. In the case of calcium aluminates, it was found that the reactivity with water increases with crystal size. Slowly cooled clinkers therefore show a faster solidification when used in cement [11, 14, 15].

1.1.7 Portland cement production

The calcium aluminate, in the ground cement clinker, reacts spontaneously with the formation of e.g. water-rich tabular tetracalcium aluminate hydrate ($4CaO \times Al_2O_3 \times nH_2O$). A solid paste that can no longer be processed is formed in a very short time. In order to be able to work with cement mortar respectively concrete for longer, calcium sulfate (gypsum and/or anhydrite) is added to the cement during grinding. In the presence of sulfate, a layer of ettringite, a sulfate-rich calcium aluminate sulfate hydrate (Figure 1.1.7), forms on the surface of the tricalcium aluminate crystals in contact with the water, which initially prevents the crystal surface from contacting the water.

During cement grinding, a large part of the grinding energy is lost as heat. As a result, the mills heat up to temperatures in some cases to more than 120 °C. At these temperatures, the gypsum dehydrates to hemihydrate ($CaSO_4 \times 2H_2O \rightarrow 2CaSO_4 \times 0.5H_2O + 1.5H_2O$). Even if the calcium sulfates used have approximately the same solubility, they differ in the speed of dissolution. Thus, hemihydrate dissolves faster than gypsum and anhydrite, this is due to its crystal structure disturbed by the drainage. The amount and the ratio of gypsum to anhydrite is adjusted to accommodate for the reactivity of the calcium aluminate in the clinker.

The complex chemical reactions of Portland cement with water have been researched for decades and are discussed in detail, for example, in the following books and articles [1, 4, 5, 7, 10, 13, 19, 21, 29–32].

Figure 1.1.7: Scanning electron microscope images of Ettringite crystals (left) and a large Portlandite crystal, i.e. $Ca(OH)_2$ (right) surrounded by Ettringite.
Source: Forschungsinstitut der Zementindustrie/VDZ, Düsseldorf.

1.1.8 Other cement constituents

In large parts of the world, today's cements are composed of various constituents. In addition to Portland cement clinker, the European cement standard EN 197-1 defines granulated blast furnace slag, pozzolana, fly ash, burnt shale, limestone and silica dust as permissible main and secondary constituents (Table 1.1.5) [8].

Table 1.1.5: Overview of common cement and concrete constituents.

Constituent	Latent-hydraulic	Pozzolanic	Availability
Granulated blast furnace slag	×		High
Silica fume		×	Low
Fly ash	×	×	High
Natural puzzolan		×	High
Oil shale [34]	×		Medium low

A distinction is made between latent-hydraulic, pozzolanic-reacting or predominantly inert materials. In the following, a brief overview will be given of the material flows currently relevant in terms of quantity, with which modern cements are produced. In the future, other by-products from other industries will probably also become interesting.

1.1.9 Ground granulated blast furnace slag

For more than 100 years, the slags from pig iron production (see also Section 6.4) have been used in granulated, glassy form as blast furnace slag for cement production [19]. Ground granulated blast furnace slags are referred to as latent-hydraulic because they form solid hydrate phases upon, for example, sulfatic or alkaline excitation with water. These are comparable to those resulting from the reaction of Portland cement clinker with water. The reactivity rate of granulated blast furnace slags is significantly lower than that of Portland cement clinker.

The slags result from the production of pig iron in a blast furnace from the secondary constituents of the ores used, the coke ash and the so-called aggregates (limestone, dolomite stone or quartz sand, etc.). The aggregates lower the iron melting point, act as fluxes, and remove sulfur and phosphors from the ore [24–26]. Due to its lower density, the blast furnace slag floats on the molten iron during the process and can flow out of the blast furnace both, continuously or discontinuously. The slag content is around 160–470 kg per t of pig iron produced [24–26].

The slag acquires the hydraulic properties, of interest for cement production, when it is cooled very rapidly from its liquid aggregate state and solidifies to a glassy consistency. Today, granulation plants (up to 720 t/h [24, 27]) are frequently used for this purpose, in which cooling takes place with 6 to 10 times the amount of water. The slag is quenched very quickly from 1500 °C to below 840 °C and decomposes into a sand-like product called granulated blast furnace slag. Good granulated blast furnace slags are usually very light in color. The color provides information about the blast furnace process and composition [24].

The blast furnace slags are stored in the open air (see also Section 6.4) and, depending on the porosity, the grading curve and the storage conditions, a residual moisture of about 10% is obtained [24]. The residual moisture is unfavorable for subsequent cement production, so that the granulated blast furnace slags generally have to be dried in an energy-intensive process before processing (heat consumption: approx. 0.1–0.5 GJ/t).

Ground granulated blast furnace slag can be used as a concrete additive or as a main cement constituent. According to the European cement standards EN 197-1 [8] and EN 15167-1 [20], granulated blast furnace slag, to be used as main cement constituent or concrete additive, must be at least 2/3 glassy solidified. In addition, it is required that granulated blast furnace slag consists of at least 2/3 calcium oxide (CaO), magnesium oxide (MgO) and silicon dioxide (SiO_2) and contains aluminum oxide (Al_2O_3) and small amounts of other compounds. The standards also stipulate a ratio of $(CaO+MgO)/SiO_2$ of greater than 1.0 [8, 20]. Because of the content of CaO slag is latent-hydraulic.

The granulated blast furnace slag content in European cements may be up to 95% according to EN 197-1 [8]. The Portland cement clinker, when mixed with the blast furnace slag, provides the alkaline excitation when water is added. The granulated blast furnace slag surface corrodes and forms calcium silicate hydrates with lower Ca/Si ratios than pure Portland cement. The reaction rate is much slower than that of Portland cement clinker. However, the hardening potential of cements containing blast furnace slag is at least equivalent.

Since granulated blast furnace slag (Figure 1.1.8) is usually harder than Portland cement clinker, the co-grinding of these two components in the cement mills leads to the Portland cement clinker accumulating in the finest fraction of the cement. Due to these very different reactivities and properties of the two components, and in order to adapt cement properties to market demand, the trend in the cement industry is towards separate grinding of granulated blast furnace slag and clinker with subsequent blending.

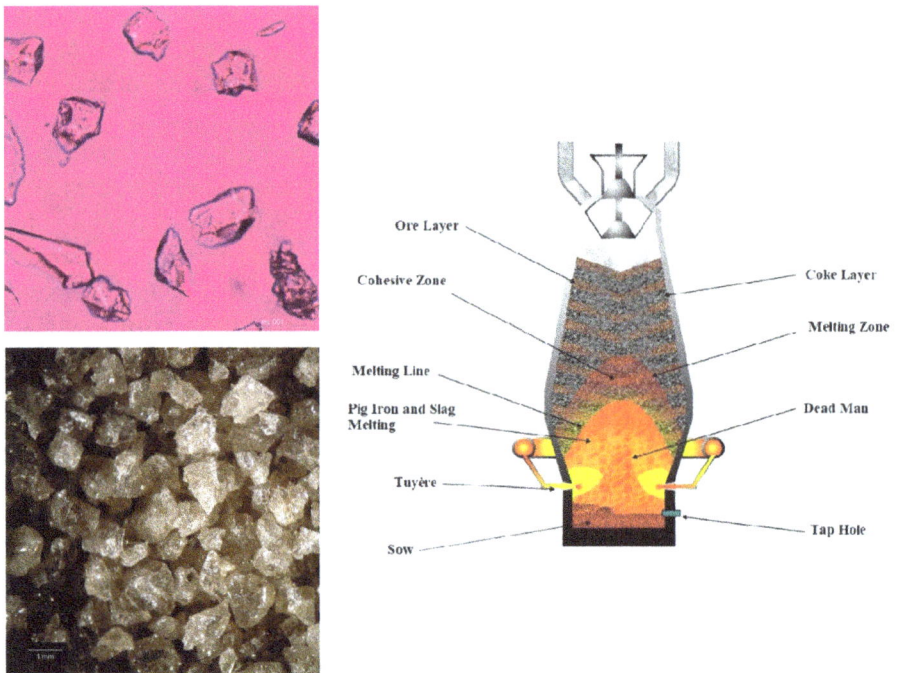

Figure 1.1.8: Fraction 40–60 µm of crushed blast furnace slag in transmitted light microscope (top left); unground granulated blast furnace slag in reflected light microscope with typical vitreous luster
(bottom left; **Source:** FEhS Institute 2017). Blast furnace schematic (right; **Source:** Tobias Broch [41]).

Large emissions of CO_2 (process-related and fuel-related) are released during the pig iron production process. Due to increasing socio-political pressure and rising CO_2 costs under the European Emissions Trading System, European industry is working on pig iron production using a hydrogen-based process known as direct reduction. This process will only produce small quantities of granulated blast furnace slag as a by-product. Moreover, these will not be comparable in quality with today's granulated blast furnace slag.

1.1.10 Artificial pozzolan: fly ash

There are various artificial pozzolans. The most relevant artificial pozzolan in terms of quantity is fly ash. Fly ash is obtained by the electrostatic or mechanical separation of dust-like non-combustible particles from flue gases of coal-fired plants (e.g. thermal or electric power plants). The chemical composition therefore depends strongly on the coal used. Fly ash can be silica-rich or lime-rich. The former exhibits pozzolanic properties; the latter, in addition to pozzolanic properties, also exhibits low hydraulic properties [7, 8, 13].

Fly ash, due to its extraction from the gas stream of the furnace, is generally spherical and largely glassy solidified, Figure 1.1.9. The average grain diameter of coal fly ash is between 10 and 30 μm. Only 10–30% of fly ash particles are generally larger than 45 μm [7, 8, 13, 28].

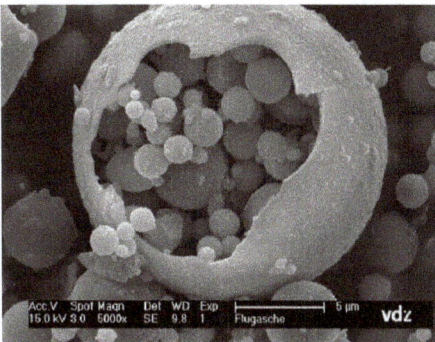

Figure 1.1.9: Scanning electron micrograph of fly ash.
Source: Forschungsinstitut der Zementindustrie/VDZ, Düsseldorf.

Fly ash is used as a main cement constituent in cement production and is added to concrete as an additive. The glassy SiO_2 portion of fly ash reacts with the $Ca(OH)_2$, (portlandite, Figure 1.1.7) which is released from Portland cement clinker hydration, and forms calcium silicate hydrate:

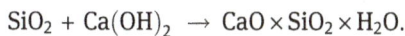

$$SiO_2 + Ca(OH)_2 \rightarrow CaO \times SiO_2 \times H_2O.$$

Other minerals formed by the reaction have since been described [28]. The use of fly ash can significantly improve the properties of concrete. Due to the spherical components, the workability of concrete is significantly improved. The heat generation, which has to be considered especially for massive concrete components (components of large dimensions), is reduced when fly ash is used, thus, lowering thermal stresses and cracking. Furthermore, fly ash improves the resistance to chemical attack and reduces the risk of alkali aggregate reaction that can disintegrate concrete [4, 7, 8, 13, 33].

It is generally assumed that fly ash availability will decrease due to the CO_2 reduction targets of many countries and the move away from coal-fired power generation.

1.1.11 Natural pozzolans

Natural pozzolans are substances with a siliceous and/or alumo-silicate composition. Pozzolans do not harden independently after being mixed with water, however, finely ground and in the presence of water; they react at ambient temperature with dissolved $Ca(OH)_2$ to form calcium silicate hydrates and calcium aluminate compounds, and these fuse to form a solid structure. These compounds are comparable to those formed during the hardening of Portland cement clinker. Pozzolans consist mainly of glassy reactive silica (SiO_2) and alumina (Al_2O_3). The mass fraction of reactive silica (SiO_2) must be at least 25.0% according to EN 197-1 [8]. Conform to ASTM C 618 [35], the sum of (SiO_2), Al_2O_3 and Fe_2O_3 must be at least 70%.

Natural pozzolans are generally substances of volcanic origin or sedimentary rocks of chemical-mineralogical composition [4]. Natural pozzolans are, for example, tuff, perlite, diatomite, zeolites, etc. The rate of formation of calcium silicate hydrates of pozzolans is much slower than that of Portland cement clinker. The strength contribution of pozzolans is often observed only after 56 to 90 days, while Portland cement reaches the final strength already in the first 28 days [4].

Pozzolans will in all likelihood once again become more important in cement production, due to climate policy developments and their enormous availability.

1.1.12 Quality assurance and process control in modern cement plants [36, 37]

Quality assurance and production monitoring in the cement industry have a long tradition, which in Germany, for example, goes back to the time of product standardization in 1877–1878. Then as now, the valid product standard specifies the minimum testing effort required to ensure that cements conform to the standard.

The tasks of today's works laboratory are diverse, must be performed in a timely manner and exceed the testing effort required by the standard. In addition to ensure that products conform to standards, the overriding goal is to provide customers with products of consistently high quality.

To achieve this quality goal, the product must be monitored in cement plants through regular sampling at every step of production, starting with the raw materials and ending with the loading of the cement.

Although the production steps in the cement plants are comparable in principle, they differ in detail, so that the focus of quality assurance is different in the individual plants.

In modern cement plants, central production and process monitoring is carried out in an automated laboratory (Figure 1.1.10). At each individual process step (raw meal, kiln feed, clinker, cement, etc.), samples are taken automatically at defined time intervals. Automatic samplers fill partial quantities of the sample into pneumatic tube containers and send the sample via a pneumatic tube system directly to the automated laboratory. In this laboratory, e.g. particle size distribution (PSA), chemical composition (XRF), mineralogical composition (XRD) and color are determined fully automatically. The analysis results are transmitted to the process control room. In addition to the materials produced at the plant itself, other raw materials such as granulated blast furnace slag, fly ash, trass, gypsum, anhydrite, etc. are also used in modern cement plants. Incoming samples are taken from all delivered raw materials and the respective plant-specific chemical-physical product requirements are checked.

Figure 1.1.10: Automated laboratory at the Ennigerloh cement plant (left); sands have a considerable influence on the development of compressive strength. Since no natural sand has the required uniformity in the grain band and in the chemical-mineralogical composition, a specifically machine-composed test sand (right) is used in the laboratories of the cement plants to determine the standard strength class according to EN 197-1 [8, 38].

The European cement standard EN 197-1 [8] specifies the properties and requirements for standard cements and their constituents. For the product to comply with this standard, it must be continuously evaluated on the basis of random samples. Samples of each cement type must be taken twice a week, during which the standard chemical and physical parameters (e.g. setting behavior and strength development) are determined, on defined mortar mixes of cement, CEN standard sand (a very uniformly produced quartz sand composed from different sand pits and produced on a large scale [38]) and water. Only complete monitoring ensures uniform production of the commodity product.

1.1.13 Energy consumption in cement production

With a relatively high energy consumption of around 30 TWh per year (fuel and electricity) and an energy cost share of more than 30% of gross value added, cement production is one of the most energy-intensive industries in Germany [39].

The production of one ton of cement requires a specific energy input of around 3.200 kJ/kg, whereby the energy input has remained relatively stable since the beginning of the 2000s (Figure 1.1.11).

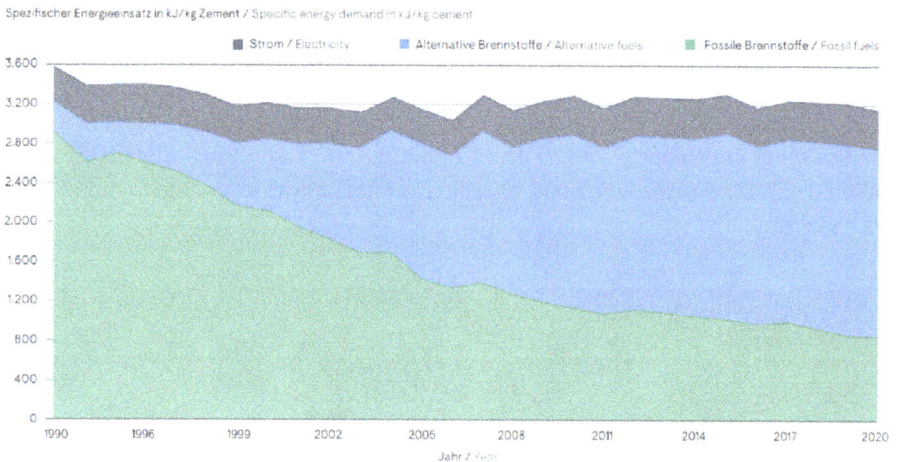

Figure 1.1.11: Development of the specific energy use in cement production [40].

As Figure 1.1.11 shows, the majority (2020: around 87%) of the energy consumption is accounted for by fuels that are used in the kiln to generate the high process temperatures for burning the cement clinker. Around 13% of the total energy consumption is for electrical energy, which is mainly used for processing raw materials (approx. 25%), burning and cooling the clinker (approx. 24%) and for grinding the cement (approx. 46%).

Although measures to increase energy efficiency have always been implemented in the cement industry, higher demands on product quality (especially more powerful, finer ground cements) as well as tightening of environmental legislation and the associated measures for exhaust gas cleaning have led to a tendency towards an increase in electrical energy consumption. When it comes to fuels, the increasing use of alternative fuels (as a substitute fuel, especially for hard coal and lignite) leads to higher heat consumption due to the higher moisture content [38].

Due to technical measures to reduce CO_2 in order to achieve the statutory climate protection targets (including carbon capture and storage (CCS)/carbon capture and utilization (CCU)), a further increase in energy and, above all, electricity requirements in cement production can be assumed in the future [38].

References

[1] Zementtaschenbuch. Verein Deutscher Zementwerke e.V., Forschungsinstitut der Zementindustrie, Düsseldorf.

[2] Design and Control of Concrete Mixtures, Cement Association of Canada, 7th Canadian edition, 2002. ISBN 0-89312-218-1.

[3] Lamprecht H-O. Opus Caementitium – Bautechnik der Römer, 5. Auflage, Beton Verlag, Düsseldorf, 1996.

[4] Hewlett PC. Lea's Chemistry of Cement and Concrete, 4th edition, Arnold, London NW1 3BH, 1998.

[5] Labahn O, Kohlhaas B. Cement Engineers' Handbook, 4th edition, Bauverlag GmbH, Wiesbaden, 1983.

[6] Hofmänner F. Portlandzement-Klinker – kleine Gefügekunde, Holderbank Management und Beratungs-AG, Jona, Switzerland, 1973.

[7] Locher FW. Zement: Grundlagen der Herstellung und Verwendung, Verlag Bau + Technik, Erkrath, Germany, 2000.

[8] EN 197-1. Teil l: Zement – Zusammensetzung, Anforderungen und Konformitätskriterien von Normalzement (Dt., Engl., Frz.), CEN, Brüssel, February 1999.

[9] Duda WH. Cement-Data-Book. Bd. l Internationale Verfahrenstechniken der Zementindustrie, 3. Aufl. (Dt. u. Engl.), Bauverlag GmbH, Wiesbaden, 1985.

[10] Stark J, Wicht B. Bindemittelbaustofftaschenbuch: Band 3: Brennprozesse und Brennanlagen, 2. Auflage, VEB Verlag für Bauwesen, Berlin, 1986.

[11] Locher FW. Einfluß der Klinkerherstellung auf die Eigenschaften des Zements. Zement-Kalk-Gips, 1975, 28, 265–72.

[12] Swayze MA. A report on studies 1. The ternary system CaO-C3A-C2F, 2. The quaternary system as modified by 5% magnesia. Am J Sci, 1946, 244, 65–94.

[13] Stark J, Wicht B. Anorganische Bindemittel – Zement – Kalk – Spezielle Bindemittel, Schriften der Bauhaus Universität Weimar, 1998. issue 109. ISBN3-86068-093-5.

[14] Sylla H-M, Steinbach V. Einfluss der Klinkerkühlung auf Erstarren und Festigkeit von Zement. Zement-Kalk-Gips International, 1975, 28, 357–62.

[15] Sylla H-M, Steinbach V. Einfluss der Klinkerkühlung auf die Zementeigenschaften. Zement-Kalk-Gips International, 1988, 41, 13–20.

[16] Hoenig V, Sylla H-M. Technische Klinkerkühlung unter Berücksichtigung der Zementeigenschaften. Zement-Kalk-Gips International, 1998, 51, 318–33.

[17] Locher FW. Einfluss der Brennbedingungen auf die Klinkereigenschaften, Cement Production and Use, Engineering Foundation, US Army Research Office, 1979, 81–95.

[18] Sylla H-M. Einfluss reduzierenden Brennens auf die Eigenschaften des Zementklinkers. Zement-Kalk-Gips International, 1981, 34, 618–30.

[19] Keil F. Zement – Herstellung und Eigenschaften, Springer-Verlag, Heidelberg, 1971. ISBN 3-540-05167-8.

[20] EN 15167-1. Hüttensandmehl zur Verwendung in Beton, Mörtel und Einpressmörtel – Teil 1: Definitionen, Anforderungen und Konformitätskriterien; Deutsche Fassung EN 15167-1, 2006.

[21] Taylor HFW. Cement Chemistry, Academic Press Limited, London, 1990. ISBN 0-12-683900-X.

[22] Rankin OA, Merwin HE. The Ternary System CaO-Al$_2$O$_3$-MgO. J Am Chem Soc, 1916, 38, 568–88.

[23] Greene KT, Bogue RH. Preliminary study on portions of the systems Na$_2$O-CaO-Al$_2$O$_3$-Fe$_2$O$_3$ and Na$_2$O-CaO-Fe$_2$O$_3$-SiO$_2$. J Res Nat Bur, 1948, 40, 225–34.

[24] Ehrenberg A. Hüttensand – Ein Leistungsfähiger Baustoff mit Tradition und Zukunft – Teil 1 und 2, Beton-Informationen 4 und 5, 2006, 35-64 and 67–95.

[25] Holleman AF, Wiberg N, Wiberg E. Lehrbuch der Anorganischen Chemie, Walter de Gruyter & Co., Berlin, New York, Auflage, 1985, 91–100. ISBN 3-11-007511-3.

[26] Shatoka V, Pliskanovskyy S, Kharakhulakh V. The present day and perspective development of the ironmaking industry in Ukraine. Stahl und Eisen, 2005, 125, 39–42.

[27] Maas H, Peters K-H. Der Einfluss der Granulation auf die hydraulischen Eigenschaften des Hüttensandes. Zement-Kalk-Gips International, 1978, 6, 300–01.

[28] Schulze SE. Zur Reaktivität von Steinkohlenflugaschen und ihrer Rolle bei der Hydratation flugaschehaltiger Zemente, Schriftenreihe der Zementindustrie, 2014. issue 81, ISBN 978-3-7640-0604-4.

[29] Locher FW, Richartz W, Sprung S. Erstarren von Zement; Teil l: Reaktion und Gefügeentwicklung. Zement-Kalk-Gips International, 1976, 29, 435–42.

[30] Locher FW, Richartz W, Sprung S. Erstarren von Zement; Teil 2: Einfluss des Calciumsulfatzusatzes. Zement-Kalk-Gips International, 1980, 33, 271–77.

[31] Sylla H-M. Erstarren von Zement; Teil 3: Einfluss der Klinkerherstellung. Zement-Kalk-Gips International, 1982, 35, 669–76.

[32] Locher FW, Richartz W, Sprung S, Rechenberg W. Erstarren von Zement; Teil 4: Einfluss der Lösungszusammensetzung. Zement-Kalk-Gips International, 1983, 29, 224–31.

[33] Schiessl P, Wiens U, Schröder P, Müller C. Neue Erkenntnisse über die Leistungsfähigkeit von Beton mit Steinkohlenflugasche – Teil 1 und 2, Beton Verlag, Düsseldorf, 2001.

[34] Feige F. Zur wirtschaftlichen Verwendung des Ölschiefers bei Rohrbachzement. Zement-Kalk-Gips International, 1992, 2, 53–62.

[35] ASTM C 618-19. Standard Specification for Coal Fly Ash and Raw or Calcined Natural Pozzolan for Use in Concrete, ASTM International, West Conshohocken, PA, USA.

[36] Boos P. Qualitätssicherung und Prozesskontrolle im modernen Zementwerk, Newsletter Technik der HeidelbergCement AG, Heidelberg, 2010.

[37] Boos P, Herrmann N, Scholz R. Quality assurance and process control in modern cement plants. Zement-Kalk-Gips International, 2013, 4, 44–52.

[38] Boos P. Standard sand as the standard for worldwide monitoring of cement quality. Cement International, 2013, 1, 72–75.

[39] Loos F. Leiter Energiemanagement Deutschland, ESG Business line coordinator Cement, HeidelbergCement AG, personal communication, 2021.

[40] Association of German Cement Works (VDZ). Cement Industry at a Glance 2020/2021, Berlin, 2020, 13.

[41] Broch T. Notwendigkeiten der gezielten Zusammensetzung des Hochofen-Möllers, Presentation at the 87th Working Group "Building Materials" of the FEhS Institute, FEhS-Institut für Baustoff-Forschung e.V., Duisburg, 2010.

1.2 Inorganic insulation materials

Rainer Pöttgen, Thomas Fickenscher

Inorganic materials in construction chemistry comprise inorganic binders and filling materials (Chapter 1.1), inorganic sealing materials and additives (Chapter 1.3), inorganic pigments (Chapter 1.4) and inorganic insulation materials [1–5]. The latter are summarized in the present chapter. These materials play an important role for thermal insulation of all kinds of buildings and technical equipment, primarily with respect to save energy, but also with an enormous impact on the gross national products. Also many organic and organic-inorganic hybrid materials find application as insulation materials; however, they are not discussed herein.

The inorganic insulation materials can roughly be subdivided into two different branches, fibrous materials (glass wool and rock wool) and cellular materials (calcium silicate and cellular glass). The quality of an inorganic insulation material is determined by its thermal conductivity (diffusivity). To give an example, typical thermal conductivity values (in units of $W\ m^{-1}\ K^{-1}$) are 97.5 for aluminum, ca. 0.75 for concrete, ca. 0.13 for wood, ca. 0.04 for foam glass, ca. 0.38 for cork and ca. 0.12 for glass and rock wool. This short comparison readily shows that low thermal conductivity materials are an important prerequisite for energy-intelligent construction purposes. The technologically most important inorganic insulation materials are summarized in the following subchapters.

1.2.1 Glass wool, stone wool and mineral wool

Both glass wool and stone wool are made of mineral fibers and generally denoted mineral (or insulation) wools (the general technical name for any kind of fiber materials made from molten minerals through a spinning process) [6]. Glass wool (also known as so-called fiberglass) is produced from a mixture of glass (>70%, mostly recycled glass), sand, soda ash and limestone with 0.5–7% of a binder (mainly phenolic resins for enhancing the strength and achieve a certain texture of the final product) and an addition of ca. 0.5% of mineral oil, in order to bind any kind of dust in the final product. The basic raw materials for rock wool are feldspar, dolomite, basalt, anorthosite and coke (as energy source). Next to raw materials, also recycled rock wool can be added to the process as well as slag residues from the metal industry.

For both glass wool and rock wool, the raw materials are first homogeneously melted. The melting temperature depends on the mixture of the raw materials and lies in the range of 1300–1600 °C. Three different processes for fiber production are known: (i) the drawing process for endless fibers, (ii) blow molding for smaller

https://doi.org/10.1515/9783110733143-002

fibers and (iii) a centrifugal process (high-speed spinning), where droplets of the melt are centrifuged on a flywheel (similar to the cotton candy production).

In the final process, the individual fibers are treated with a binder and an impregnating agent in order to strengthen the material and get a hydrophobic surface. The two wool types have different application forms: rolls, slabs and flakes (loose filling material for wall cavities or attics). The roll material (Figure 1.2.1) is typically used for wall, roof and floor insulation of buildings. Further uses are structural insulation, pipe insulation, filtration and soundproofing.

Figure 1.2.1: Cutouts of rock wool (left) and glass wool (right) batts.

The most important parameter for a highly effective insulation with mineral wool is the vigorous exclusion of humidity, since any moisture content drastically increases the thermal conductivity.

1.2.2 Calcium silicate

The source for calcium silicate, $CaSiO_3$, is the mineral wollastonite. Synthetic calcium silicate can be obtained from calcium oxide (through thermal decomposition of $CaCO_3$) and silica. Calcium silicate is also the slake forming during the synthesis of white phosphorus ($2Ca_3(PO_4)_2 + 6SiO_2 + 10C \rightarrow P_4 + 6CaSiO_3 + 10CO$). It is a white solid that is insoluble in water.

Wollastonite finds application as filler in diverse polymers, as insulation material and for fire protection (also as so-called refractory filler). Highly porous insulating boards of calcium silicate are produced with the addition of binders and blowing agents. A major advantage of calcium silicate-based insulation materials is their mold resistance, thus its broad application for the rehabilitation of damp walls. Calcium silicate boards (Figure 1.2.2) are one of the best mold-preventing insulation systems for rooms with enhanced humidity.

Figure 1.2.2: Porous calcium silicate insulation boards, kindly provided by CASIPUS GmbH.

1.2.3 Cellular glass

Foam-like insulation material made of glass is called cellular glass or foam glass. It is mainly produced from recycled glass (mainly flat glass from cars and windows). The waste glass pieces are melted with additions (in order to get a melt with low melting temperature) of feldspar, dolomite, iron oxides, manganese oxide and sodium carbonate. The cooled frozen glass is then powdered in planetary mills, mixed with a small quantity of carbon and placed in the desired form of stainless steel. The mixture is finally heated to 1000 °C. Oxidation of the carbon powder leads to the foaming process leaving a close cell structure with thin cell walls. Controlled cooling leaves the cellular structure (Figure 1.2.3). Usually, the larger foam glass blocks are processed as insulation boards, but also a continuous process leads to so-called foam glass gravel (with an extremely low density of <200 kg m^{-3}), a bulk material with excellent insulation properties.

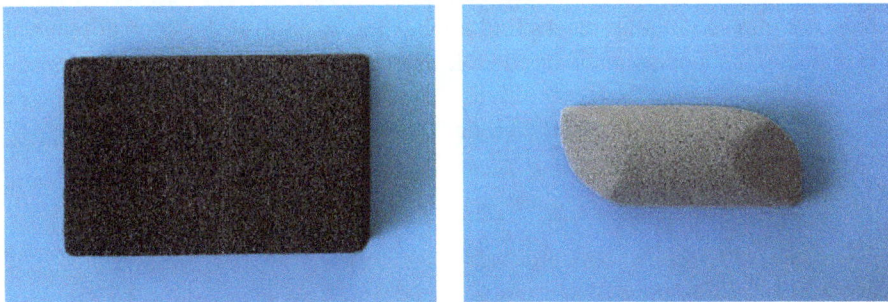

Figure 1.2.3: (left) A block of foam glass® T3 + for use as insulation board. The sample was kindly provided by FOAMGLAS. (right) A piece of foam glass for corneal removal (cosmetic application).

The large advantage of foam glass is its non-flammability and high impermeability. However, foam glass is not completely frost-proof. The surface cellular structure can be destroyed by water uptake and frost. Foam glass has a high compressive strength and excellent recycling ability. Due to its high thermal resistivity, foam glass also finds application for diverse fireproof panels.

We should also mention expanded glass granulate in this chapter. This granulate is produced from residues of glass industry. Glass flour is mixed with binders and blowing agents and the granules are obtained after an annealing process at 800–900 °C in a kiln. The granules result in mm size and have a fine internal pore structure. Different grain fractions can be obtained by sieving. Besides thermal insulation, expanded glass granulate finds also application for soundproofing. The granulate can be used for cavity filling or as filling material for low-density concrete or thermal insulation or acoustic plaster.

1.2.4 Pumice, vermiculite and perlite

Pumice [7] has volcanic origin. It forms through violent eruption during volcanic explosions. Super-heated molten rock with water and gas inclusions under high-pressure conditions rapidly cools and depressurizes during the volcanic eruption and leaves the peculiar bloated, frothy pumice habit. The final product shows a lot of (finally air-filled) vesicles that lead to the very low materials density. In analogy to the cellular glasses discussed above, pumice can be considered as a mineral foam. The composition of pumice is close to obsidian (60–75% SiO_2, 15–17% Al_2O_3, 1–3% Fe_2O_3, 1–2% CaO, 7–8% Na_2O/K_2O; density 2.5–2.6 g cm^{-3}); however, it has only about one third of its density. Thus pumice can float on water. It has a broad distribution on the earth surface and large deposits are almost available worldwide.

The low density and extremely porous nature lead to a manifold of applications. Pumice powder is used as abrasive in many cosmetics and as polishing agent in dentistry to obtain smooth tooth surfaces and also in pencil erasers. Pumice stones cut into rectangular blocks find application as abrasive for corneal removal (Figure 1.2.4). Its moderate abrasiveness is used for so-called stonewashed denims.

Figure 1.2.4: A pumice stone for abrasive cosmetic application.

The massive porous nature of pumice is used in horticulture. Pumice can absorb water and nutrients and helps to retain water. This is used for Bonsai soil mixes (soil conditioner). Pumice-enhanced soils in sports fields, parks and golf courses contain up to 10% pumice addition.

In construction chemistry pumice is widely used with respect to its low density and its low thermal conductivity, a consequence of the gas (air) inclusion. The main technological use concerns lightweight concrete and insulative low-density cinder blocks (Figure 1.2.5). Furthermore, pumice is the raw material for diverse soundproofing bricks.

Figure 1.2.5: A low-density pumice brick (left) and a pumice brick filled with insulating mineral foam. The bricks were kindly provided by Bisotherm®.

Vermiculite belongs to the class of phyllosilicates (layered silicates) with substantial water content. Its chemical formula is $(Mg,Fe^{2+},Fe^{3+})_3[(Al,Si)_4O_{10}](OH)_2 \cdot 4H_2O$; charge neutrality is achieved through a precise Fe^{2+}/Fe^{3+} vs. Al^{3+}/Si^{4+} ratio, adjusting the total cation charge. The incorporated water is released during the so-called dry heating process in the temperature range between 700 and 1000 °C. The sudden water evaporation pushes the crystal flakes apart and the vermiculite blows up in a worm-like shape, drastically increasing the volume and the absorbency. This leads to an excellent low-density insulation material with low thermal conductivity. Vermiculite flakes and a vermiculite plate (for application in a fireplace) are presented in Figure 1.2.6.

Especially the vermiculite granules find application for fire protection purposes, for soundproofing, as chemically inert and fire-resistant insulating transport bulk or as additive in fire-resistant plaster. Vermiculite flakes are also used along with Portland cement for lightweight insulation concrete.

Perlite shows similar thermal properties as vermiculite. It derives from the volcanic glass obsidian (>70% silicic acid content). The latter slightly decomposes under devitrification and gets cracks that include water (2–5 wt%). During heating at 800–1000 °C, perlite loses its water and shows a volume increase up to 15–20 times

Figure 1.2.6: Vermiculite flakes (left) and a fire-resistant plate (right).

of the original one along with a drastic reduction of the bulk density to 0.9–1.0 g cm^{-3}. The small perlite balls (Figure 1.2.7) in the order of some mm (often silicone impregnated for better pourability) are mainly used for bulk insulation and expansion insulation. Furthermore, perlite is used in composite materials with inorganic or organic binders in order to form insulating blocks.

Figure 1.2.7: Loose bulk granular perlite insulation material. The sample was kindly provided by Bisotherm®.

1.2.5 Aerogels

The latest innovation of highly effective thermal insulation materials concerns aerogels (nanogels, nanoporous foam) [8–11] which have extremely low density (<0.2 g cm^{-3}) and very high porosity. The precursor compounds for silica-based aerogels are silicon alkoxides Si(OR)$_4$, typically tetramethylorthosilicate (TMOS) or tetraethylorthosilicate (TEOS). In the case of TMOS, the (catalyzed) hydrolyses can simplified be written as Si(OCH$_3$)$_4$ + 2H$_2$O → SiO$_2$ + 4CH$_3$OH. The formed sol transforms to a gel (the so-called sol-gel process) and is subsequently aged in its mother liquor. The diffusion-controlled aging conditions, the aging time and the pH medium are the important parameters for the technological process, in order to get the desired pore structure. Besides the pure silica-based aerogels, aerogels from other metal oxide precursors (aluminum or chromium oxide) or carbon-based precursors are also known.

The decisive final step is the drying process, since the dried samples should keep the structure they had in the wet stage. This can be achieved by supercritical drying (a process that is well known from the decaffeination of coffee). The liquid within the gel is replaced by air without any volume change. Aerogels are almost transparent and extremely temperature stable. Although they are almost exclusively composed of pores, they have high compressive strength. The pore size is in the nanometer range and the inner surface of such an aerogel can be as large as $1000 \text{ m}^2 \text{ g}^{-1}$. The very small pore size reduces the gas thermal conductivity, since the mean free path length of the gas molecules is larger than the distance between the pore walls, avoiding heat transport through molecular collisions. This is the essential property for application of aerogels (as insulating mats or as plaster additive) as very efficient insulation material. The thermal conductivity is that low that already thin layers are sufficient. As an example, we present a mat (Figure 1.2.8) of the non-combustible aerogel SLENTEX®, commercialized by BASF. This highly efficient insulation material is extremely flexible and exhibits a lambda value of only $19 \text{ mW m}^{-1}\text{K}^{-1}$.

Figure 1.2.8: Slentex® insulation matt. The sample was kindly provided by BASF.

The extremely low thermal conductivity of some aerogels is impressive. In an illustrative experiment, an aerogel plate is heated with a Bunsen burner at the bottom and matches at the top do not ignite [12].

1.2.6 Conclusion

Our short summary of the most important inorganic insulation materials readily shows that they all exhibit extremely energy intensive production processes, mainly high temperature treatments. This is different to the synthetic organic insulation materials (typical polymers are styropor® or styrodur®). For any of the insulation materials, there is a delicate interplay between different parameters: (i) energy consumption during production of the insulation material, (ii) the recycling ability, (iii) the insulation properties and the workability of the material and (iv) the fire-protection properties. Especially with respect to the latter point, there is a clear advantage for the inorganic insulation materials.

References

[1]   Papadopoulos AM. Energy Build, 2005, 37, 77–86.
[2]   Pfundstein M, Gellert R, Spitzner M, Rudolphi A. Insulating Materials: Principles, Materials, Applications, Birkhäuser, München, 2008.
[3]   Adityaa L, Mahlia TMI, Rismanchi B, Ng HM, Hasan MH, Metselaar HSC, Muraza O, Aditiya HB. Ren Sust Energy Rev, 2017, 73, 1352–65.
[4]   Eshrar L, Bevan R, Woolley T. Thermal Insulation Materials for Building Applications, ICE Publishing, London, 2019.
[5]   Villasmil W, Fischer LJ, Worlitschek J. Ren Sust Energy Rev, 2019, 103, 71–84.
[6]   Sirok B, Blagojevic B, Bullen P. Mineral Wool, Woodhead Publishing, Sawston, Cambridge, 2008.
[7]   Rashad AM. Silicon, 2021, 13, 551–72.
[8]   Baetens R, Jelle BP, Gustavsen A. Energy Build, 2011, 43, 761–69.
[9]   Cuce E, Cuce PM, Wood CJ, Riffat SB. Ren Sust Energy Rev, 2014, 34, 273–99.
[10]   Du A, Zhou B, Zhang Z, Shen J. Materials, 2013, 6, 941–68.
[11]   Smirnova I, Gurikov P. Ann Rev Chem Bio Eng, 2017, 8, 307–34.
[12]   https://inhabitat.com/exciting-advances-in-insulation-with-aerogel/aerogel-insulation/

1.3 Conservation: silicon chemistry in building protection

Markus Boos

1.3.1 Weathering

Building materials are designed for durability. Nevertheless, depending on their composition and structure as well as external influences, they are subject to specific decay processes (corrosion) that can impair both their functionality and their aesthetic appearance.

This is evident, for example, in cement-bound materials: For a long time, it was thought that concrete was a building material for eternity. In the meantime, however, it has been proven that concrete is also susceptible to damage and that bridges, roads and buildings made of reinforced concrete, for example, can be so severely damaged by use and/or environmental influences, so that even their stability is at risk. The situation of the German motorway bridge infrastructure, which has been neglected for decades, can be cited as an example.

This becomes particularly clear in the case of historic buildings, which show a correspondingly long course of damage due to long weathering periods. However, the deterioration of building materials has not only a purely material-economic, but also a cultural-social significance. It is not for nothing that the preservation of spiritual and material evidence of a culture is considered an important basis for the further development of a society.

The weathering processes can be traced back to natural causes (Figure 1.3.1) and to anthropogenic influences.

Figure 1.3.1: SEM image of swelling clay minerals as an example of natural causes of weathering (left) and scaling on a sandstone containing clay minerals as a typical result of this swelling (right).

https://doi.org/10.1515/9783110733143-003

It is undisputed today that the weathering of buildings is greatly accelerated by the immission of pollutants resulting from human production. Only the extent of this contribution to the overall damage picture is judged differently.

Natural and anthropogenically caused damage processes can be traced back to physical, chemical and biological attacks. However, the boundaries between the causes of damage cannot be sharply drawn. In many processes, physical, chemical and biological processes flow into each other, concerted mechanisms take place and there are close interrelations between the processes.

1.3.2 Damage due to moisture

The presence of water is a basic prerequisite for almost all physical, chemical and biological destruction processes. Thus, pollutants dissolved with the water – especially salts harmful to the building – are carried into mineral building materials. The moisture content of a building material usually increases its pollutant absorption. Acidic hydrolytic processes not only attack carbonate minerals but also dissolve aluminosilicates (Figure 1.3.2).

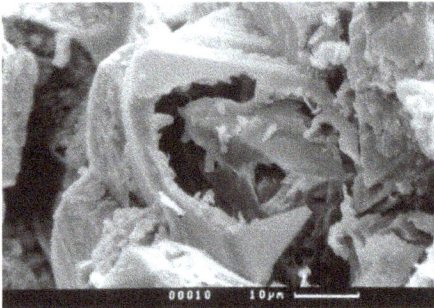

Figure 1.3.2: SEM image shows a feldspar with a hole due to hydrolytic weathering. The appearance is reminiscent of a carious molar tooth [1].

The phenomenon of façade greening, which is intensively discussed today, is also causally related to the moisture content of the building materials, as moisture is an essential prerequisite for biological colonization.

In our latitudes, the statement "water is to blame for everything" therefore applies to most of the façade damage. It should be noted that not only extreme, but also "normal", medium moisture loads lead to damage [2].

The most important water absorption mechanism of a building material is its "capillary water absorption", which describes its absorbency. If a building material is wetted with water (façade: e.g. by driving rain; unsealed area in contact with the ground: e.g. by soil moisture), the moisture is transported into the building material through its capillary pores.

1.3.3 Conservation

In the field of building conservation, conservation means taking measures to reduce the weathering mechanisms or their speed.

Without doubt, the conservation of natural stone is one of the most complex tasks in the context of building conservation. This is mainly due to the wide range of rock varieties used as natural stone.

The (protective) measures to be applied in the context of conservation can be divided into:
– structural or
– chemical processes

whereby these in turn are both
– preventive as well as of
– restorative nature.

Structural protective measures are usually aimed at keeping water away from objects to be protected through structural measures. This can be ensured, for example, by projecting roofs, curtain walls, the insertion of horizontal barrier layers or also (smaller dimensioned objects, such as sculptures) by encapsulating free-standing sculptures.

Chemical processes offer a variety of possibilities. Thus,
– water-repellent treatment of the surfaces to be protected
– the hygric swelling of rock varieties containing clay minerals that can be sustainably reduced or
– the binder matrices destroyed by weathering can be rebuilt.

1.3.4 Organosilicon active substances in building protection

With a share of more than approx. 27%, silicon is the second most abundant element in the part of the earth's crust accessible to us after oxygen (approx. 50%). However, due to its great oxophilicity, silicon is never found in elemental form. Instead, it is found in the form of silicon dioxide (SiO_2, e.g. quartz, christobalite) or in compounds of silicic acid (e.g. in the form of silicates).

The element silicon was discovered comparatively late (1824) by Berzelius. The first organosilicon compound was produced by Kipping in 1901. He also introduced the name "silicon" as an abbreviation for the "silicon ketone". Today, silicones are understood to be polysiloxanes with an organic residue.

Until the middle of the twentieth century, the use of materials containing silicon was limited to oxide ceramics, glass, and mineral building materials. This changed after Müller and Rochow simultaneously and independently developed the basic technical process for producing organohalosilanes in 1940. For this process, highly pure sand (SiO_2) is first reduced with carbon (C) to elemental silicon:

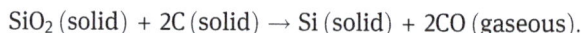

$$SiO_2 \,(\text{solid}) + 2C \,(\text{solid}) \rightarrow Si \,(\text{solid}) + 2CO \,(\text{gaseous}).$$

The silicon is then finely ground and reacted in the presence of catalysts with alkyl halides, usually methyl chloride (CH_3Cl), which has been prepared in advance from a common salt (NaCl) and natural gas (CH_4). The resulting mixture of crude silanes is separated into the individual products by means of a complex distillation process (Table 1.3.1). Of the large number of possible organohalosilanes, the methylchlorosilanes are the most economically important ones, with a production volume of more than one million tons per year as building blocks for silicon-organic compounds.

Table 1.3.1: Typical composition of the raw mixture in silane production.

Name	Abbreviation	Chemical formula	Boiling temperature	Proportion (%) in silane mixture
Dichlorodimethylsilane	Me_2SiCl_2	$(CH_3)_2SiCl_2$	70 °C	70–90
Trichloromethylsilane	$MeSiCl_3$	$(CH_3)SiCl_3$	66 °C	5–15
Chlorotrimethylsilane	Me_3SiCl	$(CH_3)_3SiCl$	57 °C	2–4
Dichloromethylsilane	$MeHSiCl_2$	$(CH_3)HSiCl_2$	41 °C	1–4
Chlorodimethylsilane	Me_2HSiCl	$(CH_3)_2HSiCl$	35 °C	< 1

The number of possible compounds increases greatly if other organic residues are attached to the silicon atom instead of the methyl group. Of the compounds that can be produced in this way, alkyltrichlorosilane ($RSiCl_3$) is of particular importance, as it is the basis of the substance classes relevant for building protection. For example, the silicones produced from it are used as injection agents in horizontal barriers that are "installed" in the event of rising damp. Longer-chain silicone resins are components of silicone resin paints. Silanes and the chemically very similar oligomeric siloxanes (made of 2–4 siloxane units) derived from them are used as water repellents (Figure 1.3.3).

1.3.5 The water-repellent adjustment of facade surfaces

The water-repellent treatment of façade surfaces is one of the oldest methods of extending the service life of building materials. While in ancient times mainly oils (e.g. linseed oil, poppy seed oil), waxes and paraffins were used [2], the classic

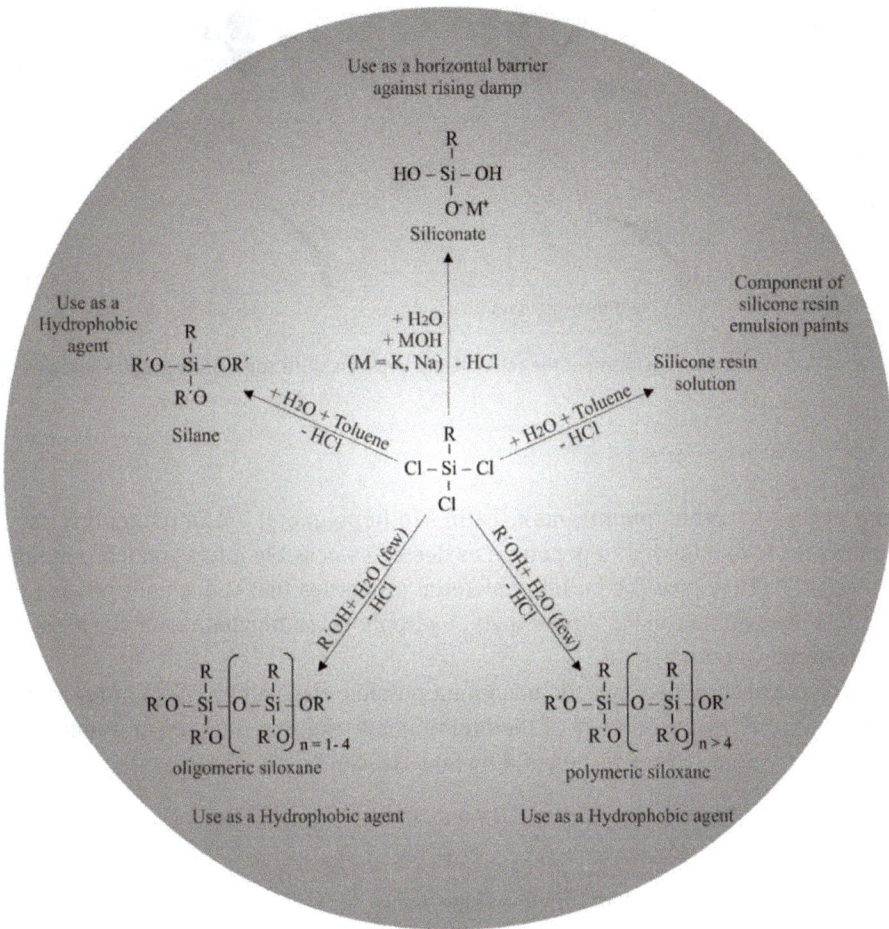

Figure 1.3.3: Alkyltrichlorosilane (RSiCl$_3$; R = organic residue) as a basis for organosilicon products relevant in building protection.

water repellency of façade surfaces has been realized since the mid-1960s with the help of silane and/or siloxane compounds. These silicon-organic substances have two different acting molecular units ("functional groups") at their central silicon atom. The alkoxy group splits off from the central silicon atom when it encounters moisture and enables both the reaction of the organosilicon molecules with each other to form corresponding larger units, as well as the chemical bonding of the remaining molecular residue to the quartzitic components of a mineral building material.

The alkyl group consists of a non-polar hydrocarbon chain that acts like an umbrella and is responsible for the water-repellent effect (Figure 1.3.4).

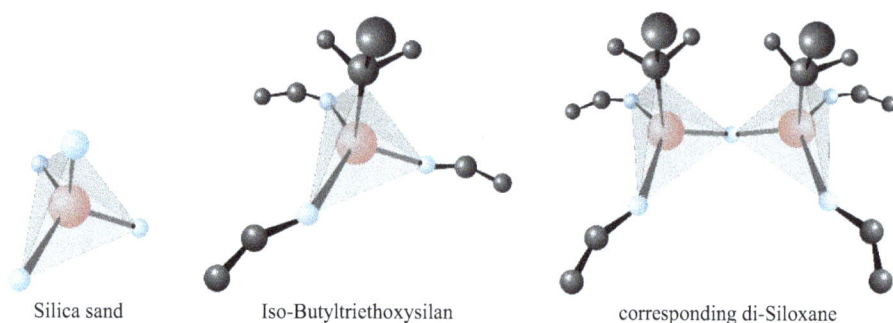

| Silica sand | Iso-Butyltriethoxysilan | corresponding di-Siloxane |

Figure 1.3.4: Model representation of the relationship between silica sand, silanes and siloxanes.

1.3.6 Effect of hydrophobic impregnations

Impregnating agents (*impraegnare*, Latin = to impregnate) are impregnating substances that are intended to penetrate as deeply as possible into a treated building material [3]. They give the building material properties that it does not possess or has lost through its use [4]. Accordingly, hydrophobic impregnations make a building material water-repellent.

The connection between the maximum possible rise height (h_{max}) of the water resulting from the absorbency of the building material and the type and nature of the capillary pore system of the building material can be calculated as follows:

$$h_{max} = \frac{2 \times \sigma \times \cos\theta}{r \times \rho \times g} \,[m]; \text{ with}$$

σ: surface tension of the water [J/m^2 = (kgm^2)/s^2],
θ: angle between water and capillary wall,
r: radius of the capillary [m],
ρ: density of the water [kg/m^3],
g: gravitational constant [9.81 m/s^2].

Wettable, hydrophilic building materials have contact angles <90° between water and surface. Accordingly, there are positive values for cos, which is equivalent to positive values for the maximum possible rise height. The water wets the building material and penetrates its capillary pores.

If the same building material is treated with a water-repellent (hydrophobic) finish, the water contracts to form drops on the building material due to the change in surface tension. The contact angle between water and surface changes to values >90°, resulting in negative values for $\cos\theta$ (Figure 1.3.5).

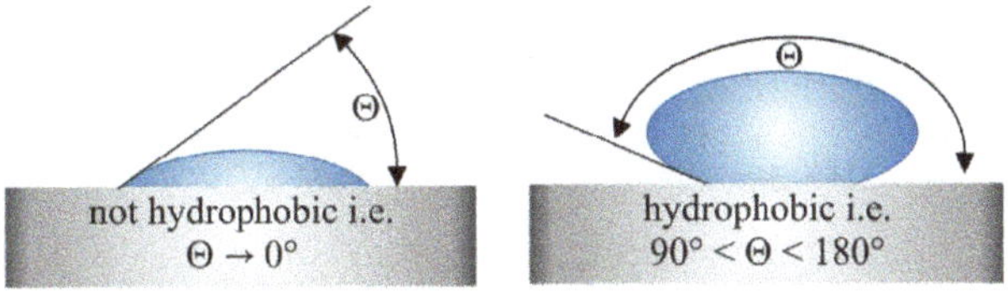

Figure 1.3.5: Change of the edge angle due to hydrophobization. Left: Small contact angle, good wetting: the water drop penetrates the substrate. Right: Large contact angle, poor wetting: the water drop lies on the surface.

The negative values for cosθ result in a negative rise height: Instead of spreading out on the surface, wetting it and penetrating the pore space, the water is pressed out of the capillary pores. This negative rise height in the capillaries is called "capillary depression". Capillary depression is often associated with a beading effect. However, contrary to what is often assumed, this is not a quality criterion for the effectiveness of hydrophobic impregnations.

Modern hydrophobic impregnations only accumulate on the surfaces of the capillary pores and only influence the capillary absorbency of the building material. Depending on the substrate and the impregnating agent used, the absorption capacity for liquid water can be reduced to almost "zero" (Figure 1.3.6).

Figure 1.3.6: Model representation (left) and photographic documentation (right) of the effect of a hydrophobic impregnation.

Since the impregnating agent only attaches to the pore walls in the form of a quasi-monomolecular film, the free cross section of the pores is hardly narrowed. Thus, the water vapor diffusion capacity (colloquially: "breathing capacity") of the building material remains almost unchanged: Water vapor can still migrate through the open cross section.

1.3.7 On the reaction of the organosilicon agents

Most products used today as water repellents in building protection can be traced back to the basic structure of the alkylalkoxysilane (Figure 1.3.7). For toxicological reasons, almost exclusively the ethoxy residue (the alcohol group "OEt"; $OR = OC_2H_5$) is used as the alkoxy group ("–OR") in facade protection.

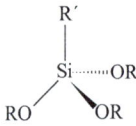

Figure 1.3.7: General structure of alkylalkoxysilanes.

Compared to all other substances still in use, these organosilicon compounds have the decisive advantage that they can establish covalent bonds to the mineral substrate by substituting their alkoxy groups (–OR) [5]. Via the alkyl radical R', which is usually a long non-polar and thus hydrophobic hydrocarbon chain (e.g. C_4H_9, C_8H_{17}), the targeted surface modification of the substrate then takes place.

Alkoxysilanes are not resistant in the presence of moisture and react in a hydrolytic polycondensation via silanols, oligosiloxanes and polysiloxanes to form silicone resins [6]. If two silanols meet, they can react to form the next larger unit, the siloxane, by splitting off water (Figure 1.3.8):

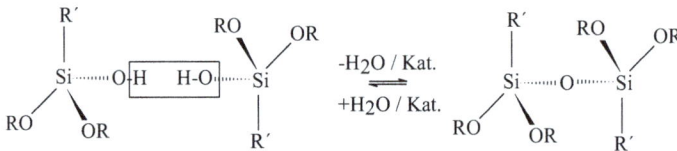

Figure 1.3.8: Schematic representation of the condensation of two alkylalkoxysilanols to form the alkylalkoxysiloxane. The siloxane owes its name to the Si–O–Si bond, also known as the siloxane bond.

Minerals, especially silicate substrates have a large number of surface OH groups (type: $HO-Si_{surf}$). Already in the sixties, the strong interaction between alkoxysilanes and these SiO_2-containing surfaces could be demonstrated [7].

The possible bonding processes for organosilicon stone protectors – starting from the alkyltrialkoxysilane – will be briefly described in the following. For steric reasons, simultaneous binding of the active ingredient via the three available alkoxy groups is not possible. Instead, the protective substance is fixed by di- or mono-functional binding to the substrate.

If the remaining alkoxy groups react intermolecularly with each other, a complex, covalently fixed network is formed. Figure 1.3.9 schematically shows the formation of the polycondensate.

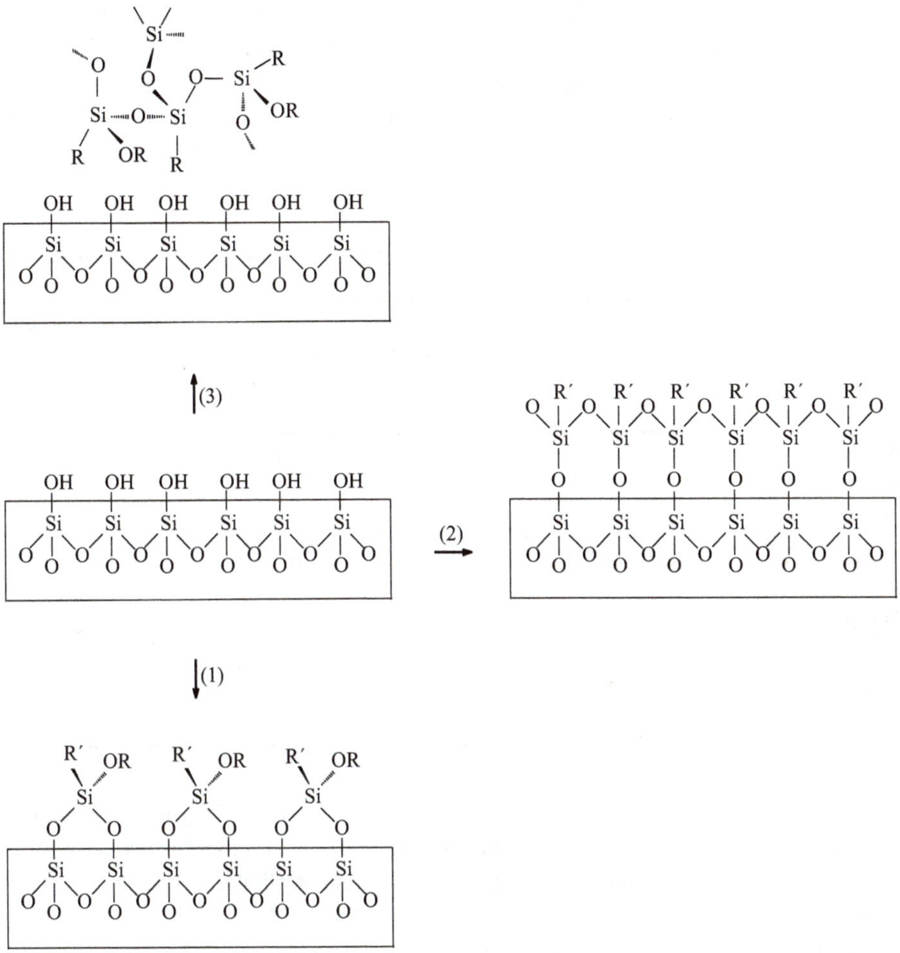

Figure 1.3.9: Schematic representation of possible forms of attachment of alkylalkoxysilanes or their condensates to silicate surfaces [1].

1.3.8 Active ingredient performance: not all impregnates are created equal

Depending on the type of organosilicon active ingredient, the organosilicon products for water repellency that are available on a large scale can be divided into
- type of organosilicon active ingredient,
- the content of the active ingredient,
- the type of solvent and
- the consistency of the product.

At present, mainly methyl- (CH_3-), isobutyl- $(i\text{-}C_4H_9-)$ and n- or iso-octyl- (n- or i-$C_8H_{17}-$) ethoxysilane are produced as raw materials on an industrial scale. Of the active substances mentioned, however, methyltriethoxysilane is not used as a monomer due to its extremely high volatility, but only in the form of a correspondingingly modified silicone resin [8]. Table 1.3.2 gives an overview of relevant technical properties of the active substances mentioned.

Table 1.3.2: Overview of silanes produced on an industrial scale that are used as water repellents.

Monomer name	Formula	Viscosity [mPa × s]	Molecular weight	Formula of the reacted molecule (theoret.)	Mass of the reacted molecule	Weight loss
Methyl-triethoxysilane "MTEO"	$(CH_3)Si(OC_2H_5)_3$	0.6	178	$(CH_3)Si(O)_{3/2}$	67	62%
Isobutyl-triethoxysilane "IBTEO"	$(C_4H_9)Si(OC_2H_5)_3$	1.0	220	$(C_4H_9)Si(O)_{3/2}$	109	50%
n- or iso-Octyltriethoxy-silane "OCTEO"	$(C_8H_{17})Si(OC_2H_5)_3$	2.1	276	$(C_8H_{17})Si(O)_{3/2}$	165	40%

The figure shows that the monomeric silanes are characterized by low viscosities, i.e. good flow properties. Therefore, they are particularly suitable for use on poorly absorbent substrates such as reinforced concrete. As the alkali resistance of the active ingredient increases with the chain length of the alkyl radical R', the iso-butyl and octyl silanes are mainly used in this field.

However, it can also be seen from Table 1.3.2 that the monomeric silanes suffer a relatively large loss of weight during the reaction at the substrate due to the splitting from their alcohol groups (from approx. 40% to approx. 62%). Since the silanes are also usually highly volatile, the weight loss in practice can be even higher than the theoretical values due to the evaporation of active ingredients, so that ultimately only a low active ingredient content remains after complete hydrolysis. To take this fact into account, the active ingredients are therefore mainly used in high concentrations or undiluted.

The loss in weight due to the splitting off of the alcohol groups is significantly lower for the higher molecular weight compounds (siloxanes, silicone resins) than for the comparable silanes. Part of the alcohol was already split off during the production of the siloxanes in the factory. At the same time, the higher-molecular compounds have no tendency to evaporate, so that high active ingredient content remains in the substrate after polycondensation. These active ingredients can therefore also be used in diluted form.

However, with increasing active ingredient size, the viscosity of the compounds usually increases (e.g. from silane to the siloxane derived from it). For this reason, the higher-molecular siloxanes are generally not used on concrete, but preferably on well-absorbent substrates.

1.3.9 Types of commercially available products

In the past, non-aqueous systems were common: the active ingredient was either used undiluted (usually silanes, "100% system") or in a form diluted with an organic solvent (usually siloxanes). For a long time, the provision of water-based, organosilicon products was considered hardly possible, because on the one hand the active ingredients – similar to oils – are not miscible with water and on the other hand they are subject to hydrolysis and react in contact with moisture (see above).

However, the problem of incorporating these moisture-sensitive active ingredients into a storage-stable aqueous product was finally solved by advances in emulsifier technology [9]. In this process, the silanes or siloxanes are finely dispersed in the water by intensive stirring. This emulsion is stabilized by emulsifiers that have oleophilic ("oil-loving") properties on one side of their molecule and hydrophilic ("water-loving") properties on the other. Due to their amphiphilic character, the oleophilic part of the emulsifier aligns itself towards the organosilicon active ingredients. The hydrophilic part stabilizes the droplets ("micelles"; Figure 1.3.10) created by stirring.

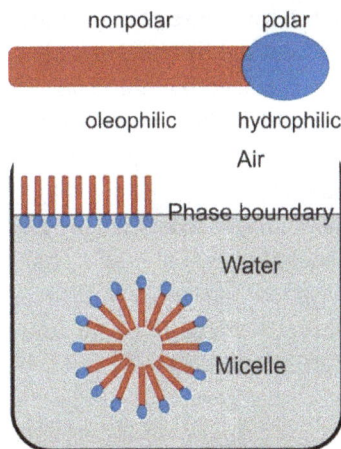

Figure 1.3.10: Model representation of the structure of surfactants and emulsifiers (left), their orientation at the phase boundary water-air (top right) and in water (bottom right).

Emulsions generally show lower penetration depths than comparable non-aqueous systems. On the one hand, this is due to the dimension of the micelles, which are naturally much larger and thus much more viscous than the individual silanes or siloxanes from which they are formed.

On the other hand, the emulsion "breaks" on its way into the mineral substrate. The two phases that are then separated again (water and active ingredient) then move on separately. Due to the lack of miscibility of the two phases, however, water-filled pores are then no longer accessible to the active ingredient [10].

Regardless of the solvent chosen, only liquid products were available until about 20 years ago. These still make up the majority of the products used today. In the meantime, however, highly viscous, aqueous systems (e.g. impregnating creams) are also available, the consistency of which results from a high concentration of active ingredient (from approx. 40 % to 80 %) in the product. These creams show a very high stability on the surface of the building material after application, so that from the combination of the two factors

- high concentration of active ingredient and
- long residence time of the active ingredient on the surface of the building material

comparatively high penetration depths can result.

1.3.10 Stone strengthening

The weathering mechanisms mentioned at the beginning ultimately lead to a loss of strength due to a weakening or cancellation of the grain contact forces. This can occur close to the surface and then leads to sanding "grain by grain". If the loss of strength occurs in a deeper zone, it can cause the detachment of large shells.

Stone strengthening is a conservation measure in the true sense of the word: it is intended to compensate for the loss of grain bonding caused by weathering [11]. Consolidation measures are required to strengthen the damaged building material matrix, i.e. one whose mechanical stability has been impaired. These measures produce a synthetic binder through chemical reactions in the cavity structure, which takes over the function of the damaged natural matrix.

The aim of strengthening is to restore the original homogeneous strength profile (uniform strength level) through a targeted supply of binder. Under no circumstances should the strength of the outer zone increase too much as a result of a consolidation measure, as a disproportionately or abruptly increasing E-modulus leads to strong stresses at the treated/untreated borderline and thus to consequential damage [12] (see Figure 1.3.11, *vide infra*).

In order to achieve optimal strengthening, that is, an increase in strength adapted to the respective rock and weathering condition, the individual weathering areas should therefore only be supplied with the amount of stone strengthener they need to achieve the original strength.

An essential prerequisite for optimal strengthening is the complete impregnation of the weathered stone zone with stone strengthener up to the unweathered core. These are applied to the building material by flooding, dipping and/or compressing.

In the flooding method, smaller areas (possibly stone by stone) are always treated wet in wet until the applied material is no longer absorbed. The compress method is often used on poorly absorbent substrates. In this process, a suitable compress material is fixed to the building material surface and soaked in such a way that the stone strengthener is available to the substrate to be treated for absorption over a long period of time.

1.3.11 Stone strengthening with water glass solutions

Water glass is the collective name for (vitreously solidified) melts of alkali silicates of different compositions as well as their solutions (general formula: $M_2O \cdot nSiO_2$). Water glasses are produced by melting together quartz sand with alkali carbonates according to

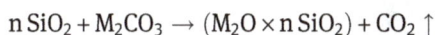

$$n\,SiO_2 + M_2CO_3 \rightarrow (M_2O \times n\,SiO_2) + CO_2 \uparrow$$

The corresponding water glass solutions are formed during solvolysis of the solidified melt with hot water under pressure [13]. With the CO_2 in the air, the solutions react accordingly

$$(M_2O \cdot n\,SiO_2)_{aq.} + CO_2 \rightarrow n\,SiO_2 + M_2CO_3$$

to form amorphous silica gel with a solidifying effect and the corresponding alkali metal carbonate. Since considerable amounts of water-soluble carbonate are precipitated with the water glass solutions and salt migration particularly is considered problematic, the use of such water glass solutions for stone strengthening has now been largely discontinued. Another disadvantage of water glass is the low penetration depths that can be achieved with them.

Potential problems associated with this low penetration depth are documented in the following figures: Figure 1.3.11 shows a detailed image of a weathered surface of a Cotta sandstone installed at the colonnade at the new Palais in Potsdam-Sanssouci. In the corresponding SEM image of the surface zone (Figure 1.3.12), a compacted surface can be seen in the right part of the image, and a clearly fissured, friable zone in the left part. It is assumed that the surface compaction is the result of a consolidation measure carried out in earlier times with a water glasses preparation. In the end, this compaction led to a peeling off the over-consolidated zone that had been saturated by the water glass.

Figure 1.3.11: Detailed image of a weathered surface of a Cotta sandstone.

Figure 1.3.12: Corresponding SEM image of the surface zone.

1.3.12 The silicic acid ester process

The collective name for the various esters of silicic acid (general formula: $SiO_2 \cdot nH_2O$) is "silicic acid ester" (general formula: $Si(OR)_4$, with R = organic residue). Silicic acid esters are the active ingredient and eponym ("silicic acid ester process") of today's most important and most common process for stone conservation. It is a consistent further development of the application of water glass and was first proposed in 1861 by August Wilhelm v. Hoffmann for the conservation of the "Houses of Parliament"/London [14].

The effect of these stone strengtheners is based on the reaction of the silicic acid ester ($Si(OEt)_4$) with water (H_2O) to form a high-molecular network of strengthening silica gel ($SiO_{2,aq}$):

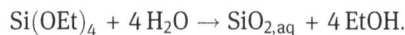

$$Si(OEt)_4 + 4\,H_2O \rightarrow SiO_{2,aq} + 4\,EtOH.$$

Silicic acid esters are produced by reacting silicon tetrachloride ($SiCl_4$) with alcohols. In the field of stone preservatives, ethanol is almost exclusively used as alcohol.

During this reaction of the monomeric $SiCl_4$, the monomeric silicic acid ester (also called tetraethoxysilane or tetraethyl orthosilicate; abbreviated form "TEOS") is formed under exclusion of moisture:

$$SiCl_4 + 4\,ROH \rightarrow Si(OR)_4 + 4\,HCl;\; \text{building protection:}\; R = Et$$

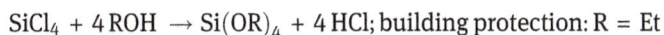

In the presence of moisture, the silicic acid ester first hydrolyzes – corresponding to the reaction of the alkylalkoxysilane (see above) – releasing ethanol.

After hydrolysis, the ester silanols $(RO)_nSi\text{-}OH_{4-n}$ ($n = 0$ to 3) condense with each other under water splitting to form larger units:

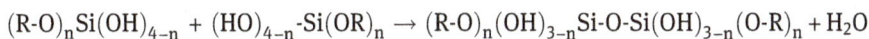

$$(R\text{-}O)_nSi(OH)_{4-n} + (HO)_{4-n}\text{-}Si(OR)_n \rightarrow (R\text{-}O)_n(OH)_{3-n}Si\text{-}O\text{-}Si(OH)_{3-n}(O\text{-}R)_n + H_2O$$

The water released in this reaction intervenes in the hydrolysis/condensation of further silicic acid ester molecules. The resulting larger molecules will (depending on the number of Si–O–Si linkages present) be called oligomeric (molar mass approx. 600–1000) or polymeric silicic acid esters (molar mass 5000). The formal end product of the hydrolytic polycondensation is a high-molecular network, the amorphous, polymeric, glass-like, and "rock-hard" silica gel (SiO_2, aq).

The following figure shows a model of the relationship between crystalline quartz sand (SiO_2), silicic acid ethyl ester ($Si(OC_2H_5)_4$) as the basic building block of stone strengtheners and amorphous silica gel ("silica gel"; SiO_2,aq; Figure 1.3.13).

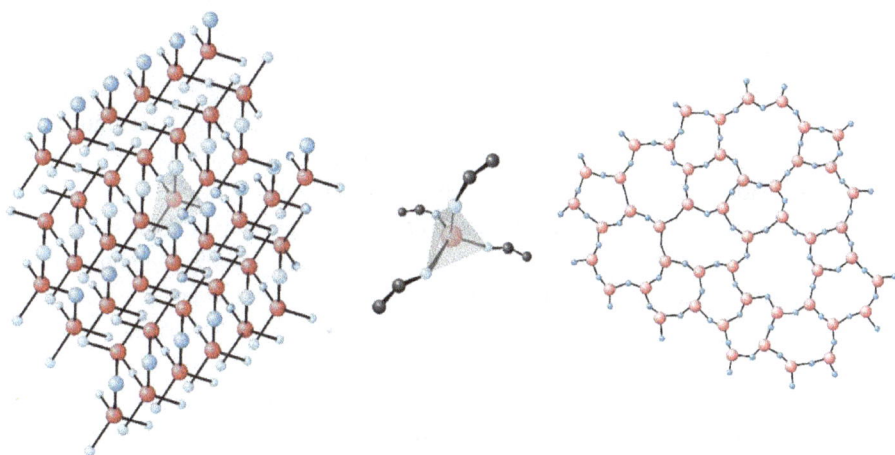

Figure 1.3.13: Model representation of the relationship between crystalline quartz sand (two-dimensional representation; left), the monomeric tetraethoxysilane (center) and amorphous silica gel (two-dimensional representation; right).

1.3.13 "Classic" stone strengtheners

The term "silicic acid ester" or "SAE" is generally used synonymously to refer to the product group "stone strengtheners". Therefore, many users believe that the silicic acid ester is a precisely defined substance (as suggested by the above formula). In fact, however, the active substance is produced in different molecular sizes during its manufacture. Therefore, all stone strengtheners available on the market contain both monomeric and oligomeric molecules. By varying the mixing ratios of small and large molecules in the finished product, the properties of the product, especially the "gel deposition rate" (amount of silica gel deposited in the pore structure of the substrate) can be varied significantly without adding "thinners".

Further variations are possible regarding
–   the reaction rate (type and quantity of catalyst) and
–   the addition of solvent(s)

Since the silicic acid ester itself is a liquid, it can in principle be used without the addition of solvents. However, the lowest possible gel deposition rate of solvent-free products is about 28%. To set an even lower rate, (a) suitable solvent(s) must therefore be added.

The hydrolytic polycondensation is followed by a gelation of the polycondensates. It can be assumed that the formal product of this reaction, the three-dimensionally completely linked SiO_2 network, is not built up. Instead, amorphous silica gels ($SiO_{2,\,aq}$) are formed, which contain ethoxy groups that have not yet been hydrolyzed and/or surface OH groups that have not yet been condensed.

In its fresh state, the silica gel contains in its pores a high proportion of small molecules from the combination of leaving groups (mainly water, but also ethanol, diethyl ether as products of the condensation reaction), which evaporate over time. The forces acting on the pores of the "rigid" gel during this process cause a "shrinkage" of the gel associated with the formation of cracks. This can be seen clearly if silicic acid esters can react with atmospheric moisture in a Petri dish. After completion of the reaction, a small-particle, hard silica gel with a clearly brittle character remains there (Figure 1.3.14).

1.3.14 Limits of "classic" stone strengtheners

In many cases, the requirements for successful stone setting can be met by the specific use of products with an adapted gel deposition rate. However, all "classical" stone strengtheners have a common characteristic property. The silica gel forming in the pore space, built up from the disordered, hydrous SiO_2 framework (see above), leads to silica gel plates of average size of approx. 10 µm. This gel plate size limits the application range of the classical silicic acid esters.

Figure 1.3.14: Silica gel remaining in the Petri dish.

For example, this brittleness causes a lack of lasting success on substrates with large cavity radii. The cause of these unfavorable cavity sizes can be, among other things, the natural pore radius distribution of the natural stone (e.g. tuff) or the formation of microcrack zones resulting from weathering (found e.g. in swellable natural stones such as reed sandstones). Modified products are therefore necessary for the solution of special problems [15].

1.3.15 The elastification of stone strengtheners

By incorporating so-called "soft segments", "elasticized" stone strengtheners could be developed. These produce a less brittle silica gel than the "classic" products, so that significantly larger gel particles are formed. This means that elasticized stone strengtheners can be used to bridge larger cracks than conventional products (Figure 1.3.15).

Figure 1.3.15: SEM image (900 × magnification) of an untreated glass frit (left) and of glass frits after treatment with a classic (center) and an elasticized (right) stone strengthener. The resulting classic gel visibly shows more and stronger cracks than the elasticized one. The latter also coats the substrate much better.

References

[1]   Boos M. Beiträge zur Natursteinkonservierung, Dissertation, Universität Münster, Münster, Germany, 1998.
[2]   Boué A. Ist Hydrophobierung heute verantwortbar? In: Wittmann FH, Gerdes A (Hrsg) 4. Internationales Kolloquium Werkstoffwissenschaften und Bauinstandsetzen, Aedificato Publishers, Freiburg, Fraunhofer IRB Verlag, Stuttgart, 1996, 245–56.
[3]   Boos M, Hilbert G, Niehoff MJ. Das Siliconharzemulsionssystem: wäßrige Alternativen zur wasserabweisenden Einstellung von Fassadenoberflächen. In: Wittmann FH, Gerdes A (Hrsg) 4. Internationales Kolloquium Werkstoffwissenschaften und Bauinstandsetzen, Aedificato Publishers, Freiburg, Fraunhofer IRB Verlag, Stuttgart, 1996, 273–82.
[4]   Weber H. Ermittlung von Kenndaten zur Beurteilung erschwerter Bedingungen. In: Remmers – Bauwerkserhaltung und Baudenkmalpflege, Bauwerksinstandsetzung unter erschwerten Bedingungen, Remmers Bauchemie GmbH, Löningen, Germany, 1996, 227–99.
[5]   Göbel T, Panster P, Albers P. Zur chemischen Fixierung von Organosilanen auf Beton, Ziegel und Natursteinen, Kongreßdokumentation zur bautec, Berlin, 09.–11.02.1994, 578–86.
[6]   Stoppek-Langner K, Grobe J. Spektroskopische Analyseverfahren – Vermittler zwischen Entwicklung und Anwendung von silicium-organischen Steinschutzstoffen. In: Snethlage R (Hrsg) Jahresberichte Steinzerfall – Steinkonservierung 1994–1996, Fraunhofer IRB Verlag, Stuttgart, 1998, 11–22.
[7]   Sterman K, Marsden S. Plast Technol, 1963, 9, 39.
[8]   Mayer H. Siliconharze – Chemismus und Eigenschaften. In: Schultze W (Hrsg) Wässrige Siliconharz-Beschichtungssysteme für Fassaden, expert-Verlag, Renningen-Malmsheim, 1997, 63–80.
[9]   Neuentwicklungen in der Bauchemie. ConChem-Journal, 1993, 2, 62.
[10] Gerdes A. Bauchemie im 21. Jahrhundert – Neue Strategien für die Entwicklung präventiver Oberflächenschutzmaßnahmen, WTA-Schriftenreihe Bauinstandsetzen und Bauphysik, 2006, 371–95.
[11] Sattler L, Snethlage R. Der Einsatz von Kieselsäureester zur Sandsteinverfestigung. In: Snethlage R (Hrsg) Denkmalpflege und Naturwissenschaft – Natursteinkonservierung I, Verlag Ernst & Sohn GmbH, Berlin, 1995, 89–104.
[12] Snethlage R. Leitfaden Steinkonservierung, Fraunhofer IRB-Verlag, Stuttgart, 1997.
[13] Holleman AF, Wiberg N. Lehrbuch der Anorganischen Chemie, 91.–100. Aufl., de Gruyter, Berlin, New York, 1985, p. 779.
[14] Weber H. Steinkonservierung in Deutschland, Bautenschutz und Bausanierung, 1987, 10(1), 6–8.
[15] Boos M, Grobe J, Meise-Gresch K, Tarlach S, Eckert H. Alterungsmerkmale unterschiedlicher Steinfestiger auf Basis von Kieselsäureestern (KSE). In: Wittmann FH, Gerdes A (Hrsg) Werkstoffwissenschaften und Bauinstandsetzung, Berichtsband zum vierten Internationalen Kolloquium, AEDIFATIO Publishers, Freiburg, 1996, 551–62.

1.4 Inorganic pigments

Sascha Broll, Rainer Pöttgen, Cristian A. Strassert

Inorganic pigments are constituted by small-sized particles that are practically insoluble. They are mainly used for their intrinsic color. Their main application areas are the automotive industry, industrial coatings, diverse construction materials, paints, besides many other areas involving coloring, anticorrosive or magnetic coatings. The broad field of inorganic pigments is well documented in the relevant basic literature [1–3]. Herein, we summarize the most important inorganic pigments while addressing their optical properties. Carbon black is discussed separately in Chapter 11.

The main advantage of inorganic pigments (as compared to organic dyes and pigments) is their high thermal stability as well as their ability to withstand external factors (weather and other atmospheric influences), electromagnetic radiation (mainly UV light) and corrosive or oxidizing influences. Also, most inorganic pigments are less expensive in their manufacturing procedure than organic chromophores.

Besides the intrinsic stability related to vast lattice enthalpies, the ability of inorganic pigments to withstand photodegradation is also related to ultrafast radiationless deactivation pathways from photoexcited metal-centered states acting as dissipative energy traps, due to geometric distortions associated to the weak ligand field splitting usually found in $3d$-block cations. Thus, their short-lived excited states prevent further photoinduced reactions and photosensitization processes (e.g., production of singlet molecular dioxygen), which would otherwise degrade the components encountered in coatings (such as fillers, other dyes and polymers). Hence, the absorbed photons are quickly converted into harmless phonons (i.e., heat). This holds true even for coordination compounds where the chromophore is an organic ligand. For instance, Cu(II) phthalocyanine is a widespread pigment owing its characteristic blue color to a ligand-centered optical transition from the doublet ground state to the excited doublet state (excited π–π* configuration). Upon intersystem crossing to the lowest metal-centered state (excited d–d* configuration), the complex quickly relaxes back to the ground state with the release of prompt heat. This is due to the fact that the low-lying metal-centered d–d* states of Cu(II) provide geometric distortion-related conical intersections with the ground state facilitating a fast radiationless relaxation. Rare-earth- and main-group-based pigments are less frequently used, but just as in the case of their transition metal counterparts, their high lattice enthalpies provide significant photochemical stabilities.

Inorganic pigments have already been used in prehistoric wall paintings. These were mainly drawn using brownish to black colors, since at that time only natural manganese oxides, charcoal and natural iron oxides were available. Since then, colored minerals and meanwhile also highly sophisticated synthetic inorganic pigments have become broadly available, making our daily life colorful – for wall paintings, pigments in polymers, construction materials, printing inks, cosmetics, etc..

https://doi.org/10.1515/9783110733143-004

Besides a sophisticated chemical synthesis, inorganic pigments deserve careful physicochemical characterization with respect to their structural properties. The most important parameters to be considered are the particle size and its distribution (polydispersity), particle shape (individual particles vs. agglomerates and aggregates), refractive index and absorption coefficient, photobleaching endurance, chemical and humidity resistance, compatibility with the binder and corrosion protection. We refer to a standard textbook on inorganic pigments [3] for this physicochemical part. The present chapter is focused on the solid-state chemistry and application of the basic inorganic pigments.

1.4.1 Titanium dioxide and other white pigments

Among all inorganic pigments, the white ones play the most important role and more than 60% of the global market of inorganic pigments concerns titanium dioxide. Although TiO_2 exhibits the three modifications brookite, anatase and rutile, only rutile and anatase find technical applications. For a detailed crystal-chemical discussion of the three modifications and their relative stabilities, we refer to basic bibliographic literature [4]. TiO_2 has a high refractive index (slightly higher for the rutile modification) and is a light-sensitive semiconductor.

The main natural raw materials for TiO_2 production are rutile, anatase and brookite, ilmenite ($FeTiO_3 \equiv FeO \cdot TiO_2$) and leucoxenes ($Fe_2O_3 \cdot TiO_2$, a weathered variant of ilmenite). Since all the naturally found raw materials contain impurity phases, the first step is a chemical digestion. The latter depends on the type of raw material. Ilmenite is treated in the so-called sulfate process with 80–98% sulfuric acid at temperatures between 170 and 220 °C:

$$FeTiO_3 + 2H_2SO_4 \rightarrow TiOSO_4 + FeSO_4 + 2H_2O$$

In this process, it is important to keep iron in the oxidation state +II. This is achieved by addition of adequate amounts of scrap iron, e.g., waste turnings. The digestion solution is finally cooled and $FeSO_4 \cdot 7H_2O$ can be separated by crystallization. With respect to waste reduction, iron(II) sulfate can either be used for the production of iron-oxide-based pigments (*vide infra*) or as a chromate substitute in cement production.

The hydrolysis of the remaining solution containing titanyl sulfate is the main step of the sulfate process. The reaction with water proceeds between 94 and 110 °C, leading to precipitation of hydrated titanium dioxide:

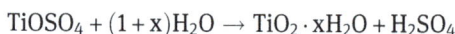

$$TiOSO_4 + (1+x)H_2O \rightarrow TiO_2 \cdot xH_2O + H_2SO_4$$

The released sulfuric acid can be recovered, concentrated and used for further processes. The next step is the dehydration of the hydrated titanium dioxide:

$$TiO_2 \cdot xH_2O \rightarrow TiO_2 + xH_2O$$

This is a calcination reaction that is carried out in a rotary kiln. The maximum temperature for this process ranges between 800 and 1100 °C and the residence time amounts up to 20 h, a heavily energy-consuming process related to the high vaporization enthalpy of the bound hydration water molecules. The operating temperature controls the anatase vs. rutile formation. As a general rule, the higher the temperature, the more favored is the relative amount of rutile in the final product. The calcination reaction in the kiln leads to TiO_2 clinkers, which are finally crushed by hammer or planetary mills. In summary, this is only a crude description of a very sophisticated industrial process. A well-documented and detailed description of the complete flow diagram of the sulfate process can be found in the literature [3].

The second important process is used for natural and synthetic rutile. It is a high-temperature chlorination process with carbon acting as a reducing agent. The products are titanium tetrachloride and carbon monoxide:

$$TiO_2 + 2Cl_2 + 2C \rightarrow TiCl_4 + 2CO$$

This reaction equation not only considers the involved titanium, but also the by-products in the raw material are chlorinated. The resulting $TiCl_4$ is then purified by distillation. Depending on the by-products, fractional distillation is eventually necessary, e.g., for a proper separation from $SiCl_4$.

In the last step, titanium tetrachloride is treated with oxygen. The temperature in the combustion reactors ranges between 900 and 1400 °C. The driving force for this transformation is the high lattice energy of titanium dioxide. The chlorine formed during combustion can be re-used for the aforementioned carbochlorination step.

$$TiCl_4 + O_2 \rightarrow TiO_2 + 2Cl_2$$

Both the sulfate and the chloride process can yield the anatase as well as the rutile modification. The outcome can be controlled by different parameters; for instance, addition of PCl_3 and $SiCl_4$ in the combustion process or doping of the titanium dioxide hydrate in the calcination process is favorable for the anatase formation. In general, higher process temperatures lead to rutile. Nevertheless, many batches of the two processes often contain both modifications. The anatase for pigment synthesis is exclusively formed by means of the sulfate process.

For their specific uses, the titanium dioxide particles require surface treatments, either with inorganic or organic substances. This so-called aftertreatment process improves the surface structure for subsequent applications, e.g., better dispersability and formation of stable emulsions, improved weather resistance, etc..

Besides the use as a white pigment, titanium dioxide finds application in many other fields: (i) as a catalyst for selective reduction and (photo)oxidation catalysis, (ii) as a precursor material for the production of complex colored inorganic pigments, (iii) for capacitors and piezoelectric applications in the field of electroceramics,

(iv) for the improvement of paper quality, (v) for all kinds of white plastics, (vi) as an UV filter pigment for cosmetic sunscreens, (vii) as a precursor for the production of lithium titanate in battery applications or (viii) for diverse technical glasses, especially self-cleaning surfaces where TiO_2 acts as the photooxidation catalyst involving atmospheric molecular dioxygen. Titanium dioxide is used as pigment in food industry under the label E171. The toxicity of titanium dioxide is under discussion, even though its bioavailability upon oral intake or dermal application is negligible [5].

Current research focusses on the photocatalytic properties of titanium dioxide [6–9], especially in the form of nanoparticles. This might open a completely new field for TiO_2 applications.

The second most important white pigment is zinc sulfide. The raw material is dissolved in sulfuric acid and addition of zinc powder reduces all metallic impurities that are more noble than zinc (electrochemical voltage series). A small amount of cobalt ions (0.02–0.5%) is then added to the filtered $ZnSO_4$ solution (stabilization against light) which is then mixed with a pure sodium sulfide solution.

$$ZnSO_4 + Na_2S \rightarrow ZnS + Na_2SO_4$$

The white precipitate is dried and annealed at 700 °C to obtain the optimal particle size. The optical properties of zinc sulfide are size-dependent. Particles with a diameter range between 290 and 300 nm show the best scattering power. The zinc sulfide obtained by this precipitation/calcination route has the wurtzite structure (h-ZnS). Hydrothermally synthesized ZnS crystallizes with the sphalerite type (c-ZnS). The cubic modification has a slightly better scattering power. Zinc sulfide is usually not used alone: the most common pigment is lithopone, a white pigment that consists of a mixture of ZnS and $BaSO_4$ (used as a filler to enhance the scattering efficiency).

The production of barium sulfate starts from the mineral baryte, which partly contains significant quantities of impurities. The raw material is thus reduced with carbon (1000–1300 °C in a rotary kiln) in order to produce water-soluble BaS. Barium carbonate plays a minor role as raw material. It can be treated with hydrogen sulfide at 1000 °C.

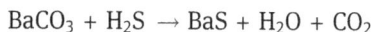

$$BaSO_4 + 4C \rightarrow BaS + 4CO$$

$$BaCO_3 + H_2S \rightarrow BaS + H_2O + CO_2$$

The filtered barium sulfide solution is then mixed with pure sulfuric acid, leading to a precipitate of white barium sulfate.

$$BaS + H_2SO_4 \rightarrow BaSO_4 + H_2S$$

Pure $BaSO_4$ finds application for stabilization of the electrodes of lead-acid batteries and as a reducing agent (reduction of whisker growth). Another important application is its use as an X-ray contrast agent. A saturated solution of barium sulfate has a Ba^{2+} concentration of only 0.005 g L^{-1} which is not toxic for the patient.

The vastest field of application of barium sulfate, however, concerns the production of lithopone. The $ZnS/BaSO_4$ material can be obtained by different coprecipitations which results in different $ZnS:BaSO_4$ ratios, for example as follows:

$$ZnSO_4 + BaS \rightarrow ZnS + BaSO_4 \ (1:1\ ratio)$$

$$ZnSO_4 + 3ZnCl_2 + 4BaS \rightarrow 4ZnS + BaSO_4 + 3BaCl_2 \ (4:1\ ratio)$$

In analogy to pure ZnS, the lithopone product is heated (650–700 °C) in a kiln for optimization of the particle size: ca. 300 nm for ZnS and ca. 1 µm for the filler $BaSO_4$. The $ZnS:BaSO_4$ ratios obtained in the above reactions are not the ones commercially used. Lithopones with ZnS contents of 30% and 60% are the most common types. They are mainly applied when high pigments concentrations are needed. A well-known application is opaque white in a color paint box. This is also the case on larger scale for primers. Lithopone is further used as white filler for plastics or putties.

Zinc oxide is one of the oldest white pigments. The raw materials for ZnO production are mainly ores and secondary sources like ashes, slags or residues from galvanization processes. All raw materials involve impurity phases. Thus, they are first converted to a raw mixture that contains around 60–75% ZnO. Since zinc forms no stable carbide, ZnO can easily be reduced with carbon (1000–1200 °C reaction temperature) leading to zinc vapor and carbon monoxide. Alternatively, it can be reduced with carbon monoxide.

$$ZnO + C \rightarrow Zn + CO \ \ or \ \ ZnO + CO \rightarrow Zn + CO_2$$

The purified zinc is then oxidized either with a mixture of carbon monoxide and oxygen, or with pure oxygen.

$$Zn + CO + O_2 \rightarrow ZnO + CO_2 \ \ or \ \ Zn + 1/2 O_2 \rightarrow ZnO$$

The resulting products are finally annealed at temperatures up to 1000 °C in order to improve the particle properties. Most ZnO particles have a surface aftertreatment, i.a. improving their dispersability and decreasing dust formation. An alternative wet-chemical route is possible: zinc carbonate can be precipitated from a pure zinc sulfate solution. The oxide is finally obtained by thermal decomposition of $ZnCO_3$.

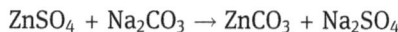

$$ZnSO_4 + Na_2CO_3 \rightarrow ZnCO_3 + Na_2SO_4$$

$$ZnCO_3 \rightarrow ZnO + CO_2$$

Besides titanium dioxide and lithopone, zinc oxide plays only a subordinary role in the field of white pigments for paints and coatings. An important application for ZnO is rubber industry (vulcanization accelerator) and its medical/cosmetic application. ZnO containing ointments and sprays are used due to their anti-inflammatory and skin-protecting effect related to the complexation and precipitation of exudates (including the inhibition of inflammatory responses and drying of wounds). Zinc oxide is also a natural UV absorbing material and used as an effective sun-blocker.

For the sake of completing the series of white pigments, we will herein shortly discuss basic lead carbonate ($PbCO_3 \cdot 2Pb(OH)_2$ or $2PbCO_3 \cdot Pb(OH)_2$), the so-called *white lead* and antimony(III) oxide (Sb_2O_3, *antimony white*). Due to their toxic character, these pigments find no longer application on a broad scale. They have been essentially replaced by titanium dioxide.

The synthesis of basic lead carbonate starts from a soluble lead salt, most suitably with lead acetate that is reacted with carbon dioxide under adequate conditions.

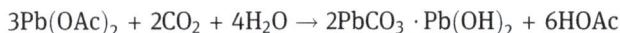

$$3Pb(OAc)_2 + 2CO_2 + 4H_2O \rightarrow 2PbCO_3 \cdot Pb(OH)_2 + 6HOAc$$

Antimony(III) oxide can be synthesized by oxidation of pure antimony powder, or by a direct combustion of the mineral stibnite at high temperature (850–1000 °C). In the latter process, the Sb_2O_3 product is recovered after vaporization.

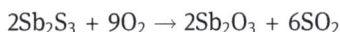

$$2Sb + 3/2O_2 \rightarrow 2Sb_2O_3$$

$$2Sb_2S_3 + 9O_2 \rightarrow 2Sb_2O_3 + 6SO_2$$

Besides their heavy metal-related toxicity, both *white lead* and *antimony white* have another severe disadvantage when used in paints: traces of gaseous hydrogen sulfide in ambient air (from fermentation or anaerobic decomposition processes) lead to the formation of PbS (brown) or Sb_2S_3 (brown) and thus to a creeping darkening of the paint.

1.4.2 Colored pigments

In the following subchapters, we will discuss the commercially available colored pigments. Carbon black is discussed in Chapter 11.1 along with the complete series of carbon-based functional solids. Natural colored pigments have already been used profusely in antique paintings [10], many of them dating back to prehistoric times. Besides the iron oxides discussed below, several mineral-based colored pigments have found applications, e.g., cinnabar (red, HgS), realgar (red, As_4S_4), orpigment (yellow, As_2S_3), azurite (blue, $2CuCO_3 \cdot Cu(OH)_2$) or malachite (green, $CuCO_3 \cdot Cu(OH)_2$). Although these pigments show excellent color properties, those containing harmful elements are no longer used for toxicological reasons. This is also the case for several synthetic pigments. Typical examples are Pb_3O_4 (orange, minium), $BaCrO_4$ (barium yellow), $Cu(AsO_3)_2$ (Scheele's green) or $Co_3(PO_4)_2$ (cobalt violet). A complete compilation of the many pigments is given in the literature [3].

In most of the cases involving relatively hard cations and anions, the absorption of light is related to localized yet parity-allowed optical transitions into excited electronic configurations. This includes changes in the occupation of metal-centered *d*-orbitals, so-called ligand-field states, which are usually observed for the lower oxidation states of the relevant cationic centers; hence, the strength of the ligand field determines the resulting color (e.g., for chromium and iron oxides). Alternatively, excitation into ligand-to-metal charge-transfer states dominates the optical properties when higher oxidation states of the metals are involved (e.g. for vanadates, chromates, molybdates, tungstates and permanganates). Charge-transfer states can be reached by absorption of light and usually display large oscillator strengths, whereas metal-centered states are only optically active if the chromophore has no inversion center (otherwise the optical transition is parity-forbidden, a.k.a. Laporte's rule, which does not apply for tetrahedral or distorted octahedral coordination environments). In fewer examples, the color is due to the particular band structure of the solid resulting from the extended overlap of both cationic and anionic orbitals, leading to semiconducting properties (e.g. II/VI and III/V semiconductors as well as other materials involving soft cations and anions from heavier elements such as Cd, Hg, Ga, In, P, As, Sb, S, Se or Te). For semiconductors, the (in)direct band-gaps are relevant.

1.4.2.1 Iron oxides

The family of iron oxides is by far the largest one within the group of colored pigments. This is due to the widespread availability of cheap raw materials as well as to their different colors. The most important pigments are α-FeOOH (goethite, iron oxide yellow), γ-FeOOH (lepidocrocite, yellow/orange), α-Fe$_2$O$_3$ (hematite, iron oxide red), γ-Fe$_2$O$_3$ (maghemite, light red/dark violet) and Fe$_3$O$_4$ (magnetite, black). Several of these pigments are available as natural materials; however, they contain impurities and frequently not the desired particle sizes. For these reasons, high-quality iron oxides are used. Iron sulfate is a suitable precursor that *inter alia* is a side product of the TiO$_2$ production (*vide supra*). After a first dehydration step, the monohydrate is oxidized in a rotary kiln and the intermediately formed iron(III) sulfate decomposes under release of sulfur trioxide (which might further decompose to the dioxide).

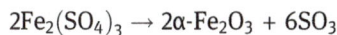

$$FeSO_4 \cdot 7\,H_2O \rightarrow FeSO_4 \cdot H_2O + 6H_2O$$

$$6FeSO_4 \cdot H_2O + 3/2O_2 \rightarrow \alpha\text{-}Fe_2O_3 + 2Fe_2(SO_4)_3 + 6H_2O$$

$$2Fe_2(SO_4)_3 \rightarrow 2\alpha\text{-}Fe_2O_3 + 6SO_3$$

An alternative is the precipitation process with the same educt. α-FeOOH can be calcinated to yield α-Fe$_2$O$_3$.

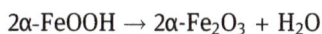

$$2FeSO_4 + 4NaOH + 1/2O_2 \rightarrow 2\alpha\text{-}FeOOH + 2Na_2SO_4 + H_2O$$

$$2\alpha\text{-}FeOOH \rightarrow 2\alpha\text{-}Fe_2O_3 + H_2O$$

Synthetic iron oxide pigments are also available from the Laux process (in fact, this process was originally developed for the production of aniline!). Nitrobenzene can be reduced with pure iron under water uptake, leading to aniline and the respective iron oxides. This process is mainly used for the obtention of Fe$_3$O$_4$ and FeOOH, but also iron oxide red is produced from pure Fe$_3$O$_4$ by combustion:

$$9Fe + 4C_6H_5NO_2 + 4H_2O \rightarrow 3Fe_3O_4 + 4C_6H_5NH_2 \text{ (black magnetite)}$$

$$2Fe + C_6H_5NO_2 + 2H_2O \rightarrow 2FeOOH + C_6H_5NH_2 \text{ (yellow goethite)}$$

$$2Fe_3O_4 + 1/2O_2 \rightarrow 3\alpha\text{-}Fe_2O_3 \text{ (red hematite)}$$

Besides paintings and fillers in plastics, the iron oxides are used in large scale as additives for concrete in paving stones and roof tiles and as yellow/red pigments in clinker bricks and roof pans. Fe_3O_4 is used as a toner pigment.

An important parameter for the iron oxide pigments is their particle shape and particle size, both depend on the production process. The particle size can be adjusted by precise milling processes. As an example, we will shortly discuss the influence of the particle size on the color appearance for α-Fe_2O_3: 0.001–0.05 µm is yellowish-red, 0.1–10 µm is yellow-red to violet and 10.0–100.0 µm is metallic brown to black.

Finally, we will briefly comment on black iron-based pigments. Carbon black (Chapter 11.1) is the most important black pigment, but also some iron-containing spinels have found applications. Typical cation combinations are Ni(II) + Fe(II, III) + Cr(III), Mn(II) + Fe(II, III) or Co(II) + Cr(III) + Fe(II).

1.4.2.2 Chromium(III) oxide

Cr_2O_3 is the most important green pigment. Trivalent chromium oxides exhibit only minor toxicity, due to their extremely low solubility related to their vast thermodynamic stability as solids. It crystallizes with the corundum-type structure with an extraordinary high lattice enthalpy. Hence, Cr_2O_3 has a high color fastness and acid resistance. This explains the hardness (Cr_2O_3 has the highest abrasion wear of all inorganic pigments) and its chemical inertness. The raw materials for Cr_2O_3 production are all Cr(VI) salts, which require careful handling, due to their intrinsic toxicity. Suitable synthesis routes involve the reduction of Cr(VI) with sulfur or carbon, leading directly to Cr_2O_3 and soluble sodium sulfate or the corresponding carbonate.

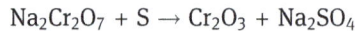

$$Na_2Cr_2O_7 + S \rightarrow Cr_2O_3 + Na_2SO_4$$

$$Na_2Cr_2O_7 + 2C \rightarrow Cr_2O_3 + Na_2CO_3 + CO$$

Wet chemical routes reduce the dichromate with a thiosulfate-containing solution. The resulting hydroxide is subsequently calcinated:

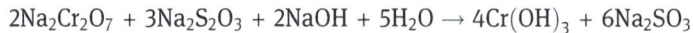

$$2Na_2Cr_2O_7 + 3Na_2S_2O_3 + 2NaOH + 5H_2O \rightarrow 4Cr(OH)_3 + 6Na_2SO_3$$

$$2Cr(OH)_3 \rightarrow Cr_2O_3 + 3H_2O$$

A very clean reaction is represented by the thermal decomposition of ammonium dichromate (the so-called volcano experiment). The decomposition starts at 200 °C and solely delivers gaseous by-products. The industrial process uses sodium dichromate and ammonium sulfate as precursors.

$$(NH_4)_2Cr_2O_7 \rightarrow Cr_2O_3 + N_2 + 4H_2O$$

$$Na_2Cr_2O_7 + (NH_4)_2SO_4 \rightarrow Cr_2O_3 + N_2 + Na_2SO_4 + 4H_2O$$

Besides the excellent mechanical properties, Cr_2O_3-based pigments exhibit strong light scattering ability and high absorption cross sections, leading to a very good hiding power. The green pigment is widely used as coating for steels. Well-known everyday examples are the green railway cars and green channel or railway bridges. Cr_2O_3 is also an excellent green pigment for plastics.

1.4.2.3 Blue pigments

Ultramarine blue and the cyanido complexes of iron (so-called iron blue) are the most relevant blue pigments. The natural mineral lapislazuli is similar to ultramarine blue. Historically, the pigment was manufactured as a ground powder obtained from lapislazuli, mainly by artists. Meanwhile, synthetic ultramarine blue is available. The chromogeneous groups in ultramarine blue are small polysulfide radical anions (clearly detectable by EPR), mainly $\cdot S_3^-$, which can be optically excited into energetically higher electronic configurations (i.e., localized excitons). The host structure stabilizing the anions in ultramarine blue is constituted by an alumosilicate with sodalithe-derived cages. The synthesis of ultramarine blue pigments is one of the most sophisticated processes in pigment production. The complete technical process is described in detail in the relevant literature [2]. The composition of synthetic ultramarine blue pigments can range from $Na_8[Al_6Si_6O_{24}] \cdot S_x$ (sodium-rich ultramarine) to $Na_{8-x}[Al_{6-x}Si_{6+x}O_{24}] \cdot S_x$ (silicon-rich ultramarine blue). The exact composition determines the color. Besides the well-known deep blue ultramarine blue pigments, green, red and violet ones can also be synthesized.

The relevant raw materials for ultramarine blue synthesis are kaolinite ($Al_4(OH)_8Si_4O_{10}$), which is first dehydrated to yield meta kaolinite ($Al_4(OH)_{8-2x}O_xSi_4O_{10}$) and later on annealed with sodium carbonate to form $Na_6[Al_6Si_6O_{24}]$. The sulfur source is provided by sodium polysulfide (NaS_x), which is produced from sodium carbonate and elementary sulfur under sulfur dioxide atmosphere. The subsequent reaction of $Na_6[Al_6Si_6O_{24}]$ with NaS_x leads to a colorless intermediate, the so-called pre-ultramarine blue, which is then slowly oxidized under formation of the $\cdot S_3^-$ radicals. Other possible radicals are $\cdot S_2^-$ and $\cdot S_4^-$.

The second group of deep blue pigments are hexacyanidoferrates(II). Their synthesis starts from yellow blood liquor salt ($K_4[Fe^{II}(CN)_6]$), which reacts with iron(II) sulfate to a white precipitate (*Berlin white*).

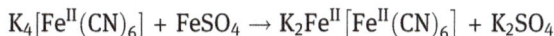

$$K_4[Fe^{II}(CN)_6] + FeSO_4 \rightarrow K_2Fe^{II}[Fe^{II}(CN)_6] + K_2SO_4$$

The white precipitate is subsequently oxidized (with hydrogen peroxide or sodium chlorate), yielding the insoluble yet dark-blue $Fe_4^{III}[Fe^{II}(CN)_6]_3$. The iron blue pigments have extremely strong color derived from the photoinduced charge-transfer from the softer C-coordinated Fe(II) centers to the harder N-coordinated Fe(III) cations (i.e., metal-to-metal charge transfer); however, their disadvantage is their low thermal stability related to the CN-ligands, which limits the application to <180 °C. Iron blue pigments are used for diverse printing applications.

1.4.2.4 Further pigments

The pigments discussed in Chapters 1.4.2.1–1.4.2.3 largely determine the pigment market. Nevertheless, there are many other interesting pigments that are used for special applications, and are sometimes called niche products. Some of them are briefly summarized in the present subchapter.

– The so-called mixed metal oxides are solid solutions, where different metal cations show mixed occupancies on Wyckoff sites of simple, stable structure types, mostly rutile and spinel (Figure 1.4.1). The color of the pigment can be adjusted by varying the metal cations. The most important rutile-based pigment is $(Ti, Cr, Sb)O_2$ (chrome antimony titanium buff rutile). Other examples are $(Ti, Ni, Sb)O_2$ (nickel rutile yellow) or $(Ti, Mn, Sb)O_2$ (manganese rutile brown). In the family of spinels, a well-known pigment is $CoAl_2O_4$ (cobalt blue) besides $Co(Al,Cr)_2O_4$ (cobalt chromium spinel, blue green), $(Co, Zn)(Ti, Ni)_2O_4$ (cobalt titanium nickel zinc spinel, green), $(Co, Li)(Ti, Li)_2O_4$ (cobalt titanium spinel, turquoise), $(Co, Fe)(Ni, Cr, Fe)_2O_4$ (iron cobalt chromium nickel spinel, black, for application in ceramics) or $CuCr_2O_4$ (copper chromium spinel, black). They can be produced by grinding and annealing or coprecipitation followed by thermal decomposition.

zinc tin titanium murataite	cobalt aluminate spinel	copper chromium spinel	antimony nickel titanium rutile	chrome antimony titanium buff rutile
cobalt chromium spinel	cobalt titanium nickel zinc spinel	cobalt titanium spinel	cobalt chromite blue green spinel	chrome tin orchid cassiterite

Figure 1.4.1: Selected colored pigments. The samples were provided by Broll-Buntpigmente GmbH & Co. KG. Fotos by Thomas Fickenscher.

- Lead chromate and lead molybdate constitute yellow, orange and red pigments. Besides pure yellow $PbCrO_4$ (chrome yellow), phases of the solid solution $Pb(Cr, S, Mo)O_4$, can be adjusted with respect to their color, namely molybdate orange and molybdate red. The three pigments are used in paints, coatings and plastics.

- Cadmium sulfide is an excellent semiconductor-based pigment (*mailbox yellow*). Also in this case, different solid solutions are known: (i) $Cd_{1-x}Zn_xS$ is greenish-yellow and $CdS_{1-x}Se_x$; depending on the selenium content, it shows orange to red colors. Due to the high costs and the heavy-metal-related toxicity of cadmium, these pigments are nowadays only rarely used.

- Bismuth vanadate, $BiVO_4$ is a yellow alternative to cadmium sulfide. It forms directly by a sintering reaction: $Bi_2O_3 + V_2O_5 \rightarrow 2BiVO_4$.

- Cerium sulfide pigments display orange to red colors. Pigments of the solid solution $Ce_{2-x}La_xS_3$ are orange, while pure Ce_2S_3 exhibits reddish color, which depends on the modification and the thermal treatment.

- Similar to the cerium sulfide pigments, oxonitride pigments have also been developed as substitutes for toxic cadmium-based pigments (yellow/orange/red pigments). Promising candidates are $LaTaON_2$, $CaTaO_2N$ and $SrTaO_2N$ [11]. Here, different solid solutions give access to color variations. The oxonitride pigments have excellent color strength and high thermal stability.

- Fe_2O_3, Fe_3O_4, CrO_2 and $BaFe_{12}O_{19}$ ($\equiv BaO \cdot 6Fe_2O_3$) are magnetic pigments. This class of materials is separately discussed in Chapter 9.8.

- Several oxidic pigments are used as anticorrosive coatings. Minium (Pb_3O_4, as a dispersion in linseed oil) was long time used for steel protection. It is no longer used nowadays, due to the heavy-metal-related toxicity of lead. A broadly used corrosion inhibitor is zinc phosphate tetrahydrate (phosphate conversion coating technique). Also, chromium and aluminum phosphate and chromates are used in such conversion processes [12].

- Metal effect pigments contain tiny metal flakes (aluminum, copper or brass phases in mm to μm size), which are oriented parallel to the coated surface upon application of a shear force. The flakes have high metallic luster allowing the light reflection. The metal effect pigments can be used with transparent binders or in combination with other colored pigments. Such colored metal effect coatings are well known from the automotive sector.

- Pearlescent pigments show multiple partial reflections of the incident light on platelet-like pigment particles at different depths of the coating. This effect is known from some clamshells, where aragonite platelets and proteins build a layer-like composite. Synthetic pearlescent pigments are represented, for example, by titanium dioxide-mica stacks or related core-shell particles. In this case, these pigments show interference colors and hence they are called interference pigments.

- Luminescent, fluorescent and phosphorescent pigments belong to the category of photofunctional materials. Many aspects of this substance class are discussed in Chapter 8.

1.4.3 Fillers

Fillers for pigments are inorganic solids that are almost insoluble in the application system. They are usually represented by white pigments with a refractive index below 1.7. Fillers are usually less expensive than the pigment itself and, for economic reasons, are used as so-called extenders. They increase the volume of the coating and furthermore lower the consumption of the binder. Depending on the pigment system and the application, the following aspects are important for the choice of the correct filler: particle size, particle shape, density, color and the interaction with the binder.

Most fillers are oxides and for economic reasons, many natural fillers are used, for example, quartz, calcite, chalk, baryte, calcium sulfate, kalin, talc, mica feldspars or nanoclays. Synthetic fillers have higher purities. Relevant compounds in this class are represented by silicon, aluminum and magnesium oxide, the alkaline earth metal carbonates, calcium and barium sulfate, as well as diverse silicates.

References

[1]   Buxbaum G, Pfaff G. (Eds.). Industrial Inorganic Pigments, 3rd edition, Wiley VCH, Weinheim, 2006.

[2]   Bertau M, Müller A, Fröhlich P, Katzberg M. Industrielle Anorganische Chemie, 4. Auflage, Wiley-VCH, Weinheim, Germany, 2013.

[3]   Pfaff G. Inorganic Pigments, De Gruyter, Berlin, 2017.

[4]   Jeitschko W, Pöttgen R, Hoffmann R-D. Structural Chemistry of Hard Materials. In: Riedel R. (Ed.) Ceramic Hard Materials, Wiley-VCH, Weinheim, 2000, 3–40.

[5]   Hartwig A. MAK Collect Occup Health Saf, 2019, 4, 857–69. 10.1002/3527600418. mb1346367d0067.

[6]   Rahimi N, Pax RA, Mac A, Gray E. Progr Solid State Chem, 2016, 44, 86–105.

[7]   Rahimi N, Pax RA, Mac A, Gray E. Progr Solid State Chem, 2019, 55, 1–19.

[8]   Haider AJ, Jameel ZN, Al-Hussaini IHM. Energy Procedia, 2019, 157, 17–29.

[9]   Verma R, Gangwar J, Srivastava AK. RSC Adv, 2017, 7, 44199–224.

[10] Coccato A, Moens L, Vandenabeele P. Herit Sci, 2017, 5, 12. 10.1186/s40494-017-0125-6.

[11] Jansen M, Letschert H-P. Nature, 2000, 404, 980–82.

[12] Sankara Narayavan TSM. Rev Adv Mater Sci, 2005, 9, 130–77.

1.5 Anodized aluminum and particle coatings

Thomas Jüstel, Konstantin Lider

The coating of a surface with an inorganic material is a commonly applied method to modify and adjust particular properties of a substrate. The hardening of drilling tools or the passivation of non-noble metals with a thin oxide layer may serve as illustrious examples.

1.5.1 Anodized aluminum

Anodized aluminum (German: Eloxal – **El**ectrolytic **ox**idation of **al**uminum) is aluminum, which is a poorly electronegative metal with a high affinity for oxygen [1–4]. If exposed to air, it is oxidized and thus coated by a thin, dense oxide layer. Under the influence of atmospheric humidity, a hydrated layer is additionally generated, which shifts the potential of its surface into a more noble range and ensures that the material has good corrosion resistance in humid air. A freshly polished aluminum surface is immediately covered with a natural thin oxide layer with a thickness of 0.005 to 0.015 μm if it is stored in the surrounding atmosphere, which protects the base material from further attacks. Once this natural protective layer is damaged, healing takes place immediately by reaction with atmospheric oxygen and formation of an oxide layer:

$$4\,Al + 3\,O_2 \rightarrow 2\,Al_2O_3$$

Aluminum is not only the third most abundant element in the earth's crust with 7.57 wt%, it is also the most abundant metal. Moreover, a continuous increase in global aluminum production is observed (Figure 1.5.1). About 100 years ago, it was considered as a rare element and was as expensive as gold, but nowadays it has become one of the most important construction metals for the global economy.

Aluminum is a material with many applications. It is a relatively soft but a tough metal. Its most important properties are probably its low specific weight of 2.7 g/cm^3, its high mechanical strength due to suitable alloying components and heat treatments, and its relatively high corrosion resistance compared to other metals.

In alloys with magnesium, silicon or other metals, strength levels are also achieved that are only slightly below those of steel. It is therefore used in applications where low mass is required, such as in the transportation sector to contribute to lower fuel consumption and, of course, in the aerospace and automotive industries. Good thermal and electrical conductivity, high reflectivity, high elasticity and thus the resulting low processing costs are additional characteristics of this metal.

https://doi.org/10.1515/9783110733143-005

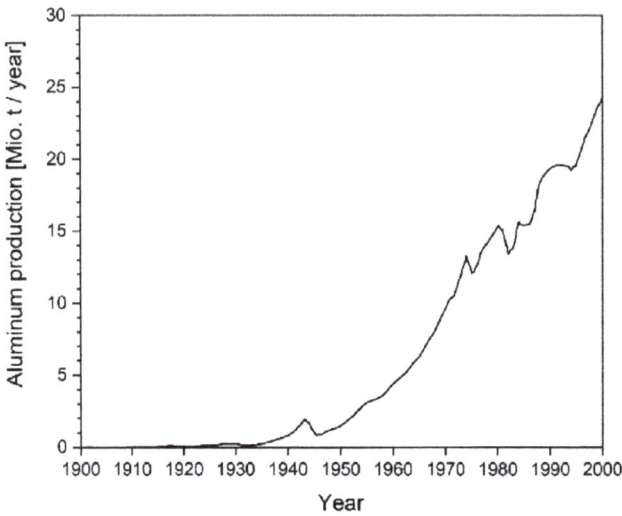

Figure 1.5.1: Development of the aluminum production between Y1900 and Y2000 [5].

Aluminum is magnetically neutral, i.e. just Pauli paramagnetic, and has non-toxic, colorless corrosion products, which makes it suitable for use in the chemical as well as the food industry. Further applications are also found in electrical engineering, electronics, packaging, containers, mechanical engineering, construction, optics, displays, light sources and many others.

The already mentioned properties of aluminum can be specifically changed and improved by surface treatments. Once aluminum is exposed to an external influence, such as mechanical stress, for example, deformation, breakage and attrition, its service life is diminished. Also chemical corrosion due to chemical or electrochemical reactions can reduce the product lifetime.

1.5.1.1 Anodic oxidation

As already mentioned above, the naturally forming oxide layer on the aluminum surface by exposure to air is responsible for the unexpectedly high corrosion resistance of the metal, despite its low electronegative nature. By chemical or electrolytic processes, synthetic oxide layers can be created and enforced, which are many times superior to the natural one.

Coatings produced by anodic reaction are superior to surface oxides produced by chemical reaction in terms of quality, mechanical stress, corrosion resistance and chemical resistance. Anodizing of aluminum is commonly referred to as "elox" and is a commonly known shortcut for electrolytic oxidation of aluminum.

In the electrolytic oxidation of aluminum or the anodizing process, unlike galvanic processes, no metallic layer is deposited on the base material. In this process, an electric current passes through an aluminum anode in a suitable electrolyte. The negatively charged anions migrate to the anode where they are discharged, converting the upper aluminum layer into an oxide or hydroxide (Al_2O_3 or $Al(OH)_3$). The surface structure and properties are thus reproduced one-to-one. The oxide layer diffuses 2/3 into the base material and grows 1/3 onto the base material. The growth is thus strongly dependent on the surface structure and the type of aluminum alloy, but these influence not only the growth of the oxide layer but also its optical appearance, inherent coloration and coloring capacity. Therefore, it is essential to carefully select the base material.

There are some theories and models on how aluminum is transformed into the oxide and thus one can describe the porous oxide layer and its growth. One of these theories was published in 1953 and is the layer structure according to the Keller-Hunter-Robinson model. Here it is described that the anodic oxide layer consists of an almost pore-free barrier layer with an upper porous layer on top. At the beginning of the electrolysis, an insulating barrier layer builds up, depending on the electrolyte type and parameters. This high-resistance layer heats up and the oxide layer is solved by the electrolyte, whereby the pores are then formed. The barrier layer in turn regenerates at the bottom of the pores and grows back.

Following the above mechanism, new pores are then formed and this process is repeated about one hundred times a minute. The current conduction in turn is mainly taken over by the electrolyte inside the pores, which heats up and promotes the regeneration at the bottom of the pores. The layer then continues to grow until equilibrium is reached between the dissolution of the oxide layer and its growth. The porous layer also has a very good adsorption capacity, which can be used to color the layer.

There are a variety of processes for anodizing a component. These vary depending on the intended use.

Electrolytes that are mainly used in practice are based on chromic acid, sulfuric acid or oxalic acid. The most commonly used anodizing processes today are those that use a sulfuric acid electrolyte. Important in the anodizing of aluminum is the electrolyte composition, bath temperature, current density and alloy composition of the aluminum used. All these factors have an influence on the subsequent growth of the layer. The oxidation reaction releases energy and takes place at the interface only.

By changing the operating parameters, the growth of the oxide layer can be modified and adjusted. The pore growth and their structure change and so do the properties of the oxide layer produced. Hardness and layer thickness as a function of the parameters can be seen in Table 1.5.1. It is also evident how the pore volume has an influence on the hardness of the oxide layer. The smaller the pore volume or diameter, the higher the hardness of the anodized layer.

1.5.1.2 Hard-anodizing of aluminum

The hard-anodizing process was developed in order to be able to produce particularly thick layers with a high hardness of an anodically produced oxide layer. As

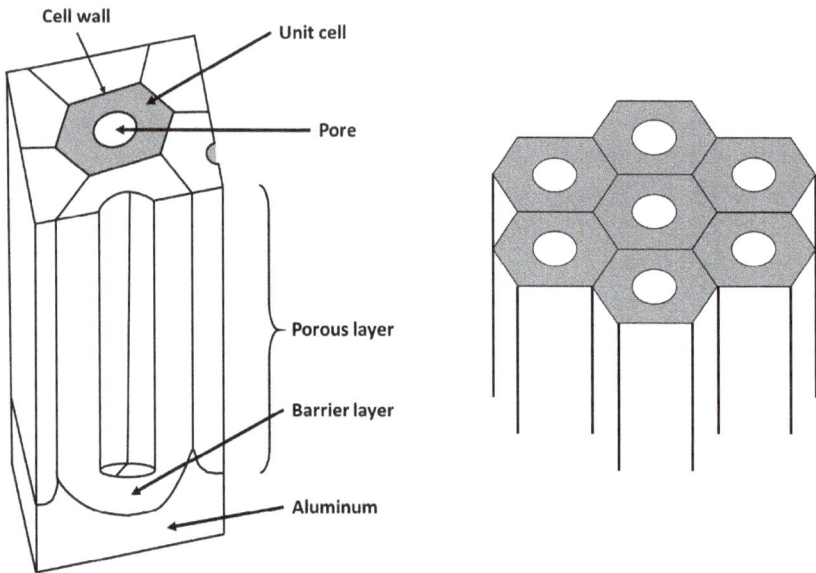

Figure 1.5.2: View of a cell of the alumina layer in section (a) and in top view (b) [6].

Table 1.5.1: Eloxation parameter and its influence.

Increase	Pore volume	Rigidity	Achievable layer thickness
Sulfuric acid concentration	↑ Increase	↓ Decrease	↑ Increase
Aluminum concentration	↓ Decrease	↑ Increase	↓ Decrease
Electrolysis temperature	↑ Increase	↓ Decrease	↓ Decrease
Anodic current density	↓ Decrease	↑ Increase	↑ Increase
Movement of the electrolyte	↓ Decrease	↑ Increase	↑ Increase

already reported, the maximum layer thickness of the oxide layer is reached as soon as the dissolving rate of the layer in the electrolyte corresponds to the formation rate; thus, when the mentioned equilibrium is established between layer dissolution and layer growth. The layer dissolution increases more and more with time and the increase of the layer. This occurs mainly in the pores (Figure 1.5.2). The layer growth, in turn, steadily decreases. This happens due to the increase of the ohmic resistance in the pores and as a result of secondary reactions taking place. Therefore, the layer thickness can only be increased if parameters such as temperature and acid concentration are reduced. In addition, the current density must be increased. As a result of the increase in current density and the decrease in acid concentration, the voltage generated increases, which leads to a local temperature rise in the anode region. For this reason, particularly good cooling of this electrolyte is an unavoidable requirement during the commissioning and during the production of thick oxide layers. The layer growth is based on the same principle as for the normal anodizing of aluminum. In hard anodizing, the electrolyte composition, current density, bath temperature and alloy composition play a very important role in producing the desired thickness of the oxide layers.

Decreasing the acid concentration and bath temperature, as well as increasing the current density in hard anodizing has an effect on the layer growth, when compared to anodizing. As already mentioned, in anodizing the oxide layer grows about 1/3 up and 2/3 into the base material. Hard anodizing, on the other hand, produces a layer that grows 50% into and 50% onto the base material.

However, the high achievable coating thicknesses, which in the hard-anodizing process can be from 25 to as much as 150 µm, are not solely responsible for the fact that hard anodizing produces a hard and wear-resistant oxide coating.

It is known that increasing the current not only increases the coating thickness, but also the hardness as a result of the reduction in porosity. Thus, the high layer thickness and the reduced pore volume in combination are responsible for the hardness. At the same time, the reduced porosity and the thicker barrier layer due to the higher current results in a higher corrosion resistance of the oxide layer. In addition, the different layer growth results in an inherent coloration, which makes the coloring of these layers more difficult or only possible to a limited extent. However, the resulting inherent colors, which are also dependent on the layer thickness and alloy composition, can be used according to individual preferences. For example, the resulting brown – gray shades have been used specifically for architectural purposes. As with the anodizing process itself, there are also different process techniques for hard anodizing. These differ in electrolyte composition and the use of direct current or alternating current superimposed direct current.

1.5.1.3 Coloring of anodic oxide layers

After anodic oxidation in sulfuric acid or oxalic acid, and to a certain extent in chromic acid, the aluminum obtains a porous oxide layer which has a strong adsorbing property. This porous oxide layer absorbs oils, greases or any coloring substances readily. There are several ways to color an anodically produced oxide layer.

In adsorptive dyeing, organic dyes are adsorbed into the pores of the oxide layer. Subsequently, densification is required to seal the pores and the color contained therein (Figure 1.5.3). Since the absorption capacity of the oxide layer is no longer given after densifying or sealing due to the closing of the pores, dyeing had to be carried out after anodizing before the densifying/sealing step.

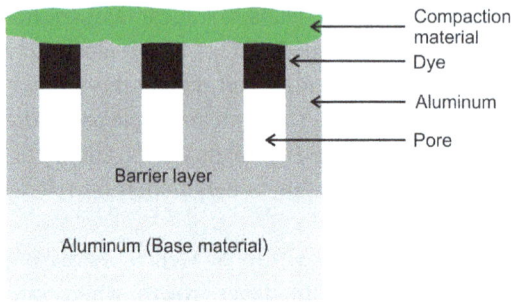

Figure 1.5.3: Schematic representation of a colored oxide layer [6].

Dyeing with inorganic dyes works in the same way in principle. However, no adsorption of organic dyes takes place here. Instead, inorganic heavy metal salts are precipitated and converted. These are then sedimented in the pores. A typical example of inorganic coloration is the gold coloration in iron ammonium oxalate. Here, Fe^{3+} is precipitated by hydrolysis as $Fe(OH)_3$ in the pores, resulting in the gold tones.

Another type of coloring is electrolytic coloring. Here, a previously anodically oxidized aluminum is treated electrolytically in a metal salt solution. First, the oxide layer is activated in a direct current phase, i.e. the barrier layer in the pore is penetrated. Then, through the use of alternating current, an electron pickup of the metal dissolved in the electrolyte takes place, for example, Sn, Zn or Cu. This results in precipitation of the metal salt/metal as a metal oxide in the eloxal pore. Depending on the amount and type of metal used, lighter to darker brown/grey shades up to black result.

1.5.1.4 Compacting/sealing

Sealing is the final step in the electrolytic oxidation of aluminum. During this process, the pores of the oxide layer are closed and the layer loses its absorption capacity. Proper sealing also ensures that the anodized work piece achieves its optimum corrosion and weather resistance. Some possible sealing processes (densification) are
- Hot water sealing
- Sealing with steam
- Cold impregnation (cold sealing)

1.5.1.5 Hot water sealing

Hot water sealing is the most commonly used form of sealing. It requires temperatures of >96 °C and dipping times of 3 to 4 min per μm layer thickness. When sealed in hot water, the anhydrous alumina absorbs water of crystallization, which is called hydration, forming boehmite-type crystals ($Al_2O_3 \times H_2O$ or $AlO(OH)$).

When anodic oxide layers are sealed in hot water, the first thing to hydrate occurs in the alumina of the barrier layer and the porous layer. Subsequently, the penetration of water into the depths of the pores is prevented and the alumina is hydrated only at the surface, which leads to the closing of the pores by volume increase.

1.5.1.6 Sealing with steam

Steam sealing behaves in the same way as hot water sealing. The advantages of sealing in steam are a faster effect, the nature and pH value of the water are secondary, as is the preservation of constant conditions. In addition, the risk of dye bleeding is minimized. Disadvantages are expensive and complex construction of the process cells and difficult operation.

1.5.1.7 Cold impregnation

Cold impregnation (cold sealing) is performed by using nickel fluoride salts and does not involve hydration. Nickel precipitates in the layer as nickel hydroxide and the fluoride reacts with the aluminum on the surface.

The advantage of this sealing process is the saving of energy, since no such high temperatures are required as with others. In addition, there is a shorter treatment time, no weathering deposits and a very good compaction quality.

Disadvantages of the cold sealing process are the susceptibility to green tinting of colorless layers, a cost-intensive disposal and that organic colorings can bleed more easily. In addition, this process may cause allergy concerns due to the use of nickel and is more complex to maintain than, for example, hot water sealing.

1.5.2 Particle coatings

The strategy of coating the surface of a substrate has been extended to microscale and even nanoscale particles for the optimization of their chemical and physical properties. Coated microscale or nanoscale particles are also known as core-shell particles, while the film or particle type shell may serve as a protection layer for the core or may act as an interface to the environment to introduce or support desired chemical or physical properties with regard to the application.

As far as protection of a surface is concerned, whether of a particle, a glass, metal or plastic substrate, solely rather few inorganic materials are nowadays in application (Table 1.5.2). The material selection is quite limited, since the coating material must be environmentally compliable and exhibit a high melting point, low solubility, chemical stability and a wide optical band gap, i.e. large optical window, in case of core particles with an optical function. Therefore, most coating materials are derived from binary or ternary oxides or from phosphates.

Table 1.5.2: Inorganic coating materials, some typical coating precursor compounds and their optical band gaps.

Coating material	Precursor compounds	Band gap [eV]	Reference
SiO_2	$Si(OR)_4 + NH_4OH$ $Na_2SiO_3 + HCl$	8.4	[7]
α-Al_2O_3	$Al(OR)_3 + NH_4OH$ $AlCl_3 + NH_4OH$	8.3	[8]
$Ca_2P_2O_7$	$Ca(NO_3)_2 \times 4H_2O + (NH_4)_2HPO_4$	8.3	[9]
$AlPO_4$	$Al(H_2PO_4)_3 + H_3PO_4$	8.1	[10]
MgO	$Mg(OR)_2 + NH_4OH$ $MgCl_2 + NaOH$	7.8	[7]
$MgAl_2O_3$	$Al(OR)_3 + Mg(NO_3)_2 + NH_4OH$	7.8	[11]
γ-Al_2O_3	$Al(OR)_3 + NH_4OH$ $AlCl_3 + NH_4OH$	7.2	[12]
Y_2O_3	$Y(OR)_3 + NH_4OH$ $YCl_3 + NH_4OH$	5.6	[7]
La_2O_3	$La(OR)_3 + NH_4OH$ $LaCl_3 + NH_4OH$	5.3	[13]
Ca/Sr-polyphosphate	$Na[PO_3]_n$ with n = 2–15 "Graham's salt"	5.3 (CaP_4O_{11})	[14]

Beyond the protection of surfaces, particle coatings are widely applied to improve physical/chemical properties such as the color coordinates of (luminescent) pigments, adhesion strength (Figure 1.5.4), specific targeting, (bio)chemical sensing, light in- or out-coupling or electron emission (R = −CH$_3$, −C$_2$H$_5$ or −C$_3$H$_7$).

Coatings can also serve for the suppression of non-radiative recombination processes caused by surface defects or high frequency phonons, which is an issue for the application of very small II/VI, e.g. (Zn,Cd)(S,SeTe), or III/V, e.g. (Al,Ga,In)(N,P, As), semiconductor nanoscale particles, also called quantum dots (QDs). Therefore, QDs are regularly coated by a shell and sometimes even by an onion shell structure in order to improve chemical and thermal stability as well as photoluminescence efficiency (Figure 1.5.5).

CdSe QDs coated by CdS and ZnS or by Al$_2$O$_3$ are applied in LED backlit LCD TVs, wherein the blue light from (In,Ga)N LEDs is converted into green and red light for the respective pixels.

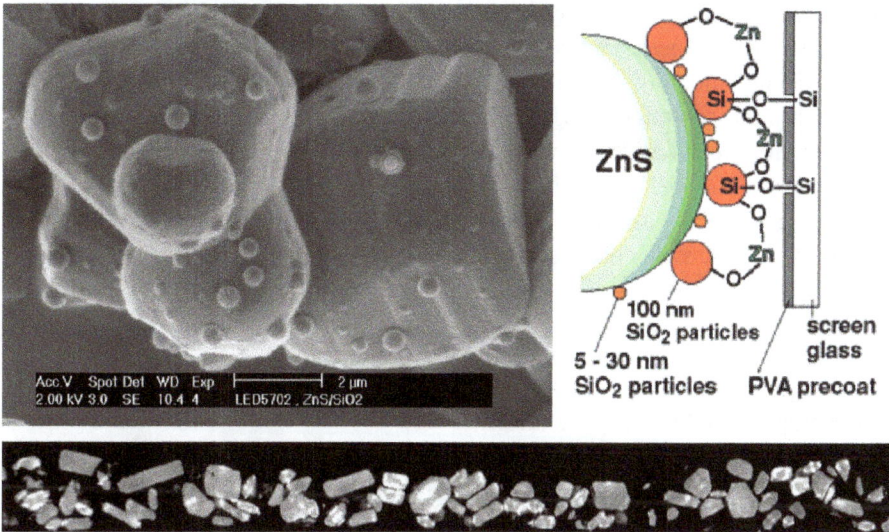

Figure 1.5.4: SEM image of ZnS:Cu,Al particles (green emitting CRT phosphor) coated by SiO₂ (top left), sketch to illustrate the adhesion of coated ZnS:Cu,Al onto screen glass of CRTs (top right) and microscopy image of a cross section of a ZnS:Cu,Al layer onto CRT screen glass (bottom).

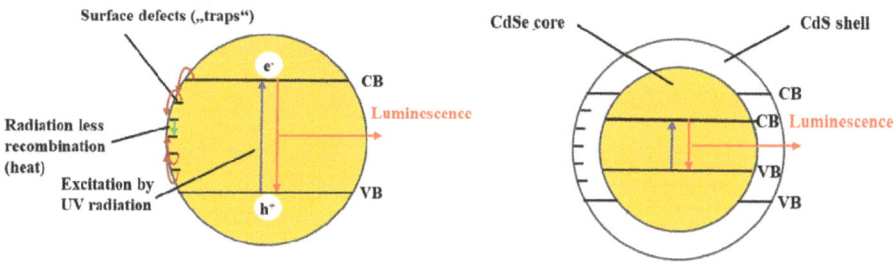

Figure 1.5.5: Sketch of a CdSe QD with illustration selected quenching processes at the surface (left) and a core-shell QD with reduced surface quenching due to an exciton reflective CdS shell (right).

1.5.2.1 Particle coatings to reduce device degradation

As mentioned above, inorganic particle coatings are widely applied on inorganic pigments and particularly on luminescent pigments since coatings reduce the degradation of industrially implemented phosphors. In most application areas phosphors operate under extreme conditions, for instance, in Hg or Xe discharge lamps, cathode-ray tubes, plasma display panels (PDPs) and phosphor converted light-emitting diodes (pcLEDs). Consequently, they undergo irreversible degradation, which determines in many cases the operational device lifetime. Moreover, the overall energy conversion efficiency of the aforementioned devices depends decisively on the absorption strength and quantum efficiency of the utilized phosphor, which is the vital point within the energy conversion chain, in which the electrical energy is converted into desired UV radiation or visible light (Table 1.5.3). Therefore, it is important to select suitable luminescent materials, which have to satisfy simultaneously a number of requirements specified for a particular device, its application and existing operating condition. It is well established that an investigation of the lifetime is of crucial importance before any industrial application can be seriously considered. As a result, an emphasis is especially put on the material stability and resistance against deterioration, which guarantee devices with a satisfactory operational lifetime.

Table 1.5.3: The overall wall plug efficiency and the most important energy loss factors of selected electrical light sources and information displays utilizing luminescent materials for image or light generation, modified after [15].

Device	Cathode-ray tube	Plasma display panel	Fluorescent lamp	Phosphor converted LED
Wall plug efficiency [%]	1–2	2–3	a) Linear 25–30 b) Compact 15	50–60
Major energy loss factors [%]	Shadow mask: 70–90 Phosphors: 80 Deflection joke: 50	Discharge: 90 Phosphors: 70	Discharge: a) Linear: 30 b) Compact: 40 Phosphors: 55	LED chip: 40 Phosphor: 25

Degradation of commercially utilized phosphors is a major concern and there are continuing research efforts to fully understand degradation mechanisms involved (particularly in luminescent materials) in order to prevent or minimize their effects on long-term stability. A wide variety of factors are responsible for the depreciation, including particle morphology and size distribution, surface defects, crystal structure, redox stability of activators, interaction with the chemical surrounding or with the discharge, reabsorption due to impurities and so on. Moreover, the degradation process is related not only to exposure to optical radiation or to the impact of highly energetic plasma components but also to thermal processes, which are unavoidable during fabrication of devices such as discharge lamps and emissive displays. Some industrially relevant examples of phosphor degradation mechanisms are the

- photooxidation of the activator of the blue emitting phosphor $BaMgAl_{10}O_{17}:Eu^{2+}$,
- Hg consumption in low-pressure Hg discharge lamps of $BaMgAl_{10}O_{17}:Eu^{2+}$, of the green emitting phosphor $Zn_2SiO_4:Mn^{2+}$, and of the UV-A emitting phosphors $BaSi_2O_5:Pb^{2+}$ or $Sr_2MgSi_2O_7:Pb^{2+}$,

- dissolution in aqueous suspension, that is, phosphor paints for lamps and display coatings, of silicate type phosphors such as $BaSi_2O_5:Pb^{2+}$, $Sr_2MgSi_2O_7:Pb^{2+}$, $Zn_2SiO_4:Mn^{2+}$ or $(Ca,Sr,Ba)_2SiO_4:Eu^{2+}$,
- and the hydrolysis by air moisture of LED phosphors such as $(Ca,Sr)S:Eu^{2+}$.

In order to prevent or at least to reduce irreversible degradation processes during device manufacturing and/or operation, a particle coating technique or other surface modification methods are frequently used.

A convincing example was the extension of the application of the blue emitting phosphor $BaMgAl_{10}O_{17}:Eu^{2+}$, which is in use in fluorescent lamps since the early 1970s, towards plasma TVs. This was solely possible by its stabilization by a particle coating, which is transparent for the 172 nm Xe_2^* excimer band, but not for the 147 nm Xe resonance line of the applied Xe/Ne excimer micro-discharges to prevent oxidation of Eu^{2+} by photoionization. This has been achieved by a particle coating on the basis of nanoscale MgO as shown in Figure 1.5.6 [16]. This measure enables plasma TVs with a lifetime of at least 30,000 h without image burn-in, which means the bleaching of the blue colors was prevented. Another advantage of an MgO coating onto plasma TV phosphors is the reduction of the firing voltage, which imposed the application towards all three RGB phosphors.

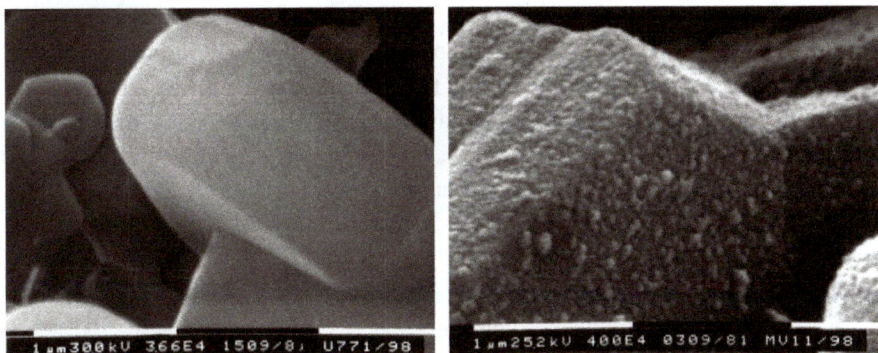

Figure 1.5.6: SEM image of the fluorescent lamp and PDP phosphor $BaMgAl_{10}O_{17}:Eu^{2+}$ uncoated sample (left) and MgO-coated sample (right), magnification ~40k.

The Hg consumption of fluorescent lamps, whether linear or compact, was a long-term issue in lamp development to enhance lamp lifetime and to fulfill the strong demand of the lamp market to reduce the lamps' Hg content for environmental reasons. Therefore, the soda lime lamp glass as well as those luminescent pigments known to contribute to Hg consumption, such as $BaMgAl_{10}O_{17}:Eu^{2+}$ and silicates were coated by Al_2O_3 to reduce their Hg take-up.

Another coated luminescent pigment in application is the lead activated phyllosilicate sanbornite $BaSi_2O_5:Pb^{2+}$, which is the mostly used phosphor in fluorescent lamps for tanning purposes. This material suffers from strong degradation in suspension due to its hydrolysis according to

$$BaSi_2O_5 + 2H_2O \rightarrow H_2Si_2O_5 + Ba^{2+} + 2OH^- \text{ and } H_2Si_2O_5 \rightarrow 2SiO_2 + H_2O$$

and also, during lamp operation by its Hg take-up resulting in a grey absorption layer due to the formation of metallic Hg onto the particles (Figure 1.5.7 top left). These problems were successfully overcome by the application of adhesive Al_2O_3 nanoparticles (alon-c) to the phosphor suspension and by a particle coating with La_2O_3 (first layer) to reduce its solubility in water and by Al_2O_3 (second layer) to further reduce its Hg take-up.

A recent application case for nanoparticle-coated luminescent pigments is motivated by the degradation of LEDs due to the efficiency loss of the phosphors, since LEDs often operate in contact to humidity, NH_3 or CO_2. The chemical reaction taking place at the surface of LED phosphors is an important degradation mechanism limiting the operational lifetime of phosphor converted LEDs. Therefore, a lot of efforts have been made to stabilize LED phosphors, such as the green emitting ortho-silicate $(Ba,Sr)_2 SiO_4$:Eu or the red emitting alkaline earth sulfide $(Ca,Sr)S$:Eu^{2+} by particle coatings (Figure 1.5.8), mostly by a thin layer of homogeneously precipitated SiO_2.

1.5.2.2 Particle coating techniques

In general, a particle coating can be deposited onto individual pigment grains by the addition of nanoscale particles or by the formation of a continuous thin layer from diluted precursors. The thin layer may be a glass phase based on an inorganic compound, e.g. SiO_2 or polyphosphates, or a polymer, e.g. polyethylene glycol (PEG). The following list summarizes most industrial relevant techniques to obtain core-shell or coated particles:

- Encapsulation by polymers
- Pigmentation by nanoparticles
    - addition to the dry pigment powder
    - addition to the pigment suspension
- Precipitation methods
    - non-homogeneous precipitation
    - homogeneous precipitation by hydrolysis
- Gas phase deposition methods
    - fluidized bed chemical vapor deposition (FB-CVD) of oxides by oxidation of metal organic compounds, e.g. $Al(CH_3)_3$ or $Mg(CH_3)_2$
    - plasma CVD of elements, e.g. diamond like carbon (DLC)

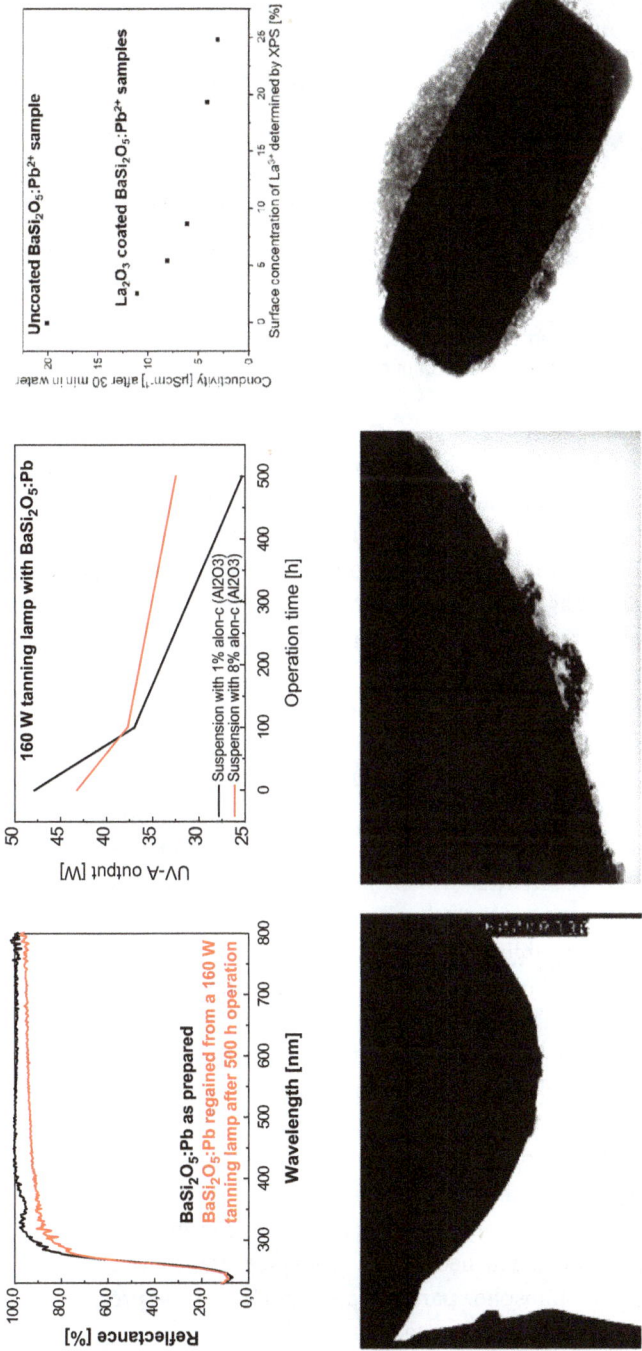

Figure 1.5.7: Reflection spectra of $BaSi_2O_5$:Pb^{2+} (top left), UV-A output of a 160 W tanning lamp up to 500 h (top middle), conductivity of a suspension of uncoated and La_2O_3 coated $BaSi_2O_5$:Pb^{2+} (top right). TEM images of uncoated $BaSi_2O_5$:Pb^{2+} (bottom left), La_2O_3 coated $BaSi_2O_5$:Pb^{2+} (bottom middle) and additional Al_2O_3 coated $BaSi_2O_5$:Pb^{2+} (bottom right), magnification ~200k.

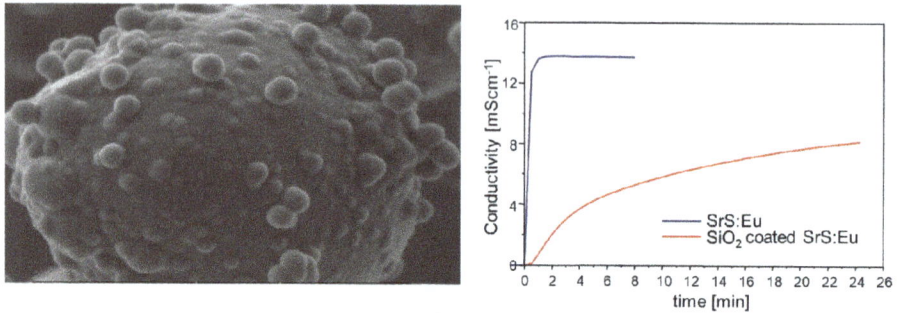

Figure 1.5.8: SEM image of a SiO_2 coated SrS:Eu particle, magnification ~30k (left) and the time-dependent conductivity of an aqueous suspension of uncoated SrS:Eu and SrS:Eu coated by SiO_2 in water (right).

Thus, particle coating techniques can be subdivided to yield organic or inorganic coatings. The former technique comprises the encapsulation with organic compounds leading to the formation of microcapsules or microspheres. This method has been found to be of special interest for the controlled release of drugs, genes, hormones and other bioactive agents. It provides the benefits of increased resistance against rapid degradation, targeted delivery, controlled release rate and prolonged duration of bioactive agents.

Fabrication of coatings by precipitation is one of the most important applications of the sol-gel process. Inorganic coatings with a wide range of chemical compositions can be deposited onto a variety of substrates including pigments using liquid precursor solutions comprising metal ions or alkoxides. Even though, such precipitation techniques are more demanding than just adding nanoscale particles, this process is a less expensive alternative to chemical vapor deposition techniques.

Therefore, nanoparticle coatings onto (luminescent) pigments are often generated by sol-gel chemistry [17] enabling homogeneous precipitation of nanoparticles from an alcoholic or aqueous solution. This process can be roughly subdivided into four steps:

1. The coating material precursors, for example, metal chlorides, nitrates or alkoxides, are dissolved in the solvent, while the pH is fixed to a certain value, e.g. by a buffer.
2. Then the pigment to be coated is dispersed in the above-mentioned solution to yield a pigment slurry.
3. This suspension is brought into contact with a hydroxide concentration enhancing species until a pH value is reached, which guarantees the complete hydroxide precipitation onto the phosphor particles and stability of the precipitate. A homogeneous pH increase can be promoted in several ways:

- Contact of the stirred slurry to an ammonia comprising atmosphere:

$$NH_3 + H_2O \rightleftharpoons NH_4^+ + OH^-$$

- Thermally induced or catalyzed hydrolysis of urea, potassium cyanate or urotropine:

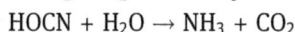

$$H_2N\text{-}CO\text{-}NH_2 + H_2O \rightarrow 2\,NH_3 + CO_2$$
$$HOCN + H_2O \rightarrow NH_3 + CO_2$$
$$N_4(CH_2)_6 + 6\,H_2O \rightarrow 4\,NH_3 + 6\,CH_2O$$

- Photochemical cleavage of NaN_3 in water [18]:

$$2\,NaN_3 + 2\,H_2O \rightarrow 2\,Na^+ + 3\,N_2 \uparrow + H_2 \uparrow + 2\,OH^-$$

To improve the coating homogeneity, the suspension can be subjected to ultrasound treatment during agitation in order to prevent agglomeration of uncoated pigment particles.

4. The coated pigment is filtered off, dried and calcinated so that the precipitated material, mostly a hydroxide, converts into the desired oxide nanoparticles or films and simultaneously the adhesion of the coating material onto the phosphor surface is enhanced, so that a dense and compact layer is obtained (Figure 1.5.6).

Many approaches for the application of Al_2O_3, SiO_2, MgO or Ca polyphosphate (CaPP) coatings onto functional pigments by sol-gel chemistry have been conducted due to the achieved performance and lifetime improvement observed for the obtained pigments [18–22].

However, the above-described process has turned out to be inapplicable to luminescent pigments based on alkaline earth or transition metal silicates, for instance $BaSi_2O_5$:Pb^{2+} (UV-A phosphor for tanning lamps) and Zn_2SiO_4:Mn^{2+} (green phosphor for fluorescent lamps and plasma TVs). These pigments are all sensitive towards hydrolysis at such low pH value at which most homogeneous precipitation processes are taking place. A low pH value of the suspension should be the starting point of a sol-gel type coating process to prevent fast and thus undesirably inhomogeneous formation of metal hydroxide. The stability of the coated luminescent material in an aqueous slurry is extremely important from the standpoint of industrial application because during a device manufacturing process, e.g. display screens or lamps, organic or water-based phosphor suspensions are used. Although an organic medium has advantages such as easiness in drying, aqueous slurries are preferred due to environmental concerns. Furthermore, leaching out of the activator during the coating performed in water can adversely affect phosphor performance during subsequent device operation. This process should also be minimized.

Therefore, methods of coating a luminescent pigment by which hydrolysis of the material and thus degradation of its optical properties, such as reflectivity and quantum efficiency are prevented are given in literature [23]. Herein, a chelating ligand, e.g. ethylenediamine tetraacetate (H_4EDTA), is claimed to be added to the metal oxide precursor. The complexation of trivalent metal cations according to

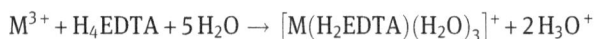

$$M^{3+} + H_4EDTA + 5 H_2O \rightarrow \left[M(H_2EDTA)(H_2O)_3\right]^+ + 2 H_3O^+$$

prevents the precipitation of $M(OH)_3$ until an alkaline pH value has been reached. As a result, the masked metal cations M^{3+}, e.g. La^{3+}, Y^{3+} or Al^{3+}, will not precipitate as a hydroxide before the pH is in the range 7 to 10, which is advantageous for $BaSi_2O_5$: Pb^{2+} or other silicate type pigments, since their hydrolysis in this pH range is much slower than in an aqueous slurry adjusted to an acidic pH value.

In case of $(Ca,Sr)S:Eu^{2+}$, which rapidly undergoes hydrolysis in aqueous solutions and which is even sensitive towards atmospheric trace components like H_2O, NH_3 or CO_2, the coating with SiO_2 has to be accomplished in alcoholic solution, preferably in ethanol. The rather low solubility, the moisture stability and the carbon dioxide impermeability of the coating material contribute to the stabilization of the luminescent material which is consequently rendered by its insensitivity to the surrounding environment. Moreover, hydrophilic properties of the coating material can increase the ability of the coated material to be homogeneously dispersed. In this particular case, nanostructured coating material deposited onto the μ-scale luminescent pigment, due to the different material properties, e.g. morphological and refractive index differences, results in anti-reflective surface, higher light outcoupling efficiency [24].

The pigmentation of microscale pigments by nanoscale particles is advantageous for industrial processes, since this process is cost effective and enables beyond the achievement of stabilization, the implementation of other physicochemical functions. This is e.g. contrast enhancement by colored coatings of display phosphors and dielectric coatings of electroluminescence phosphors. Daylight contrast enhancement through phosphor pigmentation was first achieved for cathode-ray tubes (CRTs). To this end, blue $CoAl_2O_4$ and red α-Fe_2O_3 pigments are deposited onto the blue emitting ZnS:Ag,Al and the red emitting Y_2O_2S:Eu CRT phosphor grains (Figure 1.5.9), respectively. Such inorganic pigments reduce the phosphor screen reflectance, i.e. the greyish screen turns black, which is standard for present display technologies. However, they coating layer has to be transparent to the emitted light by the core, while absorbing the remaining spectral components of incident daylight.

The red hematite pigment was also used for the pigmentation of the red emitting phosphor Y_2O_3:Eu^{3+} in plasma display panels for TV application [25]. Finally, a so-called intrinsic pigmentation of the blue emitting PDP phosphor $BaMgAl_{10}O_{17}$:Eu^{2+} by the co-doping with Co^{2+} was reported too [26].

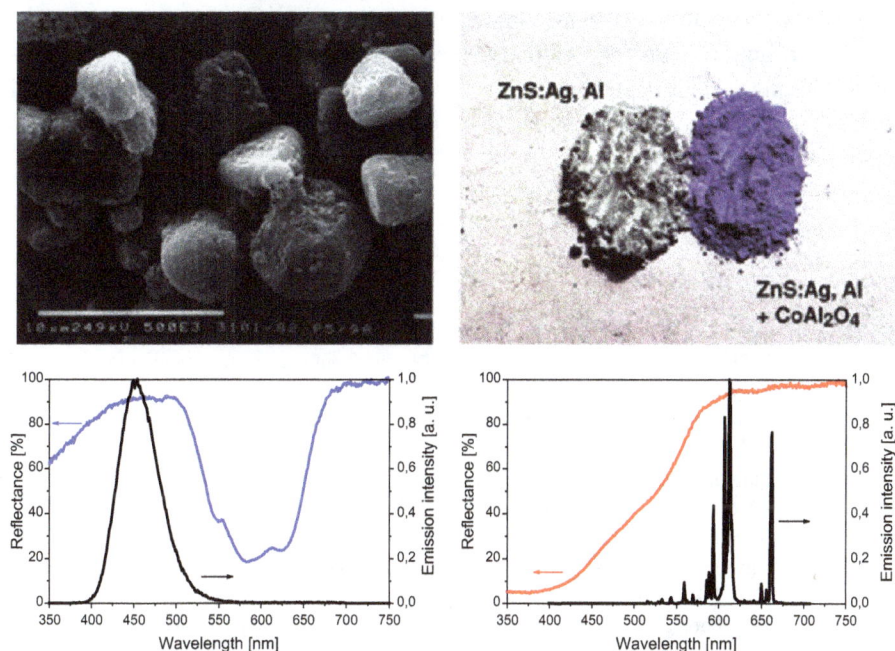

Figure 1.5.9: SEM image of CoAl$_2$O$_4$ coated ZnS:Ag,Al particles, magnification ~5k (top left) and the daylight image of coated and uncoated ZnS:Ag,Al (top right) as well as the emission and reflection spectra of ZnS:Ag,Al coated by CoAl$_2$O$_4$, Thenard's blue and Y$_2$O$_2$S:Eu coated by α-Fe$_2$O$_3$, hematite (bottom right).

1.5.3 Summary

The deposition of coatings onto substrates or particles is an effective technology to improve characteristics of functional and structural materials for practical purposes. Especially their stability and physicochemical properties can be improved to such an extent that just coated materials often become commercially viable. A convincing example is the success of the metal aluminum, which became one of the most important construction materials for the global economy during the last 100 years. This development was enabled due to its stabilization by eloxal and further surface modifications.

Many functional pigments, e.g. color or luminescent pigments, are improved by coatings with regard to stability and other physical or chemical parameters too. For instance, semiconductor nanoparticles (QDs), are coated by a wider band gap material to form a core-shell structure, which increases the quantum efficiency by means of suppressing the energy loss processes at the particle's interfaces. Coated microscale luminescent pigments, such as lamp, LED and display phosphors, enable enhanced color point consistency and an operational device lifetime to meet market requirements.

So far achieved results clearly indicate that the coating technology is an effective way to improve the characteristics of structural materials and functional pigments for many practical application areas. A thin eloxal layer onto Al metal surfaces and nanoscale particle coatings onto pigments are for many application cases truly indispensable.

References

[1]   Lider K. Technikerarbeit: Mehrfachbeschichten von Aluminium, 2013.
[2]   Wernick S, Pinner R, Zurbrück E, Weiner R. Die Oberflächenbehandlung von Aluminium, 2. Auflage, Eugen G Leuze Verlag, Saulgau, Württemberg, 1977.
[3]   Taschenbuch für Galvanotechnik, 12. Ausgabe, Langbein-Pfanhauser Werke AG, Neuss/Rhein, 1970.
[4]   Werner E. Metallische Überzüge, Carl Hanser Verlag, München, 1959.
[5]   webpage: https://commons.wikimedia.org/wiki/File:Aluminium_-_world_production_trend. svg, downloaded at 2021-12-10.
[6]   webpage: https://www.wotech-technical-media.de/womag/ausgabe/2015/11/23_koelle_ alu_11j2015/23_koelle_alu_11j2015.php, downloaded at 2021-12-10.
[7]   Strehlow WH, Cook EL. Band gaps in elemental and binary compound semiconductors and insulators. J Phys Chem Ref Data, 1973, 2, 163–99.
[8]   Fang CM, de Groot RA. The nature of electron states in AlN and α-Al$_2$O$_3$. J Phys: Condens Matter, 2007, 19, 386223.
[9]   Griesiute D, Gaidukevic J, Zarkov A, Kareiva A. Synthesis of ß-Ca$_2$P$_2$O$_7$ as an adsorbent for the removal of heavy metals from water. Sustainability, 2021, 13, 7859–67.
[10]   Dryden DM, Tan GL, French RH. Optical properties and van der Waals–London dispersion interactions in berlinite aluminum phosphate from vacuum ultraviolet spectroscopy. J Am Ceram Soc, 2014, 97, 1143–50.
[11]   Borges PD, Cott J, Pinto FG, Tronto J, Scolfaro L. Native defects as sources of optical transitions in MgAl$_2$O$_4$ spinel. Mater Res Expr, 2016, 3, 076202.
[12]   Yazdanmehr M, Asadabadi SJ, Nourmohammadi A, Ghasemzadeh M, Rezvanian M. Electronic structure and band gap of γ-Al$_2$O$_3$ compound using mBJ exchange potential. Nanoscale Res Lett, 2012, 7, 488–97.
[13]   Zhao Y, Kita K, Kyuno K, Akira Toriumi A. Band gap enhancement and electrical properties of La$_2$O$_3$ films doped with Y$_2$O$_3$ as high-k gate insulators. Appl Phys Lett, 2009, 94, 042901–3.
[14]   webpage: https://materialsproject.org/materials/mp-30983, watched at 2022, January 1st.
[15]   Ronda C. (Ed.) Luminescence, from Theory to Applications, Wiley-VCH, Weinheim, 2006, 29.
[16]   Jüstel T, Nikol H. Optimization of luminescent materials for PDPs. Adv Mater, 2000, 12527–30.
[17]   Danks AE, Hall SR, Schnepp Z. The evolution of sol-gel chemistry as a technique for material synthesis. Mater Horiz, 2016, 3, 91–112.
[18]   Bechtel H, Czarnojan W, Haase M, Mayr W, Nikol H. Phosphor screen for flat cathode ray tubes. Philips J Res, 1996, 50, 433–62.
[19]   Broxtermann M, Jüstel T. Mater Res Bull, 2016, 80, 249–55.
[20]   Seo JH, Sohn SH. Surface modification of the (Y,Gd)BO$_3$: Eu^{3+}phosphor by dual-coating of oxide nano-particles. Mater Lett, 2010, 64, 1264–67.

[21] Wang H, Yu M, Lin CK, Lin J. J Colloid Interface Sci, 2006, 300, 176–82.
[22] Yen WM, Shionoya S, Yamamoto H. Phosphor Handbook, CRC Press, Boca Raton, 2007.
[23] Jüstel T, Merikhi J, Nikol H, Ronda CR. Coating for a Luminescent Material, 2002, US 6472811 B1.
[24] Raut HK, Ganesh VA, Nair AS, Ramakrishna S. Anti-reflective coatings: A critical, in-depth review. Energy Environ Sci, 2011, 4, 3779–804.
[25] Kim BC, Lee CY, Song YH. Luminescence properties of pigment-coated Y_2O_3: Eu red phosphor with α-Fe_2O_3 by different coating methods and various exciting energy source. Jpn J Appl Phys, 2002, 41, 2066–73.
[26] Jüstel T, Bechtel H, Mayr W, Wiechert DU. Blue emitting $BaMgAl_{10}O_{17}$: Eu with a blue body color. J Luminescence, 2003, 104, 137–43.

1.6 Vitreous enamel

Jörg Wendel

Vitreous enamel is a highly effective composite material with designable properties. In general, enamel is a special glass whose thermal expansion rate is precisely matched to the substrate. It must have adhesion to the substrate in question and its melting point must necessarily be below the melting point of the substrate. It is a bulk material quenched from the melt – with a glass transition temperature (T_g). This is a general difference to amorphous materials, which are deposited by sputtering or evaporation. Vitreous enamel is not a material that can be used functionally on its own: it only becomes a material component when combined with metal or glass. The enamel is used for surface refinement and the formation of new properties in terms of material, technology and function for the entire system. The enamelled material is given new optical, hygienic, chemical and mechanical properties.

1.6.1 The material enamel, terms and definitions

The beginning of industrial enamelling can be dated to about 1750. In the period of classicism, we find the first industrial jewelry enamelling in Battersea. The beginning of a scientific work and description is made by Erbe's book "Enamel" in 1837. The enamel works are much older: 2500 BC; we find first cloisonné works in Egypt, the first recipes for colored glazes are found in Assyria about 1700 years before the beginning of our era [1, 2].

The definition of the material goes back to Adolf Dietzel, it can be found under RAL RG 529 A3 [3]: "Vitreous Enamel (American – porcelain enamel) is a glass-like material which is produced by the complete or partial smelting primarily of oxidic materials. The inorganic preparation produced in this way is fused onto workpieces made of metal or glass at temperatures of over 480 °C, in one or more coats, partially with additives".

The definition is well done, as it uses "inorganic" to clearly demarcate the material from so-called "enamel" paints and varnishes. The "partial smelting" includes partially crystalline enamels. The "primarily oxidic" composition does not exclude sulfides and fluorides as enamel raw materials. The "additives" denote clays, quartz, opacifiers and pigments, which may also be needed for full preparation. The application on "metal or glass" distinguishes the enamels from the glazes, which are fused on a ceramic carrier material.

The enamelling is then the process of applying and firing an enamel to an object. The result is a glossy inorganic coating, that is interlocked with the substrate. The enamel is a composite enamel/metal (or enamel/glass).

https://doi.org/10.1515/9783110733143-006

Enamel frits are the base materials for enamelling. A raw material mixture of quartz, feldspar, borax, soda ash, potash, aluminum oxide, titanium dioxide and adhesion-forming metal oxides is melted at temperatures of around 1400 °C. This melting process takes place in rotary furnaces or continuously in tank furnaces almost exclusively at the enamel frit manufacturers. Depending on the quenching method of the melt, granules form when rapidly cooling and solidifying the molten glass in water or flake-like products form when quenched dryly between water-cooled rollers. Granules and flakes are referred to as enamel frits.

Enamel slurries are produced as ready-to-apply suspensions by wet grinding of enamel frits and additives (e.g. quartz, clay, zirconium silicate), set-up salts, and, when needed, colorants. Their rheological behavior is adapted to the application process. Powders for electrostatic application or for application onto hot castings are produced by dry-milling the enamel frits.

Enamels must be much less viscous than glasses. It is self-explanatory that you cannot, for example, melt a soda-lime glass coating onto aluminum at 600 °C.

Worldwide, about 450 10^3 tons of enamel are produced today. The greatest demand is for enamelling steel (approx. 85%), followed by cast iron (approx. 10%), aluminum/magnesium (approx. 3%) and stainless steel (2%). The enamelling of the precious metals copper, silver, gold and platinum is done almost exclusively for jewelry production and does not play a role in terms of quantity, yet extraordinarily beautiful things are sometimes created here.

1.6.2 Durability, sustainability, physiological harmlessness and circular economy

The most common reason for enamelling is to make steel corrosion-resistant. Enamelling is the classical protection against corrosion. Corrosion not only destroys the material in question, one must also think of the consequential damage (loss of production, product contamination, safety and environmental hazards). Enamel is also suitable as a material for highly corrosive areas. Normal steels and also stainless steels can no longer do without a protective enamel layer. The symbiosis of both materials enables the production of vessels, reactors and piping with excellent suitability for the application. The strength of the structure is provided by the steel. Corrosion resistance is provided by the enamel. Enamel is gas-tight. This allows corrosion-sensitive surfaces to be protected from the attacking medium, even if it is in gaseous form. Diffusion through the enamel is impossible.

In principle, enamelled products can be recycled without much effort: the used materials can easily be returned to the material cycle. Cast iron is recycled in the cupola or medium-frequency electric furnace during cast iron production. As a bulk material, steel is in principle available in large quantities for a wide variety of applications. The recyclability of steel is very good. The use of steel as a recycling material helps to conserve resources. Unlike polymers, metals can have their properties fully preserved [4]. The thin enamel layer of the surface goes into the slag. The slags can be used, for example, in road construction for the subgrade (see also Chapter 6.4). The raw materials for enamel production – as with other types of glass – are also available in large quantities as minerals for the most part occurring naturally in the earth's crust. Thus, the principles of recovery, disposal and prevention are implemented in the best form in the waste management objective. The metals are almost fully recycled. The slag content is low and can be disposed of or also recycled. The principle of avoidance applies in full to air pollution control. Enamel is not combustible, so no further air pollution occurs during recycling. Enamel is chlorine-free – dioxin formation is ruled out during both production and recycling. Process waste generated during processing – such as overspray – can be recovered without difficulty. The conservation of resources already begins during processing. Through the use of powder technologies, utilization rates of 98% can be achieved.

The economic loss due to corrosion damage amounts to about 4% of the gross national product [5, 6]. Accordingly, the costs in Germany for this amount to approx. 64 billion euros. These costs do not include the indirect costs caused by the release of pollutants into the environment. If we consider only the chemical industry in Germany, the corrosion-related damage amounted to 3.3 billion euros in 1991 [7]. These costs are considerable for an industry that already attaches great importance to corrosion protection. About 10–15% (330 to 500 million euros) could be saved by more effective corrosion protection measures. This is certainly a good reason to consider enamelling. Components that were previously made of stainless steel can often be replaced by components made of mild steel.

A guideline from the German Federal Ministry for the Environment, Nature Conservation and Nuclear Safety (BMU) for determining the CO_2 footprint refers to product-related climate protection strategies [8]. Since enamel hardly uses any CO_2 polluting precursors in its production, its pollutant footprint in the manufacture of coatings is significantly lower. Enamel slurry in the process does not contain any solvents and therefore contributes to air pollution control.

Enamelled surfaces are environmentally friendly. Enamel finishes help manage disposal and end the throwaway practice. Enamels are enormously durable and can be recycled without much effort.

Architectural enamel finishes last over many decades even under extreme weather conditions. Furthermore, unwelcome graffiti may be removed effortlessly, even of dried-on varnish [9, 10].

Enamels for cooking and frying utensils are physiologically harmless. Today, all enamels that come into contact with food are tested for their leaching of metal ions according to EN ISO 4531 [11]. In the absence of published legal limits, the limits laid down in this document correspond to limits published by the Council of Europe in Resolution CM/Res(2013)9 on metals and alloys used in food contact materials and articles [12]. The specified limit values for materials in contact with food are complied with and thus these enamels are also harmless to health in the long term. For additional information the series of lectures, "Limits for Metals – Trace Elements or Poison" aims to acquaint the reader with the occurrence, the technical use, the content in food and the physiological effect of metals on the organism. It is astonishing how close sometimes the occurrence of the metals gets to the limits where first adverse effects on the human organism may be detected. Therefore, most of the limits act in concentration levels of µg/L (ppb) [13–17]. Likewise, the German Federal Environment Agency has published a rationale according to which all enamels in contact with drinking water must be tested [18, 19]. There is also a toxicological rationale for the reference concentrations of the evaluation criteria for the required test values [20, 21]. The assessment basis has been legally binding in Germany since August 2021. All enamels that meet these requirements guarantee the user physiological harmlessness even in long-term use.

1.6.3 Chemistry of vitreous enamel

The basis of glass formation is the network formed, for example, by SiO_4 tetrahedra. These tetrahedra are joined only at their corners by bridging oxygen atoms. Other oxides are used for glass production, all of which influence the properties of the glass melt and the finished glass. A distinction is made between network formers such as SiO_2, GeO_2, *BeF$_2$*, B_2O_3, *As$_2$S$_3$*, *As$_2$O$_5$* and P_2O_5 (arsenic is no longer used today and is mentioned only for the sake of completeness), network transformers (modifiers) such as Li_2O, Na_2O, K_2O, *Rb$_2$O*, *Cs$_2$O*, CaO and BaO, and intermediate oxides such as Al_2O_3, ZnO, MgO, *BeO*, Nb_2O_5 or Ta_2O_5. These groups are formed due to fundamentally different properties of the oxides. The oxides that are rarely or never used in vitreous enamel are printed in italic letters.

Network formers, for example, are acidic oxides and form the basic framework of the glass. They are the carriers of the property to solidify the glass amorphously. They have low coordination numbers (CN 3–4). The conditions Zachariasen set for the network formers are well known from glass (see also Chapter 3.1) and shall not be repeated here [22].

In the next lines, the effects of the network formers on the glass properties are briefly noted (CTE: coefficient of thermal expansion; TSR: thermal shock resistance; T_m: melting temperature; ↑ increase, ↓ decrease; double arrow: a lot):

SiO_2	CTE ↓, TSR ↑, T_m ↑, mechanical strength ↑, chemical resistance ↑
B_2O_3	CTE ↓, T_m ↓, viscosity ↓, mechanical strength ↑
P_2O_5	CTE ↑, UV transparency ↑, IR transparency ↓, viscosity ↓, chemical resistance ↓
Sb_2O_5	CTE ↑, chemical resistance ↑, viscosity ↓, opacifier for enamels

Network transformers are alkaline oxides and break down the network of the formers, thus changing their density. They have high coordination numbers (CN ≥ 6). Due to the small number of linking sites, the viscosity of the glass melt also decreases, which means that the glass can then be melted at lower temperatures.

Effects of network transformers on glass properties:

Li_2O viscosity ↓↓, CTE ↑
Na_2O viscosity ↓, CTE ↑↑
K_2O glass becomes "longer" (i.e. it becomes harder and the melting temperature increases)
CaO   chemical resistance ↑, viscosity ↑
MgO   glass becomes "longer", viscosity ↑, chemical resistance ↑
BaO   viscosity ↓, chemical resistance ↓, refractive index n ↑

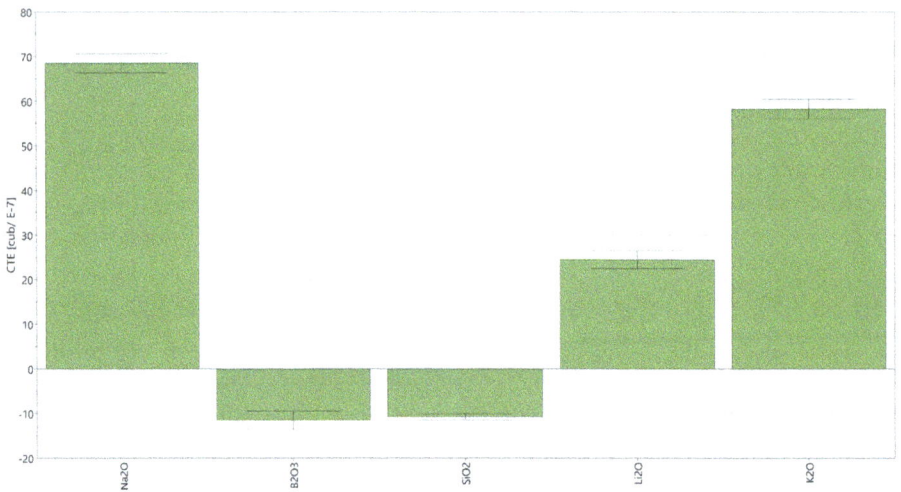

Figure 1.6.1: Influence of the alkali oxides, boron oxide and silica on the coefficient of thermal expansion in alkali silicate glasses (100–300 °C)/cub E-7 based on data from [23].

Fluegel collected thermal expansion data from scientific publications and generated a database of 900 glasses [23]. From this database, 43 glasses of composition SiO_2 65–87 mol%, B_2O_3 0–15 mol%, Li_2O 0–17 mol%, Na_2O 0–25 mol% and K_2O 0–19 mol% were selected. Within this composition, the effects of the individual components behave approximately linearly and the influence of the individual components can be calculated using partial least square regressions (PLS). Figure 1.6.1 shows the coefficients of these components within a 95% confidence interval. The expansion of the glasses is then calculated from the linear combination of these coefficients multiplied by the respective mol percentages of these components in the glass and a constant (245 E-7). From this plot, one can easily compare the influence of the oxide components. Thus, replacing the lithium oxide with the other alkali oxides will always lead to an increase in the coefficient of expansion of the glass. The influence of SiO_2 as a network former is similar to the influence of B_2O_3. Increments for calculating thermal expansion can be found in the literature – for example, by Appen [24]. The factors (or increments) found in Figure 1.6.1 (*vide infra*) for calculating thermal expansion differ from the increments given by Appen by the constant. The constant of the PLS calculation represents the center of the plane of the model. This makes changes more visible and the overall error of the calculation is smaller.

Intermediate oxides are oxides that can act either as network formers or network transformers, depending on their proportion in the glass. The coordination numbers of the intermediate oxides are between those of the network formers and those of the transformers (CN: 4–6). The glass properties can be changed by intermediate oxides without changing the basic composition of the network formers and transformers. In the case of oxides used as opacifiers, above a certain concentration the oxide is no longer used for glass formation, but then acts as an intermediate oxide. Examples are TiO_2 and ZrO_2.

Effects of intermediate oxides on vitreous enamel properties:

Al_2O_3	glass becomes "longer", mechanical strength ↑, chemical resistance ↓
PbO	T_m ↓, refractive index n ↑, el. resistance ↑, absorption of X-rays (40–80 wt%)
ZnO	viscosity ↑
CdO	absorption of thermal neutrons (30–60 wt%)
MnO	viscosity ↓, chemical resistance ↓
TiO_2	refractive index n ↑, acid resistance ↑, opacifier for enamels
ZrO_2	chemical alkaline resistance ↑, opacifier for enamels

Replacing SiO_2 in a binary glass with other oxides affects the coefficient of thermal expansion. By skillfully combining the exchanged oxides, the original properties of the original glass can be maintained or changed in a desired direction.

The classification of oxides into network formers, network transformers and intermediate oxides is an approach from a structural point of view. In general, an enamel must be easier to melt than a glass so that it can be fused onto a metallic substrate, which should not scale too much in the process. This lowering of the melting temperature can easily be achieved by alkali oxides, but the chemical resistance and thermal expansion must be adjusted by further modifications. Compared to glasses, an enamel consists of significantly more components.

Dietzel proposed a division of the oxides into Resistants I and Resistants II, as well as into fluxes (Table 1.6.1). The group Resistants I makes the enamels chemically, thermally and mechanically resistant as well as significantly more viscous. The group Resistants II does not increase the viscosity to the same extent as Resistants I, but increases the chemical resistance [25].

Table 1.6.1: Classification of oxides into nine groups.

	Network formers	Network transformers	Intermediate oxides
Resistants I (Viscosity is increased)	SiO_2, GeO_2, BeF_2	CaO, Y_2O_3	Al_2O_3, Ta_2O_5, Cr_2O_3, Co_3O_4, ZrO_2
Resistants II	As_2S_3, As_2O_5, Sb_2O_5	MgO, SrO, BaO, NiO	BeO, Ni_2O_3, Nb_2O_5, MoO_3, WO_3, SnO_2, ZnO, TiO_2
Fluxes	B_2O_3, P_2O_5, Bi_2O_3, Fe_2O_3, V_2O_5	Li_2O, Na_2O, K_2O, NaF, LiF, CaF_2, Cu_2O, PbO	MnO, CuO

The properties of the enamels can be obtained in quite different ways of composition. Some properties can be influenced directly by the components. However, there are also properties for which interactions with other components are of central importance.

1.6.4 The stratified layer construction

Enamels differ in terms of the layer structure. Conventional enamels are produced in two or more layers and direct-on enamels in only one. There are also combination processes, e.g. 2-layer/1-fire process (2c/1f).

The typical process sequence of conventional enamelling is pre-treatment of the metal, application of the ground coat with subsequent drying and firing of the coat. The ground coat creates the bond to the metal workpiece, in other words it ensures adhesion. In steel sheet enamelling, the ground coat enamel must largely dissolve the iron oxides formed during firing so that they do not penetrate into the upper enamel layer and lead to defects. Cobalt and nickel oxides are inevitable in ground coat enamels when enamelling workpieces that are only degreased and not pickled. Then, a cover coat with a slightly lower thermal expansion coefficient, slightly lower viscosity and slightly lower surface tension is applied, subsequently dried and then fired as well. The function of the cover coat is to provide the desired properties for the particular application. Some typical properties are chemical resistance, color stability and color, a smooth, glossy or matte surface and scratch resistance.

Direct enamelling processes (Table 1.6.2) are characterized by a single-layer enamel application. Direct-on enamelling must combine the properties of ground and cover coat enamelling. Due to the dark color of the adhesive oxides and the iron oxides to be dissolved during firing, light colors can hardly be achieved. In the case of direct-on white enamelling, it is necessary to nickel-plate the steel sheet beforehand. The nickel layer (Ni) is responsible for the enamel adhesion. A nickel coating of about 2 g m^{-2} is required, and the steel must also meet higher requirements. Dark colors, such as black, dark grey, brown or blue, can be easily solved with direct-on enamelling. Chemically resistant direct-on process corresponds to conventional enamelling, in which only one enamel coat is applied.

Combination processes are characterized by the fact that the cover coat (CC) is applied to the unfired ground coat (GC). This 2-coat/1-fire process (2c/1f) is beneficial as it saves a firing process and thinner coat thicknesses are possible. The enamels have the same function as in the conventional process. However, the ground and cover coat enamels must be precisely adjusted to each other because of the simultaneous firing. For this purpose, the cover coat enamel must have a slightly higher surface tension than the ground coat enamel in order to prevent the ground coat enamel from penetrating into the top coat enamel. The combination can be executed in the process sequences wet/wet, wet/powder and powder/powder. The first wet application represents a ground coat application by means of a spraying process. The powder enamel is applied electrostatically.

Table 1.6.2: Comparison of the enamelling processes (the (+) is put when the process is needed) [1, 2].

	Ni	GC	CC
Conventional enamelling	–	+	+
Direct-on white enamelling	+	–	+
Direct-on enamelling, colored	–	–	+
2c/1f enamelling	–	+	+

In case higher coating thickness is required, the 2c/1f enamelling may be combined with an additional cover coat. Then, the process results to 3c/2f. Enamels that are used for highly corrosive concentrated acids, one or two ground coats are applied and later four or more cover coats are applied to guarantee the lifetime expected for the chemical reactor.

1.6.5 Adherence to the substrates and development of tensions

Just as mentioned in the last chapter, the effects of the adhesive oxides is without question. They increase the adhesion on steel. Also copper, antimony and molybdenum oxides have a positive effect on the adhesion of enamel on steel. The most supported theory for the adhesion of enamel to steel was proposed by Dietzel: an electrochemical corrosion reaction of cobalt (or nickel) and iron that promotes adhesion. In this process, the iron dissolves locally, resulting in intense roughening. Due to the deposition of cobalt and the selective dissolution of neighboring iron, dendrite-like structures were observed, which are predominantly in direct contact with the base metal and form a large number of the so-called anchor points between enamel and base metal [26, 27]. A recent study describes the formation of a Fe-Co-Ni-Cu microalloy in the interface between enamel and steel, which allows adhesion by direct bonding to the steel substrate. The alloy precipitation appears to be a liquid–liquid type phase separation where droplets form preferentially at the outer edge of a dissolving oxide layer (fayalite) where supersaturation of iron in the enamel glass drives the immiscibility reaction. Subsequently, coalescence of the alloy particles was observed, but only after their bonding to the base metal was a high degree of mechanical interlocking and thus adhesion of the enamel layer to the steel substrate achieved. It is believed that the microalloy is a product of two partial reactions in which the iron in the glass oxidizes, while the divalent cations of Co, Ni and Cu change to the metallic state and iron(II) disproportionates to iron(III) and metallic iron, resulting in alloying of the metallic compounds [28]. EN 10209 [29] describes the quality of cold rolled steel suitable for enamelling. Regarding hot rolled steel, steels conforming to the requirements of EN 10111 [30], EN 10025 [31] and EN 10149 (HSLA steel) [32] (including grades DD 11, S235, S420 and S460) can be used successfully for vitreous enamelling with appropriate pretreatments. The usual pre-treatment processes for steel are degreasing, pickling, passivation, nickel plating or blasting, depending on the requirements.

To understand the formation of the oxide layer on cast iron during the firing procedure, we have to view the phase diagram of iron and iron oxides depending on the oxygen pressure (Figure 1.6.2). Starting at a temperature of approximately 560 °C the iron starts to form oxides.

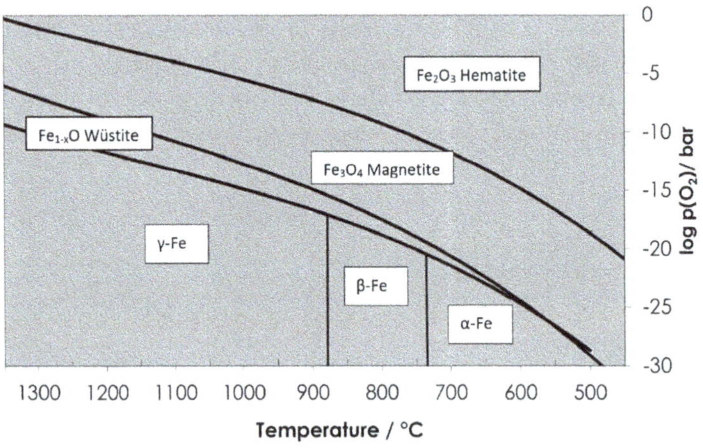

Figure 1.6.2: Phase diagram Fe–Fe$_2$O$_3$, calculated with data from [37, 38].

Depending on the amount of oxygen, it can form three different oxides: Wüstite ($Fe_{1-x}O$), magnetite (Fe_3O_4) or hematite (Fe_2O_3). The β-iron nowadays is seldom named as it has still the same body centered cubic structure of α-iron but due to the fact of exceeding the Curie temperature (768 °C) the magnetic properties are changing. Concerning the adherence of the oxide layer to the cast iron it is very important which one of the oxides is formed. We know that hematite reduces the adherence very much and magnetite increases the adherence. The wüstite is mainly a transition phase and will later, when cooling the part, disproportionate to magnetite and α-iron. Therefore, also in cast iron the formation of iron and alloying to the surface is possible. Mainly important for the adherence on cast iron is the formation of wüstite and later the formation of magnetite which leads to good adherence [33, 34]. Due to the higher weight in cast iron enamelling, longer firing times are required and the adhesion can be achieved also without the help of adhesive oxides. Cast iron suitable for enamelling needs to have a defined chemical composition [35, 36]. The usual pretreatment process for cast iron is blasting with an angular abrasive, either corundum or steel grit.

PROTHECAST® is a cobalt–chromium–molybdenum alloy according to ISO 5832–4 [39]. This alloy is used to cast prostheses for humans. Due to the cobalt and molybdenum content, the system for the adhesion of the enamel is given in the same way as for steel.

When enamelling **precious metals** like copper, silver, gold and platinum, the adhesion is created differently. For example, copper has a high solubility for oxygen in the structure. The oxygen can also diffuse easily in the structure, unlike iron. During firing, the oxygen also diffuses into the interior. Ultimately, an increasing oxygen gradient to the surface is formed and a smooth transition from the metallic to the oxidic system occurs. Enamelling precious metals in a vacuum will therefore not work [25].

Stainless steel, austenitic or ferritic, can be enamelled as well. We believe that the adhesion is caused by a thin chromium oxide layer that adheres to the substrate. The enamel adheres onto the chromium oxide. Scientific work proving this is still pending.

Among the light metals, the enamelling of **aluminum** is the most important so far. Complex aluminum profiles can be produced cost-effectively by extrusion. Enamelling ensures durable coloring. Even colors that glow in the dark can be produced with light-storing pigments. Magnesium and titanium as well as various alloys can also be enamelled [40].

Another important point for good adhesion is the balance of the thermal expansions of the substrate and the enamel. Figure 1.6.3 shows the development of tensions in the metal/enamel system during cooling: Starting with the temperature where the enamel becomes solid (T_m), in the range of T_g (transformation area (+)), the enamel will initially be under tensile stress. The neutral point (T_N) is stress-free and below T_N compressive stress (-) is built up. Given that the tensile strength of the enamel is low and the compressive strength is about 10 times higher, the

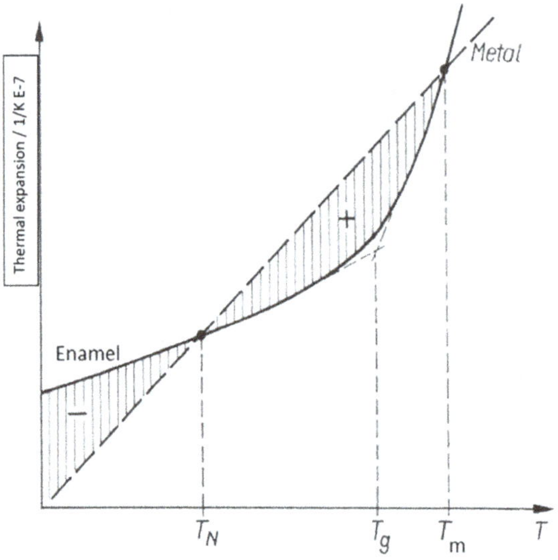

Figure 1.6.3: Schematic depiction of the forming of tensions in the system metal/enamel during cooling [41].

enamel layer must always be under a low compressive stress. This is achieved by a thermal expansion coefficient that is lower than that of the metal. Under load, some compressive stress is then first relieved before the enamel comes under tensile stress. In steel and cast iron, the stresses are far below the yield point. Ductile metals with lower yield point (e.g. Cu, Al), on the other hand, are plastically deformed. For these enamels, the stress is set correspondingly lower.

1.6.6 Opacifying and coloring enamel

The opacity of a transparent medium is caused by the light being reflected by another medium with different refraction index. The ratio of the intensity of diffuse scattering to the intensity of incident light is a measure of the opacity of the second medium. It is greater than the difference in refraction index between the opacifying particle and the medium. Fresnel's law applies here. Zirconia (n = 2.4) has one of the highest differences in refraction index compared to enamel. It is only surpassed by titanium dioxide in the two technically important modifications "anatase" (n = 2.5–2.7) and "rutile" (n = 2.7–2.86).

The particle size is also important for the reflection: if the particles are too large, their number in the given volume is statistically too small and the haze is too weak. In addition, the surface becomes rough and matte. If the particles are too small, diffraction dominates and the effect becomes wavelength dependent: now Rayleigh's law applies and the enamels become bluish and opalescent. Anatase nevertheless opacifies better than rutile because its particle size is more uniform. The particles are rather isometric while rutile likes to crystallize needle-shaped.

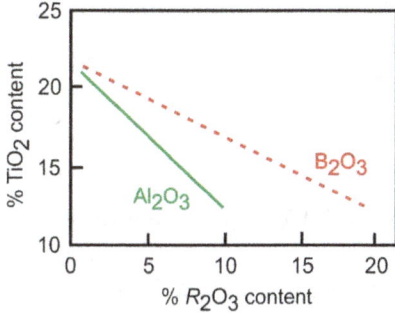

Figure 1.6.4: Onset of TiO_2 crystallization as a function of the Al_2O_3 and B_2O_3 content in the base glass 16 Na_2O / 84 SiO_2. Redrawn and adapted from [42].

Therefore, anatase is the best opacifier. The ideal particle size is between 0.1 and 0.4 μm. The crystallization behavior of anatase also depends on the composition of the vitreous enamel.

Figure 1.6.4 shows that the amount of TiO_2 required for crystallization to begin depends on the amount of Al_2O_3 and B_2O_3 present in the glass. In the absence of both components, about 22% TiO_2 is needed for crystallization. With 10% Al_2O_3 or 20% B_2O_3, about 12% is sufficient. The amount of anatase and rutile can also be modified by P_2O_5 addition (Figure 1.6.5).

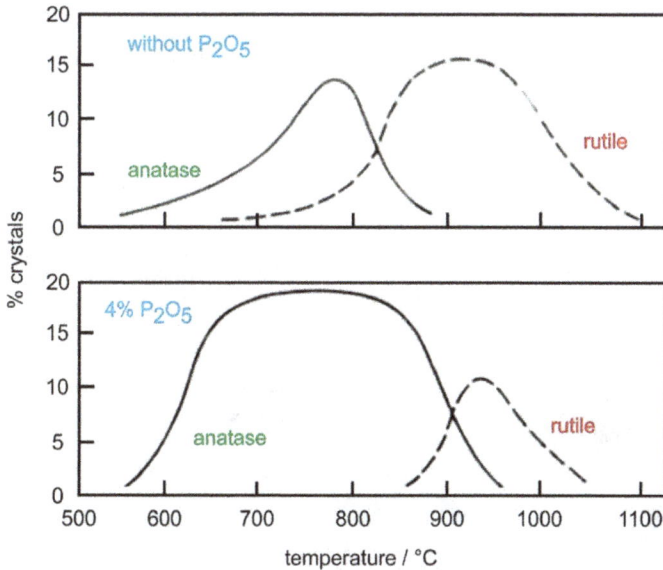

Figure 1.6.5: Dependence of the modification of the precipitated TiO_2 crystals on P_2O_5 content. Redrawn and adapted from [42].

In this way, semi-opaque and opaque white frits are obtained. For coloring, transparent frits (for dark colors) and semi-opaque frits (for lighter colors) are preferred. Color oxides are added to the mill, which should not dissolve in the melt during the firing process. Alternatively, the melt can also be colored directly. This process is used today almost exclusively for majolica frits.

1.6.7 Applications of enamel

When applying enamel, wet and dry methods are used. Wet application involves dipping, flooding (also by vacuum) and spraying. These techniques can also be combined with an electromotive force. This results in electrophoretic dipping and electrostatic wet spray application. In dry enamelling, the powder is applied through a sieve onto the red-hot part. The red-hot part can also be dipped into the enamelling powder. If the enamelling powder is fluidized, the dipping process results in a fluidized powder bed application. If the dry application is combined with an electromotive force, this leads to an electrostatic powder application. In this case, the part to be enamelled remains cold. Figure 1.6.6 shows a schematic summary of these techniques.

Using dry enamel applications, hardly any additives are used and the frit is applied almost pure. The electrostatic powder application also leaves little room for the use of color pigments, and therefore the color range of these enamels is very limited.

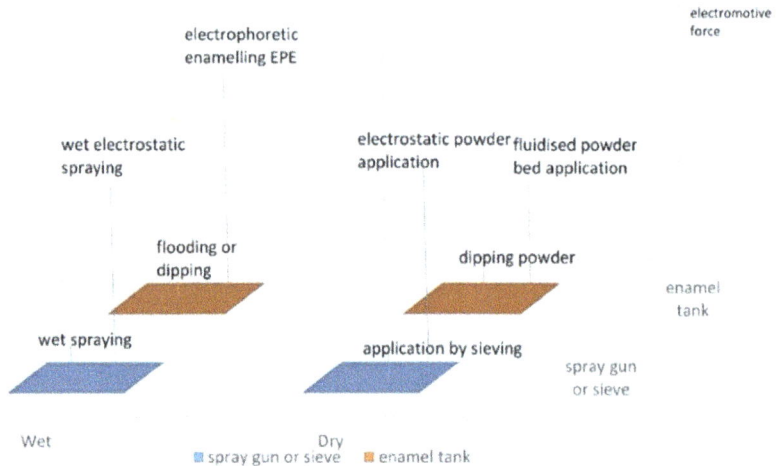

Figure 1.6.6: Application techniques for enamelling.

When preparing aqueous suspensions for wet application (the slurry), the frits are ground with additives and water. This opens a wide room for modifications. With additions of quartz, feldspar and zirconium silicate, the firing intervals can be extended or the firing behavior at difficult edges can be improved. As a result of crushing quartz or frit, breaks of chemical bonding appear within the siloxane ($\equiv$ Si–O–Si $\equiv$) groups, forming the radicals $\equiv$ Si· and ·O–Si $\equiv$ respectively. As the suspension is made in water, the recombination of those radicals plays only an inferior role. Rather than that, the free bondings are saturated forming silanole ($\equiv$Si–OH) and another silanole (HO–Si $\equiv$), respectively. Accompanied by hydroxide (OH$^-$ ions) the weak base silanole ($\equiv$Si–OH) gives away its proton easily. Due to the formation of silicate ($\equiv$Si–O)$^-$ groups on the quartz surface, the particles are negatively charged. Compensating this charge, positively charged ions are located close to the surface, forming a diffuse electrical double layer at the surface of the particles. This mechanism is valid similarly also for enamel frit particles. The thickness of the double layer determines the hydration ability of the particles and hence their tendency to sediment. The thickness of the double layer is reciprocally proportional to the field strength of the cations within the double layer. The stronger the cation is, the more the double layer is drawn tight. Strong cations lead to a coagulation of the particles forming bigger aggregates and therefore precipitation occurs. By using set-up salts, the rheology of the slurry can be easily adapted to the specific requirements of the application. Therefore, different types of clay, bentonite, polysaccharides, carboxymethyl celluloses, boron salts, sodium, potassium, magnesium and aluminum salts are used [43, 44]. For the dipping process (also vacuum process), the slurry is used with a high density so that a uniform flow and sufficient layer thickness can be achieved. The dipping process is particularly suitable for formed parts such as cooking pots. The vacuum process is used for water heaters, but in detail difficult to handle. The spraying process is particularly suitable for large flat parts such as silo panels, shower trays and bathtubs. Robots can also be easily used here. The slurry is set with less viscosity and lower density than for dipping or flooding. The use of an electromotive force for this wet application is recommended, as the layer thicknesses become more uniform and the edge coverage improves. In electrophoretic application, the part to be enamelled is connected as an anode. The negatively charged enamel particles in the slurry are guided onto the workpiece at current densities of about 3–5 A dm^{-2}. Flat workpieces are more suitable as otherwise auxiliary electrodes are needed. Complicated three-dimensional parts are not suitable for electrophoresis. Due to the deposition of the enamel particles, the conductivity at this point decreases and the next particle is deposited at a thinner point with a stronger field strength. This makes the application very uniform. At the same time, the slurry dehydrates electro osmostically and the density must therefore be controlled and kept constant. Before firing, all wet-coated parts must be dried. Only in the case of aluminum enamelling are slightly damp parts placed in the kiln.

1.6.8 Enamelled products with advantageous combinations of material properties

Advantageous combinations of material properties of enamelled products are:
- highly corrosion-resistant surface,
- strength of the base material combined with glass properties,
- scratch and abrasion-resistant surface,
- non-flammable and thermally stable,
- color-stable and non-fading surface; resistant to UV radiation,
- hygienic and easy to clean surface,
- surface with excellent sliding properties,
- environmentally friendly production and processing,
- durable, sustainable and fully recyclable product.

Enamel is used for numerous applications in household and industry. The best known household applications are listed below. The used base material is indicated in parenthesis (steel – S, cast iron – CI, aluminum – Al, titanium – Ti, stainless steel – iS, copper – Cu, silver – Ag, gold – Au, PROTHECAST® – PC, dental metal inlay – mi, ceramics – c, glass – g).
- cookware and frying utensils (S, CI, Al, iS),
- cookers (S, CI),
- oven cavities and baking trays, microwave ovens (S),
- grills, grates, indoor and outdoor fireplaces (S, CI, Al),
- radiant panel heaters with a high emission factor (S),
- hot-water tanks and instantaneous water heaters (S, iS),
- sinks, shower trays and bathtubs (S, CI),
- exhaust pipes for heating systems or corrosive air (S),
- blackboards for schools and whiteboards for conference rooms (S),
- ironing soles, hair straightener (Al),
- profile moldings for stairs and transitions (Al),
- furniture (CI, S) and
- jewelry (S, iS, Cu, Ag, Au).

Industrial applications are:
- heat exchangers for power plants (S),
- architectural panels for tunnel cladding and facades as well as for interior design (S),
- signs and placards that years later still look the same as they did on the day they were installed (S),
- cladding for petrol pumps (S),
- vessels for molten metals (S, iS),

- bolted tanks and silos for the storage of drinking water, wastewater/sewage treatment, process water, industrial effluent and agricultural waste product storage (S),
- glass-lined tanks, reactors and pipes for chemical and pharmaceutical industry (S, iS),
- valves, pressure fittings and water hydrants for the hygienic supply of drinking water and untreated water (CI, S),
- washing fountains (CI),
- watering holes for cattle (CI),
- sonotrodes or tuning forks (Ti, Al, S, iS),
- labeling for bottles (g, Al).

Medical applications:
- Artificial hip joints (PC),
- dental prostheses (mi, c).

The enamelling industry has been very active in the development of industrial standards. The committees are ISO TC 107 WG2 and CEN TC 262 WG5 as well as the national mirror committees. In 2013, the German Institute for Standardization DIN published a 509-page book of standards used in the enamel industry for the first time [45].

References

[1] DEV Deutscher Email Verband. Email – Emaillieren – Emaillierung: Einführung in Die Technologie, IBE, Hagen, 2013.
[2] DEV Deutscher Email Verband. An Introduction to Enamelling Technology and the Application of Vitreous/Porcelain Enamel, IBE, Hagen, 2015.
[3] RAL Deutsches Institut für Gütesicherung und Kennzeichnung. RAL-RG 529 A3: Email(le) und emaillierte Erzeugnisse Begriffsbestimmungen/ Bezeichnungsvorschriften, Beuth, Berlin, 2007.
[4] Wernick L, Themelis NJ. Recycling metals for the environment. Ann Rev Energy Environ, 1998, 23, 465–97.
[5] Bayer AG. Rosten ist kein Schicksal. Bayer Res, 2002, 16, 74–77.
[6] Fraunhofer-Institut für Fertigungstechnik und Angewandte Materialforschung (IFAM). Broschüre Oberflächen Korrosion Erkennen-verstehen-vermeiden, Bremen, 2005.
[7] Wendler-Kalsch E. Korrosionsschadenkunde, Springer, Berlin, 1998.
[8] Bundesministerium für Umwelt (BMU). Naturschutz und Reaktorsicherheit, Produktbezogene Klimaschutzstrategien – Product Carbon Footprint verstehen und nutzen, 1. Aufl, Berlin, 2010.
[9] Wendel J. Email ist Graffiti-prohibitiv. Mitt DEV, 2003, 51, 18–21.
[10] Wendel J. Enamel is graffiti-prohibitive. Porcellain Enamel Digest, 2005, 6–8.

[11] EN ISO 4531. Vitreous and porcelain enamels – Release from enamelled articles in contact with food – Methods of test and limits, CEN TC 262 WG5, 2018.

[12] Council of Europe. Resolution CM/Res (2013) 9 on Metals and Alloys Used in Food Contact Materials and Articles, Strasbourg, 2013.

[13] Wendel J. Grenzwerte für Metalle, Spurenelemente oder Gift?, Teil 1. Mitt DEV, 2015, 63, 38–47.

[14] Wendel J. Limits for metals, part 1. In: 23rd IEC, Florence, Italy, 2015. https://ieiworlddotorg. files.wordpress.com/2020/04/23-wendel.pdf

[15] Wendel J. Grenzwerte für Metalle, Spurenelemente oder Gift?, Teil 2. Mitt DEV, 2020, 68, 34–43.

[16] Wendel J. Limits for metals, part 2. In: 24th IEC Chicago, USA, 2018. https://ieiworlddotorg. files.wordpress.com/2020/05/pei_24thenamellerscongress_papers.pdf

[17] Wendel J. Grenzwerte für Metalle, Spurenelemente oder Gift?, Teil 3. Mitt DEV, 2021, 69, 24–32.

[18] Umweltbundesamt. Bewertungsgrundlage für Emails und keramische Werkstoffe im Kontakt mit Trinkwasser (Email/Keramik-Bewertungsgrundlage). https://www.umweltbundesamt.de/ sites/default/files/medien/5620/dokumente/anlage_1_-_email-bwgl_1._aenderung_de_0. pdf Bad Elster, 2021. Accessed February 09 2022.

[19] German Environmental Agency. Evaluation Criteria Document for enamels and ceramic materials in contact with drinking water (Enamel and Ceramics Evaluation Criteria Document). https://www.umweltbundesamt.de/sites/default/files/medien/5620/dokumente/bewer tungsgrundlage_fur_email_und_keramik_1.anderung_en.pdf Bad Elster, 2021. Accessed February 09 2022.

[20] Umweltbundesamt. Begründung der Prüfwerte der Bewertungsgrundlage für Emails und keramische Werkstoffe. https://www.umweltbundesamt.de/sites/default/files/medien/ 3521/dokumente/begruendung_pruefwerte_email-bewertungsgrundlage_rev01_final.pdf Bad Elster, 2019. Accessed February 09 2022.

[21] Environmental agency. Rationale for the reference concentrations of the evaluation criteria for enamels and ceramic materials that come into contact with drinking water (enamel/ ceramic evaluation criteria). https://www.umweltbundesamt.de/sites/default/files/medien/ 3521/dokumente/begrundung_prufwerte_email-bewertungsgrundlage_rev01_final_englisch. pdf Bad Elster, 2019. Accessed February 09 2022.

[22] Rawson H. Oxide glases. In: Cahn RW, Haasen P, Kramer EJ. (Eds.) Glasses and Amorphous Materials, vol. 9, Wiley VCH, Weinheim, 1991, 279–330.

[23] Fluegel A. Thermal Expansion Calculation of Silicate Glasses at 210 °C: Based on the Systematic Analysis of Global Databases. glassproperties.com – December 2007. Accessed February 09 2022.

[24] Scholze H. Glas: Natur, Struktur und Eigenschaften, Springer, Berlin, Heidelberg, 1988.

[25] Dietzel AH. Emaillierung: Wissenschaftliche Grundlagen und Grundzüge der Technologie, Springer, Berlin, Heidelberg, 1981.

[26] Dietzel AH. Die Aufklärung des Haftproblems bei der Eisenblechemaillierung – explanation of the adhesion problems in enamelling iron sheets. Sprechsaal, 1935, 68, 3–6.

[27] Dietzel AH. Theory of the adherence enamel to iron. Sprechsaal, 1945, 78, 8–9.

[28] Striepe S, Bornhöft H, Wendel J, Deubener J. Microalloy precipitation at the glass-steel interface enabling adherence of porcelain enamel. Int J Appl Ceram Technol, 2016, 13, 191–99.

[29] EN 10209. Cold rolled low carbon steel flat products for vitreous enamelling – Technical delivery conditions. ECISS/TC 13, 2013.

[30] EN 10111, Continuously hot rolled low carbon steel sheet and strip for cold forming – Technical delivery conditions. 2008.
[31] EN 10025-1, Hot rolled products of structural steels – General technical delivery conditions. 2005.
[32] EN 10149-1, Hot-rolled flat products made of high yield strength steels for cold forming – Part 1: General delivery conditions, 2013.
[33] Wendel J. Haftung von Email auf Gusseisen. Mitt DEV, 2005, 53, 61–66.
[34] Wendel J. Adherence on cast iron. In: Proceedings of the 20th International Enamellers Institute Congress, Istanbul, Turkey, May 15-19 2005.
[35] Wendel J. Die EEA-Empfehlung für emaillierfähiges Gusseisen. Mitt DEV, 2011, 59, 42–47.
[36] Wendel J. The EEA recomendation for enamelable cast iron. Porcellain Enamel Digest, 2012, 12–19.
[37] Darken LS, Gurry RW. The system iron-oxygen, The Wüstite field and related equilibria. J Am Chem Soc, 1945, 67, 1398–412.
[38] Darken LS, Gurry RW. The system iron-oxygen, equilibrium and thermodynamics of liquid oxide and other phases. J Am Chem Soc, 1946, 68, 798–816.
[39] ISO 5832-4, Implants for surgery – Metallic materials – Part 4: Cobalt-chromium-molybdenum casting alloy, NA 027-02-06 AA, 2014.
[40] Kühn W. Möglichkeiten und Grenzen der Emaillierung von Leichtmetallen. Oberflächen Polysurfaces, 2009, 2, 6–9.
[41] Wendel J, Hellmold P. Vitreous Enamel – A Highly Efficient Material Compound. 2007. In: Proceedings of Electron Cloud Clearing – Electron Cloud Effects and Technological Consequences "ECL2", CERN, Geneva. Switzerland, 2007, 16–19.
[42] Petzold A, Pöschmann H. Email und Emailliertechnik, Dtsch Verl Grundstoffindustrie, Leipzig, 1986, 2. Aufl, 2001.
[43] Wendel J. Rheology – Complex by interaction. In: 21th IEC Shanghai, China, 2008, 223–33, https://ieiworlddotorg.files.wordpress.com/2020/04/j_wendel.pdf. Accessed February 09 2022.
[44] Wendel J. Rheologie – kein Buch mit sieben Siegeln. Mitt DEV, 2009, 57, 18–26.
[45] DIN-Taschenbuch 486. Emails und Emaillierungen, Beuth, Berlin, 1. Aufl, 2013.

1.7 Flame retardants

Andreas Termath

1.7.1 Introduction

Destructive fires have been a thread for mankind for centuries. While the medieval urbanization and centralization of living has increased the fire danger due to the reliance on inherent combustible construction materials like wood, the industrialization age has brought additional challenges for the fire protection of lives and assets.

The urbanization culminated during the twentieth and twenty-first centuries in the creation of megacities like Dhaka and Kinshasa reaching population densities of 30,000 inhabitants/km^2. The space requirements led to the construction of high-rise buildings with limitations on evacuation capacities and firefighting measures (Figure 1.7.1). Furthermore, the demands on safety in public transportation were fueled by the upcoming globalization and the increased mobility resulting in strict fire-safety regulations in aviation, marine and railway.

Figure 1.7.1: A fireworks accident at the Address Hotel near Burj Khalifa (Dubai, UAE) on January 2016. With written permission from Shutterstock (foto 781093054; transaction number 169909).

In addition, the use of non-natural materials based on combustible raw materials lead to different ignitability and fire scenarios that need to be answered by using modern flame-retardant technologies.

https://doi.org/10.1515/9783110733143-007

1.7.2 Fire development and FR-mode of action

Three key criteria need to be fulfilled for the development of a fire, as indicated by the "fire triangle" (Emmons [1, 2] triangle) as shown in Figure 1.7.2.

Note: Newer conceptions consider the radical chain reaction mechanism required for the advancing combustion process as a fourth pillar, forming a "fire tetrahedron" [3]. For additional details on the reaction mechanisms involved in the flame itself, kindly refer to 1.7.2.2 radical-inhibiting flame retardants/gas-phase active flame **retardants**.

– Ignition source
– Heat, radiation sources (UV, IR, µw, gamma rays), flash, spark
– Combustible (oxidizable) material
– Oxidative atmosphere (mostly air, oxygen, in rare cases, non-oxygen containing atmospheres may also contribute to a fire)

Figure 1.7.2: Fire triangle representing the requirements for the evolvement of a fire.

If at least one of the three components is not present in a sufficient concentration or fully absent, the start of a fire is not possible. It shall be mentioned that the bulk material is not directly participating in the combustion process, but rather the gaseous pyrolysis products formed by thermal stress, impacting the surface (*vide infra* for details).

However, the ubiquitous availability of combustible materials and oxygen (air) renders the development of fires rather easy (Figure 1.7.3). In most cases, conflagrations are caused by the initial formations of small fires that are able to quickly develop. The formation of sufficient, flammable pyrolysis gases may lead to a flash-over, causing the fire to rise in temperature and fully penetrate structures and buildings. For a detailed review on different types of fire, we refer to Wakefield et al. and references therein [4].

The above-mentioned pyrolysis gases like carbon monoxide, alkanes, alkenes, etc. are further oxidized in the presence of oxygen, to final combustion products including hazardous asphyxiant chemicals like carbon oxides, hydrogen cyanide and others [5, 6]. Besides organic irritants (acrolein, formaldehyde, etc.), inorganic irritating combustion products like nitrogen oxides, sulfur oxides and hydrogen halides pose a significant thread when being exposed to combustion products. In fact, the large majority of fire casualties perish due to the exposure to toxic fumes [7–9].

Development of a fire

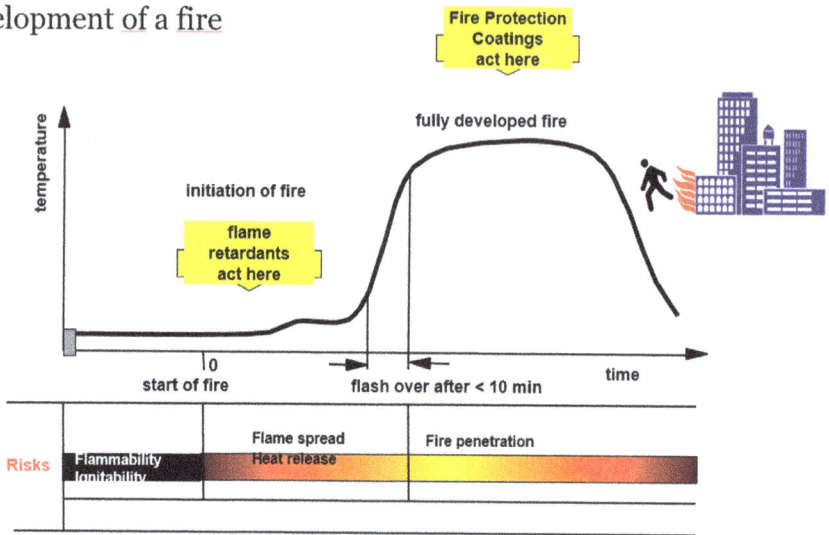

Figure 1.7.3: Heat-development of a fire as a function of time.

1.7.2.1 Ignition sources

The initial heat transfer from a commonly external source might be generally separated into four groups: chemical reaction energy, electrical heat energy, mechanical heat energy and heat generated via nuclear decomposition [10, 11].

Flame retardants can retard or even prevent the formation of developed fires by various modes of action. In contrast to fire retardants or fire-stopping technologies, flame retardants are designed for dealing with non-developed fires and low-energy threads like

– smoldering fires (cigarettes, non-burning combustion),
– electrical shortages,
– open flames (candles, matches, etc.) and
– residential heating.

The above-mentioned ignition sources represent the large majority of fire causes [11, 12 and references therein].

While the selection of bulk materials is often limited due to various demands on mechanical properties, costs, sustainability and others, ignition sources cannot be fully excluded during their life cycle (Figure 1.7.4). Therefore, flame retardants must act during the combustion process itself.

The exothermic combustion process is creating a thermal feedback, affecting the surface of the flammable material and therefore is responsible for the generation of additional pyrolysis products that will contribute to an even more severe combustion process. Here, the cycle starts over again.

ACTION IN THE GASPHASE

Thermal feedback

Oxygen

Heat

Interruption of radical chain mechanism:
Br, Cl, P

Heat
Combustion products
Smoke

Combustible gases

Dilution by water formation:
ATH, MDH

Carbonaceous layer by
P, N, B, intumescence,
nanocomposites

Decomposition area

Carbonization

Cooling (endothermic) and substrate dilution: ATH, MDH

ACTION IN THE CONDENSED PHASE

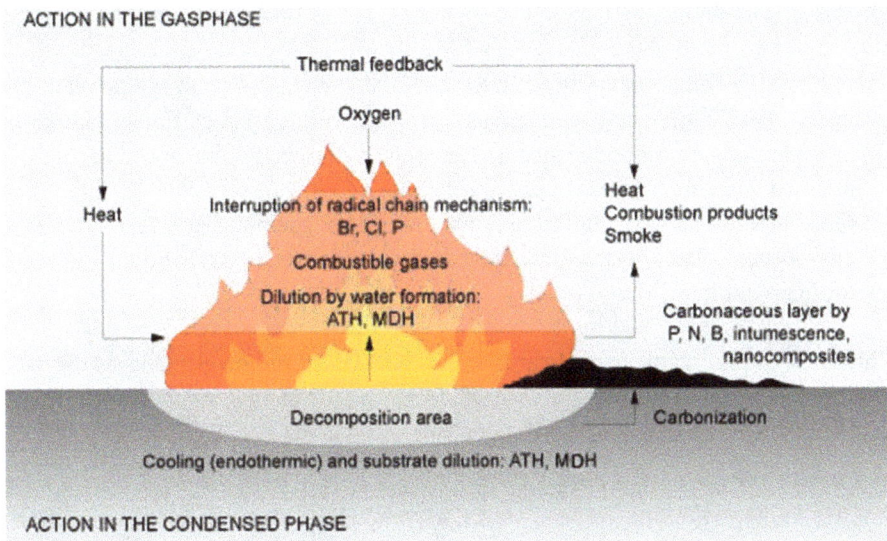

Figure 1.7.4: Detailed combustion cycle describing the various points of attack of the flame retardants.

The thermal feedback itself might be of radiative, conductive, or convective nature [13]. With increasing temperature difference in fully developed fires, the radiative heat transfer is the major contributor

$$\dot{Q_{1\rightarrow2}} = \sigma A_1 F_{1\rightarrow2}\left(T_1^4 - T_2^4\right) = \sigma\varepsilon\left(T_1^4 - T_2^4\right) \ in \ kW/m^2$$

where A is the surface area, F the view factor from surface 1 to 2, σ the Boltzmann constant and ε the emissivity.

1.7.3 Flame retardants

As their name suggests, flame retardants might not be able to fully suppress the generation of a fire but to prolong the time to ignition (flaming combustion) and the formation of developed fires allowing additional time for firefighting measures and/or for evacuation efforts.

However, to name only a few, the time-to-ignition (TTI), the heat-release rate (MARHE), the flame spread, the smoke generation and smoke toxicity are measurable parameters of interest for assessing the performance of materials containing flame retardants.

1.7.4 Modes of action

Flame retarding compounds can act during different stages, all with the target of retarding or even interrupting the combustion cycle. In this respect, flame retardants can be separated into different categories, depending on their mode of action [14]:
- ablative flame retardants,
- radical chain-reaction inhibiting flame retardants (gas-phase active),
- char-forming flame retardants (active in the condensed phase) and
- intumescent flame retardants.

While most of the organic flame retardants do contain non-carbon elements as their active structure moiety, in this chapter we will focus on pure inorganic compounds only.

1.7.4.1 Ablative flame retardants

Ablative flame retardants are heat-sensitive compounds that decompose at comparably low temperatures. Therefore, they can consume substantial amounts of reaction heat which is not available for the fire circle as described above. Prominent and applied examples are **a**luminum **t**ri-**h**ydroxide (ATH) (hydrargillite and gibbsite; decomposition temperature 170–300 °C) and magnesium dihydrate (decomposition temperature 340–450 °C), while ATH is globally the most important inorganic flame retardant (Figure 1.7.5).

Ablative (*lat.: ablatio*, removal, taking away) flame retardants usually need to be added in high loads (>100 phr) to be sufficiently active. Their filler-like properties limit the application to those, where the mechanical impact on the substrate can be tolerated. While the heat consumption through endothermic decomposition represents the major flame-retardant effect, ablative compounds like ATH in addition emit substantial amounts (approx. 0.4 L/g) of (gaseous) water in the case of fire. The oxygen-containing atmosphere is diluted, which also reduces the severity of the fire evolvement. In addition, the smoke generation in general is significantly suppressed. The remaining aluminum oxide has a high internal surface area allowing the incorporation of sooty chemical compounds and shielding the underlying areas from further heat transfer.

Borates and ammonium salts of strong inorganic acids like phosphates and sulfates may also be regarded as ablative flame retardants, due to their ammonia release at comparably low temperatures. However, due to the remaining strong acid, those compounds will also have a charring effect on the substrate and therefore will be discussed in Chapter 1.7.4.6 (Char-forming flame retardants).

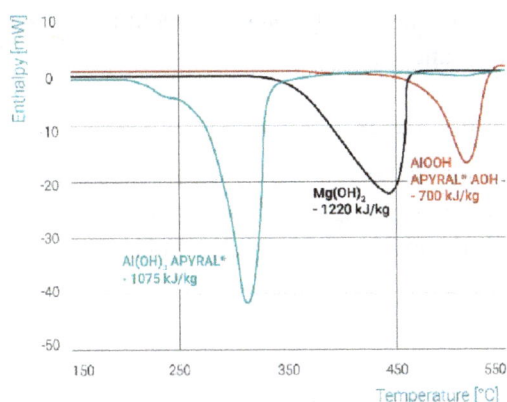

Differential scanning calorimeter data (DSC) of the mineral flame retardants **APYRAL®, APYRAL® AOH** and magnesium oxide **Mg(OH)₂**

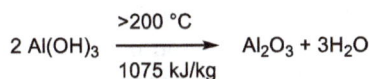

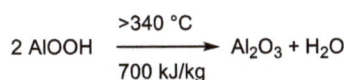

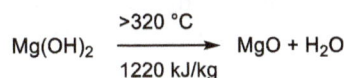

$$2\,Al(OH)_3 \xrightarrow[\text{1075 kJ/kg}]{>200\,°C} Al_2O_3 + 3H_2O$$

$$2\,AlOOH \xrightarrow[\text{700 kJ/kg}]{>340\,°C} Al_2O_3 + H_2O$$

$$Mg(OH)_2 \xrightarrow[\text{1220 kJ/kg}]{>320\,°C} MgO + H_2O$$

Figure 1.7.5: DSC data and decomposition reactions of metal hydroxides. With courtesy and copyright agreement of Nabaltec AG.

1.7.4.2 Radical-inhibiting flame retardants/gas-phase active flame retardants

The formation of incomplete combustion products and radicals is essential for the formation of flames and the propagation of the combustion. A sufficient reactivity is required to maintain a flame. Hydroxyl radicals, hydrogen radicals and methylidyne radicals within others can be observed as the most radical propagating species in flames [15, 16]. The strongly exothermic oxidation of CO to CO_2 is facilitated by the participation of the OH-radical.

$$\cdot\,OH + CO \longrightarrow CO_2 + H\cdot$$

Chemical action of gas-phase active flame retardants interferes with the combustion processes by eliminating the high-energy H and OH radicals.

Note: If sufficient energy is present, those reactive intermediates might be excited and relax to emit light in the visible spectrum:

$\dot{C}$ – H 432 nm and

:C = C: 436 nm, 475 nm, 520 nm (Swan – bands) [17].

Halogen halides emitted from halogenated flame retardants, metal halogen compounds formed from antimony trioxide, and phosphorous-containing fragments from phosphorous flame retardants are the most common radical-inhibiting flame retardants ("flame poisoning").

1.7.4.3 Halogenated flame retardants

The utilization of purely inorganic halides like ammonium halides has been widely reduced due to concerns on stability and toxicology. Over the last decades, even the industrial use of organic halide compounds of low molecular weight has been reduced in favor of the higher stable, non-migrating polymeric halides [18]. An advantage of those molecules is the often, trouble-free incorporation into organic polymer matrices like polypropylenes, polystyrenes and others due to their similar chemical and mechanical properties.

While the reactivity drops as follows, HI > HBr > HCl > HF, also stability concerns of the C–X bond need to be considered when selecting the right flame retardant. The practical use is therefore often limited to the use of bromide and chloride compounds.

Halogen acids (HX) are believed to react with H- and OH-radicals and deliberating the less reactive halogen radicals. Further H-abstraction of C–H bonds leads to even less energetics C-radicals where the hydrogen halides are regenerated.

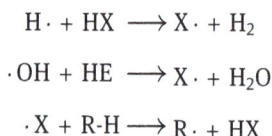

$$H\cdot + HX \longrightarrow X\cdot + H_2$$

$$\cdot OH + HE \longrightarrow X\cdot + H_2O$$

$$\cdot X + R\text{-}H \longrightarrow R\cdot + HX$$

While not being an active flame retardant itself, antimony oxide (Sb_2O_3) might be added as a potent synergist (Scheme 1.7.1) allowing the dosing reduction of the FR-compound or for the FR-boosting of halogen-containing polymers like, PVC or additivated PE or PS [19–21].

Most organic halogen-containing flame retardants break down in the case of fire while omitting hydrogen halides. Antimony(III) oxide reacts under the emission of water to the volatile antimony trihalide which represents the active flame retardant [19]. In addition, the fine SbO particles formed in the process, promote the recombination of radicals in a so-called wall effect [14, 22–24].

$$Sb_2O_3 + 6\,HX \rightleftharpoons 2\,SbX_3 + 3\,H_2O$$

$$SbX_3 + H^{\bullet} \longrightarrow SbX_2 + HX$$

$$SbX_2 + H^{\bullet} \longrightarrow SbX + HX$$

$$SbX + H^{\bullet} \longrightarrow Sb + HX$$

$$Sb + O \longrightarrow SbO$$

$$SbO + H^{\bullet} \longrightarrow SbOH$$

$$SbOH + H^{\bullet} \longrightarrow SbO + H_2$$

Scheme 1.7.1: Synergistic reaction of antimony oxide in the presence of hydrogen halides.

1.7.4.4 Red phosphorous

Providing a maximum concentration of P-atoms, red phosphorous turns out to be a highly efficient flame retardant for a variety of applications [25–30]. In the oxidative fire environment, red phosphorous passes its various oxidation states up to P(V):

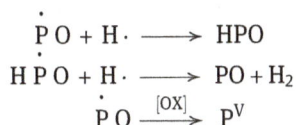

$$\overset{\bullet}{P}O + H\cdot \longrightarrow HPO$$
$$HP O + H\cdot \longrightarrow \overset{\bullet}{P}O + H_2$$
$$\overset{\bullet}{P}O \xrightarrow{\;[OX]\;} P^V$$

While the PO-radical has been as the most efficient gas phase active radical [22, 31, 32], further oxidation products will be active in the condensed phase promoting a charring of the substrate (*vide infra*).

Despite its maximum efficiency, red phosphorous has some limitations due to its naturally occurring red color. Additionally, the poor stability towards moisture accompanied by the release of toxic phosphine gas requires the addition of trapping agents (e.g. $AgNO_3$, MoS_2 and others) [33, 34]:

$$P_4 + 6H_2O \longrightarrow PH_3 + 3\,H_3PO_2$$

1.7.4.5 Aluminum hypophosphite (AHP)

The partially oxidized P compounds also do find application as gas-phase active flame retardants. A prominent inorganic member of this family is aluminum hypo-phosphite $Al(H_2PO_2)_3$ (Scheme 1.7.2). It is widely used as a flame retardant for a variety of thermoplastic polymers like PBT [35], PS [36], PA [37, 38], PU [39, 40], PP [41] and others [42].

The mostly insoluble compound provides large gas phase activity while the aluminum ions promote char formation in the condensed phase. The compound is readily

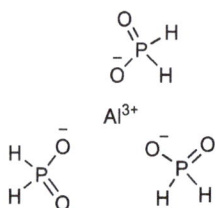

Scheme 1.7.2: Aluminum hypophosphite.

available by precipitating from hypophosphoric acid or disodium hypophosphite solution [43].

1.7.4.6 Char-forming flame retardants/active in condensed phase

The carbonization of combustible materials to less-volatile compounds is catalyzed mostly by strong acidic compounds. Due to compatibility issues with many compound matrices, they are often released upon thermal stress, only.

1.7.4.7 Ammonium (poly)phosphates

The ammonium salts of polyphosphoric acid are frequently used char-promoting agents. The strong acid formed upon decomposition dehydrates the (polymer) matrix and the vitreous layer.

Long-chained ammonium polyphosphates (APP) of crystal structure II exhibit a comparably low water solubility. However, the low thermal stability of APP-II (weight loss $T_{2\%} = 300$ °C) limits the application to non-engineering plastics.

1.7.4.8 Borates

The comparably low decomposition temperature of borax pentahydrate ($Na_2O \cdot 2B_2O_3 \cdot 5H_2O$), boric acid ($B(OH)_3$), ammonium pentaborate (($NH_4)_2O \cdot 5B_2O_3 \cdot 8H_2O$) and many other boric acid salts limit their use to applications without thermal processing involved [44]. In cellulosic applications, a major FR contribution comes from the endothermic release of water. The acidic compounds formed, additionally catalyze the dehydration of carbohydrates. The glass formation of boric oxides thermally shields the surface, and significantly reduces the smoke emission [45–47]; oxygen adsorption of reactive pyrolysis products on the surface is reduced via electron transfer processes [44].

An exemption to the low thermal stability mark zinc borates [48, 49], where $2ZnO \cdot 3B_2O_3 \cdot 3.5H_2O$ represents the most prominent compound being widely used in the flame retardancy of engineering plastics, especially polyamides [50–52].

In the combination with alkyl phosphinate salts and others, the addition of zinc borates enhances the processing stability of synergistic combinations [53, 54].

1.7.4.9 Zinc stannate

Zinc stannate, as well as zinc hydroxy stannate are used in combination with other flame retardants as efficient, less toxic antimony oxide alternative [55, 56]. In the presence of halogenated polymers or hydrogen-chloride-releasing flame retardants, Tian et al. and Horrocks et al. propose the formation of the strong Lewis-acid $SnCl_2$ and $SnCl_4$, promoting the polymer cross-linking and increasing charring [57, 58].

1.7.4.10 Carbon-based nanomaterials

The substantially larger interface surface area of nano-sized flame retardants, associated with a magnitude lower filler-to-filler distance when incorporated into polymer matrices facilitates the formation of dense char structures in the case of fire. The utilization of carbon black in polymer matrices, apart from the diluting effect (reduction of combustible material), shows synergistic effects in the combination with a variety of FRs [59–61]. Apart from char formation, the delocalized pi-electron systems are able to efficiently quench single-electron processes.

Layered carbon structures like graphene form a labyrinth-like structure offering a unique protection mechanism as quoted from Araby et al. [62].
(i)   The interlocking carbonaceous structures of the char layer act as a thermal barrier, insulating the virgin polymer underneath from heat feedback [63].
(ii) The char layer slows pyrolysis of virgin materials and reduces the overall volume of combustible elements.
(iii) The char layer formed on the surface creates "tortuous pathways" providing lengthy exchanging pathways between the volatiles and the gas phase and therefore minimizing gas permeability [63, 64].

1.7.4.11 Clay nanocomposites

The utilization of nanoclays based on various minerals like bentonite, montmorillonite and cloisite was subject to extensive research in the last decade [65, 66 and references therein]. By intercalation and exfoliation into the polymer matrix, comparably low dosing levels of the filler-like components can be achieved. However, for the efficient performance the presence of conventional FRs might be required [67–69].

1.7.4.12 Intumescent flame retardants (formulations)

Intumescent (*lat. intumere*, swelling, foaming) flame retardants often consist of a variety of chemical compounds all fulfilling single or combined functions to build-up a foam or char-like structure with an increased ability to protect underlying substrates by a heat-shielding effect (Figure 1.7.6).

The protection of materials by intumescent systems is in general superior to the above-mentioned approaches, allowing the protection of underlying structures in fully developed fires. If the incorporation of intumescents into the bulk material is not

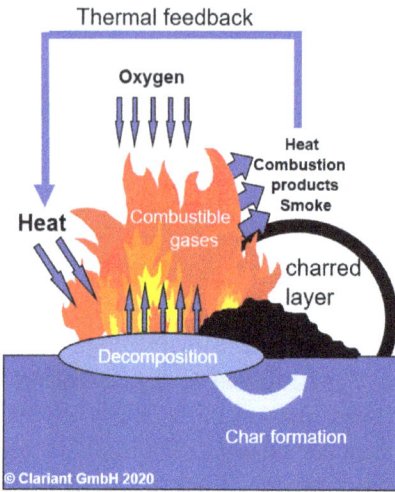

Figure 1.7.6: Decomposition sketch of intumescent flame retardants.

possible (e.g. for natural materials like wood or steel) or would alter the mechanical properties, intumescent coatings (Figure 1.7.7) can be applied onto the materials [70].

Figure 1.7.7: The intumescent foam reaches a thickness of 10 to 100 times of the originally applied coating and insulates the substrate through its low thermal conductivity.

These approaches are widely employed in the areas of public constructions and public transport (public buildings, energy sector, electromobility and others).

The following key functionalities need to be covered within an intumescent formulation to achieve an intumescent effect:

- carbonization agent (e.g. polyalcohols, starch and others),
- acid donor (e.g. ammonium polyphosphate, piperazine pyrophosphate, borates, sulfates, melamine polyphosphate and others) – dehydrates the polyalcohol to form a carbonaceous char and
- spumific (e.g. melamine, urea) – decomposes to release volatiles required for the char formation.

For an optimized protection, performance must be chemically and quantitatively optimized. Further inorganic additives may be added to stabilize the char. If incorporated in polymer matrices, the reaction of the core materials may be described as follows:

1. softening of the (thermoplastic) polymer (by melting and/or subsequent decomposition),
2. release of acidic components,

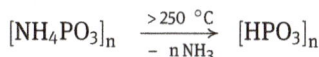

$$[NH_4PO_3]_n \xrightarrow[-\,n\,NH_3]{>250\ °C} [HPO_3]_n$$

3. dehydration of the polyalcohol component forming carbonaceous structures,

$$[HPO_3]_n + C_x(H_2O)_m \longrightarrow [HPO_3]_n \cdot mH_2O + ["C"]_X$$

4. release of non-combustible gases and expansion by the endothermic decomposition of the spumific (synchronized process) and

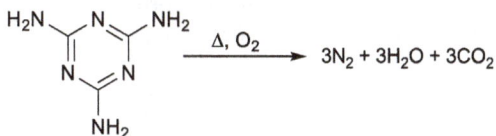

$$\text{(triazine structure)} \xrightarrow{\Delta,\ O_2} 3N_2 + 3H_2O + 3CO_2$$

5. solidification by crosslinking reactions and development of an inorganic char (optional).

$$H-[PO_3]_n-OH + TiO_2 \longrightarrow TiP_2O_7 + H_2O + H-[PO_3]_{n-2}-OH$$

Since the carbonaceous char may be oxidized in the further progress of the fire, additives like TiO_2 or other metal oxides might be used for increased stability by forming an inorganic layer consisting of titanium pyrophosphate structures [71–73]. Malucelli et al. have described DNA as a naturally occurring intumescent system incorporating all three key components in one structure moiety [74–76].

1.7.4.13 Expandable graphite

Melamine and its salts currently are the most important spumifics being employed in intumescent formulations due to an optimized thermal stability accompanied with comparably low water-solubility [70].

As a fully inorganic alternative forming a carbonaceous char [77], expandable graphite (EG) is successfully employed for flame-retarding polyurethane foams (PU-rigid insulation panels as well as soft foams) [78]. It is also used in the combination with red phosphorous or ammonium polyphosphates. For the formation of EG (Figure 1.7.8), naturally occurring graphite flakes are partially oxidized using potassium permanganate, chromium(VI) oxide or electrochemical methods and intercalated with sulfuric or nitric acid forming the graphite intercalation compound (GIC)

[79]. In a recent development, Hou et al. describe the formation of EG at comparably low temperatures using $K_2S_2O_8$ [80].

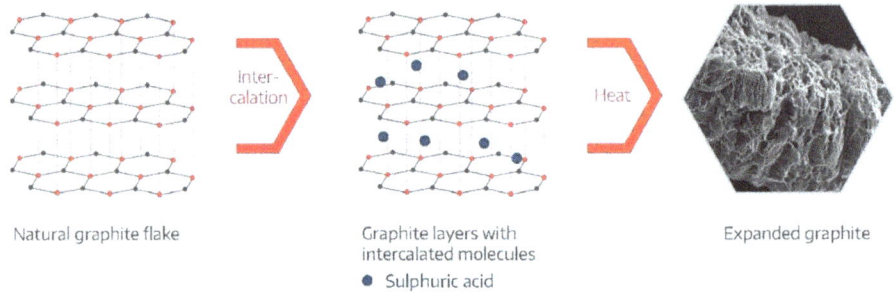

Natural graphite flake   Graphite layers with intercalated molecules   Expanded graphite

● Sulphuric acid

Figure 1.7.8: Synthesis of EG by intercalation of sulfur- or nitrogen-containing compounds. With courtesy and copyright agreement of Graphit Kropfmühl GmbH, Hauzenberg, AMG Graphite GK.

The low on-setting temperature of EG-compounds starting at approx. 180 °C allows the protection of heat-sensitive substrates, while the fluffiness of the char itself might limit the application to non-abrasive fire scenarios.

The layered structure makes EG vulnerable to applications containing processing steps with higher shear-forces (like mixing for coating applications or high-pressure polyurethane applications).

1.7.5 Outlook and challenges

1.7.5.1 Incorporation and compatibilization

The often filler-like character of flame retardants frequently renders their incorporation challenging within the different bulk materials. Balancing the influence on processability, stability and other mechanical properties as well as the flame-retardant performance, remains a constant challenge for modern flame retardants. Various approaches towards compatibilization and incorporation has been undertaken to facilitate the use of FRs and minimize their impact on bulk materials. Approaches include surface treatments, encapsulation, additivation, reactive incorporation and others.

1.7.5.2 Sustainability, persistence and recyclability

Flame retardants often do have to remain within their substrates for decades without deterioration, but still need to remain active and potent in the case of fire to ensure optimal protection. The obvious discrepancy between persistence and chemical reactivity remains a challenge for FR producers throughout.

1.7.5.3 Exposure

While flame retardants are present in a variety of consumer goods, the release of chemicals into the environment needs to be minimized, not only to maintain the highest possible fire protection, but also to reduce the exposure of chemicals in general on the user. Therefore, flame retardants are under special scrutiny concerning their toxicological and ecotoxicological profile. Low mobility, either by good matrix incorporation or reactive immobilization, combined with a low hazard profile are basic needs for modern flame retardants.

References

[1]   Emmons HW. Ann Rev Fluid Mech, 1980, 12, 223–36. 10.1146/annurev.fl.12.010180.001255.
[2]   Emmons HW, Atreya A. Sadhana, 1982, 5, 259–68.
[3]   Lazar ST, Kolibaba TJ, Grunlan JC. Nature Rev Mater, 2020, 5, 259–75. 10.1038/s41578-019-0164-6.
[4]   Wakefield JC. A Toxicological Review of the Products of Combustion. Health Protection Agency, Centre for Radiation, Chemical and Environmental Hazards, Chemical Hazards and Poisons Division, Chilton, Didcot, Oxfordshire O11 0RQ, 2010, pp. 44.
[5]   Stec AA, Hull TR. Fire Toxicity, 1st edition, Woodhead Publishing, Elsevier, Amsterdam, 2010, 728.
[6]   Gann RG, Bryner NP. Combustion Products and Their Effects on Life Safety, Vol. 1, Editor National Fire Protection Assoc, Quincy, MA, 2008.
[7]   Levin BC. Nature, 1982, 300, 18. 10.1038/300018a0.
[8]   Terrill JB, Montgomery RR, Reinhardt CF. Science, 1978, 200, 1343–47.
[9]   Beritic T. British Med J, 1990, 300, 696–98.
[10]   McKinley C. Dangerous Properties of Industrial Materials, J Wiley & Sons, Hoboken, NJ, 1979.
[11]   Stellman JM. Encyclopaedia of Occupational Health and Safety, 4th edition, International Labor Office, Geneva, Switzerland, 1998.
[12]   Troitzsch JH. J Fire Sci, 2016, 34, 171–98. 10.1177/0734904116636642.
[13]   Hurley MJ, Gottuk DT, Hall Jr JR, Harada K, Kuligowski ED, Puchovsky M, Torero JL, Watts Jr, JM, Wieczorek CJ. (Eds.) SFPE Handbook of Fire Protection Engineering, Springer-Verlag, New York, 2016.
[14]   Camino G, Costa L. Polymer Degr Stab, 1988, 20, 271–94. https://doi.org/10.1016/0141-3910(88)90073-0.
[15]   Crosley DR, Jeffries JB, Smith GP. Isr J Chem, 1999, 39, 41–48. https://doi.org/10.1002/ijch.199900004.
[16]   Wilson JWE. Fristrom RM. APL Techn Digest, 1963, 2, 2–7.
[17]   Pretty WE. Proc Phys Soc, 1927, 40, 71–78. 10.1088/0959-5309/40/1/313.
[18]   Xiong P, Yan X, Zhu Q, Qu G, Shi J, Liao C, Jiang G. Environ Sci Technol, 2019, 53, 13551–69.
[19]   Lum RM. J Polymer Sci: Polymer Chem Ed, 1977, 15, 489–97. https://doi.org/10.1002/pol.1977.170150223.
[20]   Yang YP, Brewer DG, Venart JES. Fire Mater, 1991, 15, 37–42. https://doi.org/10.1002/fam.810150107.
[21]   Lu H, Wilkie CA. Polymer Degr Stab, 2010, 95, 564–71. https://doi.org/10.1016/j.polymdegradstab.2009.12.011.

[22] Cullis CF, Hirschler MM. The Combustion of Organic Polymers, Clarendon Press, Oxford, 1981.
[23] Kuryla WC, Papa AJ. (Eds.) Flame Retardancy of Polymeric Materials, Vols 1–5, Dekker, New York, 1973–79.
[24] Cullis CF. (Ed.) Developments in Polymer Degradation, Vol. 3, Applied Science Publishers, London, 1981.
[25] Braun U, Schartel B. Macromol Chem Phys, 2004, 205, 2185–96.
[26] Zhuo JL, Dong J, Jiao CM, Chen XL. Plastics Rubber Comp, 2013, 42, 239–43. 10.1179/146580113x13650753142022.
[27] Ballistreri A, Montaudo G, Puglisi C, Scamporrino E, Vitalini D, Calgari S. J Polymer Sci: Polymer Chem Ed, 1983, 21, 679–89. https://doi.org/10.1002/pol.1983.170210304.
[28] Chen Z, Du J, Li X, Xie Z, Wang Y, Wang H, Zheng J, Yang R. RSC Adv, 2019, 9, 24935–41. 10.1039/c9ra04027g.
[29] Jiang W-J, Li Z-Z, Zhang C-Z, Fang J, Yang X-J, Lu L-D, Pu L-J. Spectr Spectral Anal, 2010, 30, 1329–35. 10.3964/j.issn.1000-0593(2010)05-1329-07.
[30] Rabe S, Chuenban Y, Schartel B. Materials, 2017, 10, 455. 10.3390/ma10050455.
[31] Hastie JW. J Res Nat Bureau Standards. Sect A. Phys Chem, 1973, 77A, 733–54. 10.6028/jres.077A.045.
[32] Granzow A. Acc Chem Res, 1978, 11, 177–83. 10.1021/ar50125a001.
[33] Laoutid F, Bonnaud L, Alexandre M, Lopez-Cuesta J, Dubois P. Mater Sci & Eng R: Rep, 2009, 63, 100–25.
[34] Fu S, Song P, Liu X. In: Fan M, Fu F. (Eds.) Advanced High Strength Natural Fibre Composites in Construction, Woodhead Publishing, Elsevier, Amsterdam, 2017, 479–508. ISBN 9780128210840.
[35] Brehme S, Köppl T, Schartel B, Altstädt V. e-Polymers, 2014, 14, 193–208. DOI: 10.1515/epoly-2014-0029.
[36] Yan Y-W, Huang J-Q, Guan Y-H, Shang K, Jian R-K, Wang Y-Z. Polymer Degrad Stab, 2014, 99, 35–42. https://doi.org/10.1016/j.polymdegradstab.2013.12.014.
[37] Cheng X, Wu J, Yao C, Yang G. J Fire Sci, 2019, 37, 193–212. 10.1177/0734904119836208.
[38] Guo C, Zhao Y, Ji G, Wang C, Peng Z. Polymers, 2020, 12, 2323.
[39] Wu S, Deng D, Zhou L, Zhang P, Tang G. Mater Res Expr, 2019, 6, 105365. 10.1088/2053-1591/ab41b2.
[40] Savas LA, Hacioglu F, Hancer M, Dogan M. Polymer Bull, 2020, 77, 291–306. 10.1007/s00289-019-02746-7.
[41] Feng C, Zhang Y, Liu S, Chi Z, Xu JR. Polymer Degrad Stab, 2012, 97, 707–14.
[42] Fan M, Feng N, Zhang Y, Wang Z, Qu M, Zhang G. Plastics, Rubber Comp, 2019, 48, 270–80. 10.1080/14658011.2019.1606580.
[43] Xiaochun H, Yonglin L, Zhihan P, Zhiquant P, Liu S. Shanghai Meilaipo Chemicals CO LTD, Method for preparing aluminum hypophosphite, patent CN103496681A, China, 2013.
[44] Shen KK, O'Connor R. In: Pritchard G. (Ed.) Plastics Additives: An A–Z Reference, Springer Netherlands, Dordrecht, 1998, 268–76.
[45] Lu N, Zhang P, Wu YN, Zhu D, Pan Z. Int J Polymer Sci, 2019, 2019, 2424531. 10.1155/2019/2424531.
[46] Yan L, Xu Z, Wang X. J Thermal Anal Calorim, 2019, 136, 1563–74. 10.1007/s10973-018-7819-1.
[47] Wu Z, Hu Y, Shu W. J Appl Polymer Sci, 2010, 117, 443–49. https://doi.org/10.1002/app.31969.
[48] Shen KK, Kochesfahani S, Jouffret F. Polym Adv Techn, 2008, 19, 469–74. https://doi.org/10.1002/pat.1119.

[49] Tugrul N, Bardakci M, Ozturk E. Res Chem Interm, 2015, 41, 4395–403. 10.1007/s11164-014-1538-4.

[50] Polat O, Kaynak C. Int Polymer Proc, 2019, 34, 59–71. DOI: 10.3139/217.3579.

[51] Savas LA, Dogan M. Polymer Degrad Stab, 2019, 165, 101–09. https://doi.org/10.1016/j.polymdegradstab.2019.05.005.

[52] Tai Q, Yuen RKK, Yang W, Qiao Z, Song L, Hu Y. Comp Part A: Appl Sci Manufact, 2012, 43, 415–22. https://doi.org/10.1016/j.compositesa.2011.10.012.

[53] Horrocks AR, Smart G, Hörold S, Wanzke W, Schlosser E, Williams J. Polymer Degrad Stab, 2014, 104, 95–103. https://doi.org/10.1016/j.polymdegradstab.2014.03.027.

[54] Braun U, Schartel B, Fichera MA, Jäger C. Polymer Degrad Stab, 2007, 92, 1528–45. https://doi.org/10.1016/j.polymdegradstab.2007.05.007.

[55] Cusack PA, Hornsby PR. J Vinyl Additive Technol, 1999, 5, 21–30. https://doi.org/10.1002/vnl.10302.

[56] Horrocks AR, Smart G, Price D, Kandola B. J Fire Sci, 2009, 27, 495–521. 10.1177/0734904109102025.

[57] Xu J, Zhang C, Qu H, Tian C. J Appl Polymer Sci, 2005, 98, 1469–75. https://doi.org/10.1002/app.22282.

[58] Horrocks AR, Smart G, Nazaré S, Kandola B, Price D. J Fire Sci, 2010, 28, 217–48. 10.1177/0734904109344302.

[59] Liu J, Zhang Y, Guo Y, Lu C, Pan B, Peng S, Ma J, Niu Q. Polymer Comp, 2018, 39, 770–82. https://doi.org/10.1002/pc.23998.

[60] Liu L, Zhao X, Ma C, Chen X, Li S, Jiao C. J Thermal Anal Calorim, 2016, 126, 1821–30. 10.1007/s10973-016-5815-x.

[61] Babu K, Rendén G, Afriyie Mensah R, Kim NK, Jiang L, Xu Q, Restás Á, Esmaeely Neisiany R, Hedenqvist MS, Försth M, Byström A, Das O. Polymers, 2020, 12, 1518. 10.3390/polym12071518.

[62] Araby S, Philips B, Meng Q, Ma J, Laoui T, Wang CH. Comp Part B: Eng, 2021, 212, 108675. https://doi.org/10.1016/j.compositesb.2021.108675.

[63] Dittrich B, Wartig K-A, Hofmann D, Mülhaupt R, Schartel B. Polymer Degrad Stability, 2013, 98, 1495–505. https://doi.org/10.1016/j.polymdegradstab.2013.04.009.

[64] Wang Y, Zhao J. Coatings, 2019, 9, 94.

[65] Morgan AB. Polymers Adv Technol, 2006, 17, 206–17. https://doi.org/10.1002/pat.685.

[66] Kausar A. In: Cavallaro G, Fakhrullin R, Pasbakhsh P. (Eds.) Clay Nanoparticles, Elsevier, Amsterdam, 2020, pp. 169–84.

[67] Ozkaraca AC, Kaynak C. Polymer Comp, 2012, 33, 420–29. https://doi.org/10.1002/pc.22165.

[68] Kaynak C, Gunduz HO, Isitman NA. J Nanosci Nanotechnol, 2010, 10, 7374–77. 10.1166/jnn.2010.2768.

[69] Dujkova Z, Merinska D, Slouf M. J Thermoplastic Comp Mater, 2013, 26, 1278–86. 10.1177/0892705712445301.

[70] Weil ED. J Fire Sci, 2011, 29, 259–96. 10.1177/0734904110395469.

[71] Friederich B, Laachachi A, Ferriol M, Ruch D, Cochez M, Toniazzo V. Improvement of thermal stability and fire behaviour of pmma by a (metal oxide nanoparticles/ammonium polyphosphate/melamine polyphosphate) ternary system. In: Fathi M, Holland A, Ansari F, Weber C. (Eds.) Integrated Systems, Design and Technology, Springer, Berlin, Heidelberg, 2010. https://doi.org/10.1007/978-3-642-17384-4_5.

[72] Horacek H. J Appl Polymer Sci, 2009, 113, 1745–56. https://doi.org/10.1002/app.29940.

[73] Yang Z, Xiao G, Chen C, Chen C, Wang M, Zhong F, Zeng S, Lin L. Coll Surf A: Physicochem Eng Aspects, 2021, 621, 126561. https://doi.org/10.1016/j.colsurfa.2021.126561.

[74] Alongi J, Carletto RA, Di Blasio A, Carosio F, Bosco F, Malucelli G. J Mater Chem A, 2013, 1, 4779–85. DOI: 10.1039/c3ta00107e.

[75] Carosio F, Di Blasio A, Alongi J, Malucelli G. Polymer, 2013, 54, 5148–53. https://doi.org/10.1016/j.polymer.2013.07.029.

[76] Alongi J, Bosco F, Carosio F, Di Blasio A, Malucelli G. Mater Today, 2014, 17, 152–53. https://doi.org/10.1016/j.mattod.2014.04.005.

[77] Wang Z, Han E, Ke W. Corrosion Sci, 2007, 49, 2237–53. https://doi.org/10.1016/j.corsci.2006.10.024.

[78] Duquesne S, Bras ML, Bourbigot S, Delobel R, Vezin H, Camino G, Eling B, Lindsay C, Roels T. Fire Mater, 2003, 27, 103–17. https://doi.org/10.1002/fam.812.

[79] Yakovlev AV, Finaenov AI, Zabud'kov SL, Yakovleva EV. Russ J Appl Chem, 2006, 79, 1741–51. DOI: 10.1134/s1070427206110012.

[80] Hou B, Sun H-J, Peng T-J, Zhang X-Y, Ren Y-Z. New Carbon Mater, 2020, 35, 262–68. https://doi.org/10.1016/S1872-5805(20)60488-7.

2 Metals and intermetallics

2.1 Resources: ores, recycling and urban mining

Martin Bertau

Everything gets started with mining (German: Alles kommt vom Bergwerk her) is an old motto coined by the miners in the late seventeenth century in Freiberg/Saxony (Germany), one of the most active mining sites in history. In fact, any chemical process has its starting point in raw materials, irrespective whether they originate from primary (mining) or secondary (recycling) raw materials. Prior to manufacturing a good, there must be some material that can be used to mount such a product. And these materials themselves originate from a feedstock, which after processing is suitable for production purposes. This ensemble is known as the value chain (Figure 2.1.1).

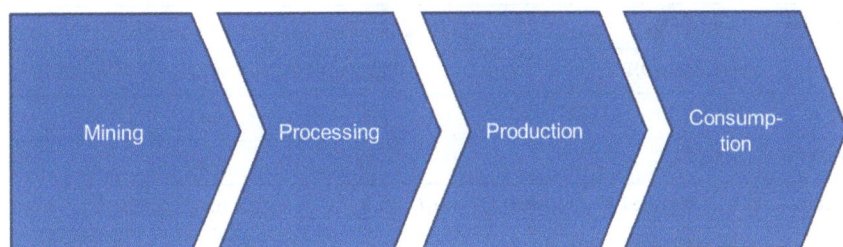

Figure 2.1.1: The value chain.

To give an example, lithium is an alkali metal, which is not existent in nature. It is accumulated in deposits which are made up from ore minerals. They are of utmost different composition and vary widely in their Li_2O content (Table 2.1.1).

It is easy to conceive that each of these minerals requires some specialized individual treatment in order to obtain lithium. And, it is conceivable, too, that secondary raw materials may be even more diverse in their composition, thus necessitating even more sophisticated recovery technology.

End-of-life products (EoL): Disused consumer goods that are no longer functional.

End-of-use products (EoU): Consumer goods that are still functional but are replaced by more up-to-date goods, e.g. smartphones, tablet PCs and televisions.

Production waste: Arises during production and is recycled wherever possible, e.g. indium tin oxide (ITO) in the manufacture of touch screens.

Dissipation: Dilution of elements (metals) in consumer goods. If a critical threshold value is not reached, recycling is no longer economically feasible and can no longer be justified in terms of energy and resource consumption.

https://doi.org/10.1515/9783110733143-008

Table 2.1.1: Lithium ore minerals [1].

Mineral	Composition	Lithium content [% Li_2O]	Average lithium content in ore [% Li_2O]
Spodumene	$LiAl[Si_2O_6]$	4 . . . 8	2.9 . . . 7.8
Lepidolite	$K(Li,Al)_3[(Al,Si)_4O_{10}](F,OH)_2$	1.4 . . . 3.6	3.0 . . . 4.1
Zinnwaldite	$KLiFeAl[(Al,Si)_3O_{10}](F,OH)_2$	1 . . . 3	2.6 . . . 3.9
Petalite	$LiAl[Si_4O_{10}]$	3.4 . . . 5.0	3.0 . . . 4.7
Eukryptite	$LiAlSiO_4$	4.5 . . . 11.8	4.5 . . . 9.5
Hektorite	$Na_{0.3}(Mg,Li)_3[Si_4O_{10}](OH)_2$	1.2	n.a.
Amblygonite	$LiAl[PO_4](F,OH)$	7.3 . . . 10.1	n.a.
Kryolithionite	$Li_3Na_3[AlF_6]_2$	~12	n.a.
Triphylite	$Li(Fe,Mn)PO_4$	2 . . . 6	n.a.
Lithiophyllite	$Li(Mn,Fe)PO_4$	2 . . . 6	n.a.
Jadarite	$LiNaSiB_3O_7(OH)$	15.7	n.a.
Bikitaite	$LiAl[Si_2O_6]·H_2O$	7.3	n.a.
Kryolithionite	$Li_3Na_3[AlF_6]_2$	~12	n.a.
Swinefordite	$Li(Al,Li,Mg)_4[(Si,Al)_4O_{10}]_2(OH,F)_4 · nH_2O$	3.7	n.a.
Salitolite	$(Li,Na)Al_3[AlSi_3O_{10}](OH)_5$	1.7	n.a.
Li ion battery	80 mg $Li·Wh^{-1}$ [2]	3.0[a]	–

[a]Calculated for a Nikon ENEL-15b cell with 1,900 mAh and 14 Wh.

If minerals are concentrated locally by nature, it is called a deposit. A similar situation is given for certain elements when being concentrated locally in an anthropogenic or technical deposit. This is what is understood by urban mining. The term originally was coined by Hideo Nanjyo (Tohoku University, Japan) and comprised the stockpile of rare metals in the discarded waste electrical and electronic equipment (WEEE) [3]. Today, however, one rather speaks of secondary raw materials addressing the sum of recyclable material. The term secondary raw materials clearly is more accurate and more helpful, since it serves to differentiate between primary raw material, i.e. feedstock from natural deposits recovered by mining activity. In public perception as well as in media coverage, the term secondary raw materials is equal to recycling, which does not hold true. In analogy to Figure 2.1.1, secondary raw material recovery equals mining in the primary raw material field. Recycling however is the processing step which comes after. It is necessary, though, to emphasize that the true meaning of recycling is different from what it is typically perceived of. Recycling means to restore primary product quality. In other words, recycling products originate from waste material, but they are indistinguishable from mining products. Secondary products however do not exhibit primary product quality. For decades, this has been the case with the few exceptions of noble metals and copper. Only in the last decade this list has incorporated phosphoric acid, indium, gallium, tin and lithium carbonate. And since some few years it is also neodymium and samarium that have found their way into this eventually growing list [4–6]. What is meant is to keep secondary materials in the economic cycle. Today, one also discriminates between different types of secondary raw materials, which makes sense, since metal recovery is greatly dependent on the nature of the waste material. End-of-life products (EoL) as well as end-of-use products (EoU) comprise manufactural goods that have exceeded their lifespan either for being unfunctional (EoL) or by having been discarded for the sake of being replaced by a more recent version (EoU). Household equipment such as refrigerators, washing machines and so on are normally EoL products, while others, such as mobile phones are EoU products. The latter case dominates wherever a social status may be derived from owning a good.

2.1.1 Reserves, resources, criticality: do we need all these terms?

There are quite some expressions, terms and definitions used in the raw material sector. It is not such a plethora that is not manageable. In fact, they are few, but useful.

When talking about natural deposits, one speaks of **reserves**, **resources** and **geopotential**. Reserves comprise those ores that can be exploited economically with existing technology. For reserves, it lacks either economy or technology, or even both, while the geopotential is the sum of everything abundant in the upper Earth's crust the existence of which is imaginable, and where recovery technology is missing, for which reason an economic assessment truly cannot be made (Figure 2.1.2).

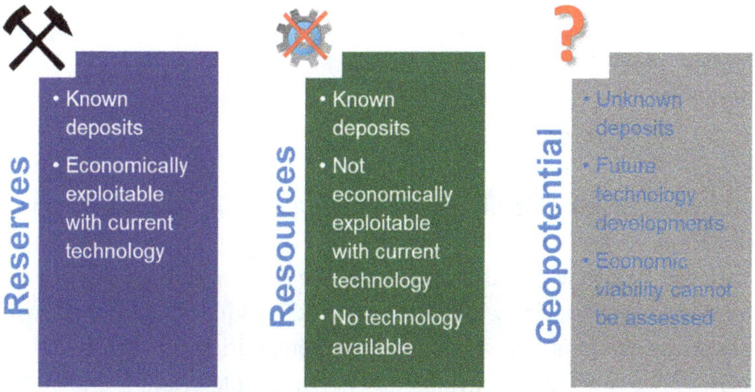

Reserves
- Known deposits
- Economically exploitable with current technology

Resources
- Known deposits
- Not economically exploitable with current technology
- No technology available

Geopotential
- Unknown deposits
- Future technology developments
- Economic viability cannot be assessed

Figure 2.1.2: **Reserves, resources and geopotential.**

The assessment of the requirement of a raw material is based on its **function**. A deposit is not being exploited for a metal, just because its color is generally recognized as beautiful. Even in the case of gold, the color of which fascinates mankind since its early days, there are other properties that render it an economically valuable good, for instance material properties, or securing monetary value by the national banks. In fact, it is the function which makes someone digging for gold or whatever raw material. In the case of copper, this can be the function of electrical conductor, or scandium is mined for producing Al-based light alloys. This function is not about the specific metal or the raw material itself.

Another term is **substitution**. What is meant here is the ability of a raw material or material to perform the same function. It has to be emphasized that one here specifically means *the same*, not something similar.

Essentiality denotes properties that cannot be substituted. Essential elements can be required by a material, process or system. In this context, bioessential elements mean those chemical entities which are indispensable for an organism and/or food production: N, P, K, Na, Ca, Mg, S, Se, Mo, Fe, Co, Ni, Mn and Zn. Useful elements, also known as beneficial elements, support the growth of organisms (plants, biomass, etc.).

Criticality stands for the supply risk for elements, the **scarcity** of which has hazardous effects on an economic system, such as a national economy or an industrial branch. The example of energy helps to distinct criticality from the concept of essentiality: Energy is essential for every organism and the functioning of society. An energy source can be critical from the user's point of view, for instance uranium or coal. But energy sources are substitutable (Figure 2.1.3).

As a thumb rule may serve that essential raw materials are *eo ipso* indispensable. They are not substitutable. Critical raw materials have a hazardous effect for the user if they become scarce. They may be substitutable.

At present, there are more and more reports on shortening raw material supply for certain elements. Often these are the rare-earth elements (REE) which have become synonymous for the raw material supply crisis which had its onset in 2010 upon China's announcements to strictly adhere to given export quota. In fact, there is hardly an element which is not becoming scarce. Mostly unnoticed by the Western countries, the entire supply situation has become increasingly critical during the last decade, with the Western governments far from being capable of responding adequately.

But why has the raw material issue become so intriguing? We are facing an unprecedented growth in world population accompanied by a shift of power to Asia. The world is changing, and new global players enter the scene watching in disbelief how established power centers, above all Europe, clear the square without opposition. As a consequence, there is a never seen before competition for raw materials on the global scale, since raw material availability at reasonable prices is crucial for the functioning of industrial production. We are presently witnessing such a situation. Producing countries withhold raw materials or intermediates for the benefit of

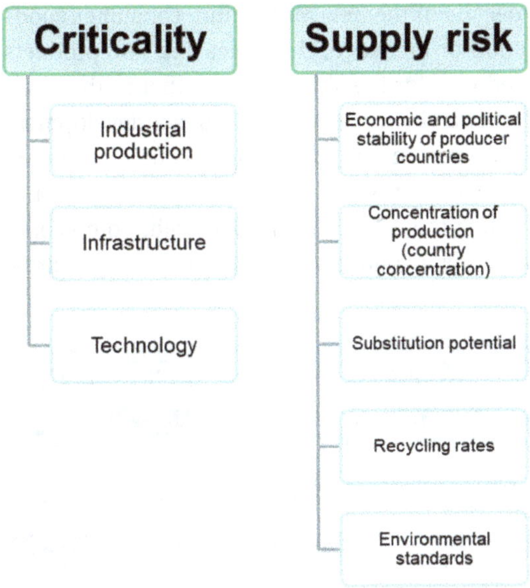

Figure 2.1.3: Criticality and supply risk. Criticality affects different areas of a national economy (left) and has its causes in different supply risks (right).

their own national economy. This shift situation that we observe is mainly attributed to the so-called BRICS countries: Brazil, Russia, India, China and South Africa where ~3.1 billion people live, roughly 40% of the world's population, with a purchasing power-adjusted share of global GDP of 32%. In addition, the MIST countries (Mexico, Indonesia, South Korea and Turkey) with ~0.5 billion people or ~7% of the world's population have a purchasing power-adjusted share of global GDP of 10%. In sum, these nine countries concentrate 42% of the world's economic power, while the G7 countries with ~0.9 billion people or ~11% of the world's population sum up to a purchasing power-adjusted share of global GDP of no more than 10%. Already at the latest the focus of global economy has shifted to East Asia and Latin America. Africa is expected to enter the scene with a development boost as of 2030. In other words, the great hunger for raw materials is yet to come.

One may be inclined to complain on that situation but creating and establishing dependences on producer countries has always been part of Western business models before, and it is too logic that other countries fill the gap that the West leaves. In response, particularly Europe is calling for intensified recycling. While the concept is correct, it will not be easy to realize, as modern products become even more complex (Figure 2.1.4). With any new generation of an electronic device or whatever, it gets even harder to separate and recycle their components. The reason is clear: the competition for raw materials has fueled developments aiming at steadily reducing raw material input. Another important issue is product cycles becoming shorter. While in the first decade of the new Millennium consumer goods were bought in response to malfunction or breakage, thus releasing EoL products to the market, it is now often EoU products, for the simple reason that in the wealthy states certain goods are conceived of as being outdated. This is a development that deliberately has been brought into action, since industry needs to generate needs in societies that have few real needs only. Again, one may be inclined to have a critical view thereupon, but in fact this is what has been fueling our economies the last decade(s), not to forget the tax income of the national treasury boards.

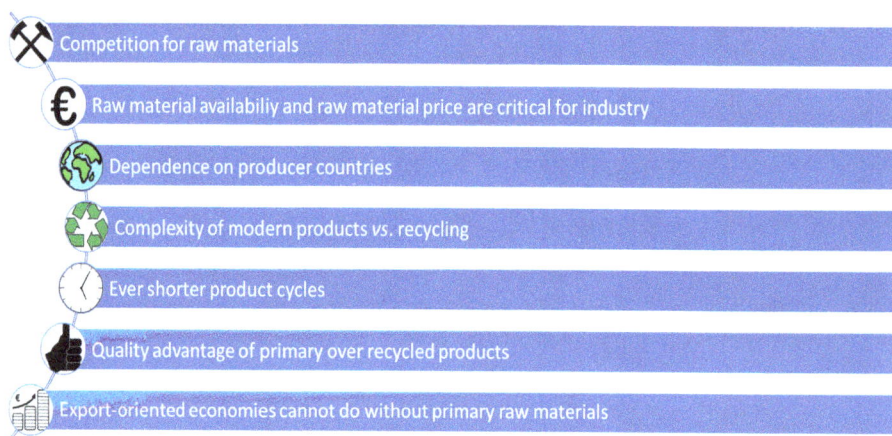

Figure 2.1.4: Factors determining raw material supply.

One issue that is being regarded extremely critical in the Western countries is mining. Mineral extraction through mining is generally regarded as dirty and environmentally harmful, while recycling appears as the clean alternative. As a result, markets for the respective products develop in parallel, because in many cases the reprocessed product does not reach primary product quality. These markets are not interconnected, what consequently multiplies efforts and CO_2 emissions while reducing energy and resource efficiency. This, however is mainly down to ideological reasons rather than for factual reasons. Decision-makers have to accept, though, that export oriented countries cannot do without a reliable primary raw material supply. That goes without saying.

Clearly, the present situation is challenging, and the much-vaunted term circular economy is nothing else but a fig leaf to cover inactivity of the past, while ignoring the developments that had their onsets in the early 1980s. Complaining will not help, though, and new concepts are needed that overcome this situation. The twenty-first century is clearly a period of time in which chemistry will play a key enabling role to solve these challenges. With natural mineral deposits on the one hand and anthropogenic resources on the other becoming more and more complex, classical methodology will get outdated increasingly with unwitnessed speed. Classical disciplines as process engineering or metallurgy already now are no longer capable of giving the right answers to everything. Not to be misunderstood, the disciplines are indispensable parts of the backbone of industrial production. However, one has to concede that recent developments have not been given the necessary recognition, and the intercourse between the disciplines is underrepresented. In fact, the complexity of raw materials inevitably requires the chemist's knowledge of substances. Together with the other disciplines these new tasks may be mastered.

One such approach in this direction is circular resources chemistry (German: Wertstoffchemie), also referred to as advanced resources chemistry [7]. This concept sets a full stop to what has been practiced in the last centuries and is a complete renunciation from the traditional. Instead, this future-directed approach focuses less on individual target elements. It is holistically oriented and attempts to extract the maximum amount of raw materials from the ore once it has been mined. The term circular resources chemistry describes the entirety of the origin-independent processes and methodologies to produce chemical raw materials. The concept eliminates the boundaries between primary raw materials (mining), secondary raw materials (recycling) and renewable raw materials (biomass). Holistic production techniques shift the boundaries of resource availability, as no more distinction is made between raw materials and residual materials (Figure 2.1.5).

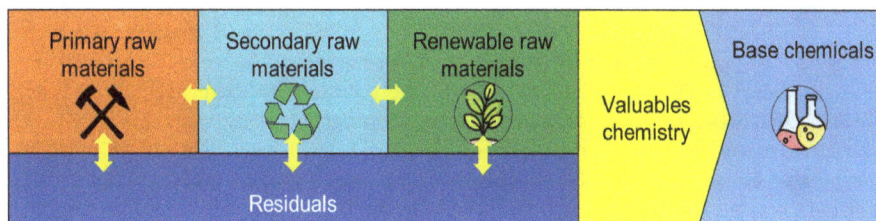

Figure 2.1.5: Circular resources chemistry. Graphics sources: author's own work, www.flaticon.com (free images).

Why origin-independent? Classical industrial plants are designed to process material from one or some few specific deposits. At present however, the inorganic raw material sector commences to repeat a development that has onset in the organic, i.e. petrochemical sector in the 1970s, the creation of value in the producing countries. Like petrol exporting countries that have successfully established pre-processing of crude oil to intermediates in their own countries, metal producing countries have initiated comparable developments. The consequence is less material available on the markets that complies with the needs of established industries. Chemistry, in turn, is capable of focusing on the product irrespective of where the starting materials originate from.

Therefore, a characteristic feature is that both mine products and recyclable materials are processed along one single process chain. In other words, material that already has undergone one or more product cycles is inevitably regenerated to primary product quality, as there is one single technology chain. Only the entry points can differ. As a consequence, circular resources chemistry (Figure 2.1.6) implements the zero-waste principle as far as possible. In this context, it should be kept in mind that completely waste-free production is an ideal goal that cannot be fully achieved in commercial production. However, minimizing waste product amounts to the greatest possible extent, as offered by this concept, comes very close to this objective [8].

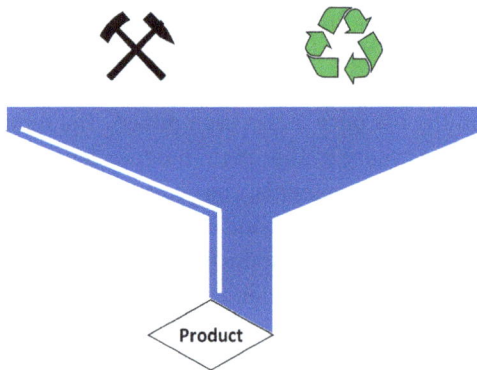

Figure 2.1.6: Circular resources chemistry profits from a broad input material variability. Since the process is suited to digest and convert (almost) all ore minerals as well as recyclable material to one single product quality, this type of raw material processing is origin independent. There is no more distinction between primary, secondary or renewable origin of raw materials. The funnel model serves to illustrate the functioning. The opening is wide enough to accept a broad variety of incoming ore material and recyclable material, while the chimney is the process chain along which the material is converted to give one uniform product quality.

Already today there is a series of processes that work commercially according to the circular resources chemistry principle. Examples are the Sepselsa process for the recovery of rare earth elements (REE) from fluorescent lamp waste, or the MagnetoRec process which with a half-ton-per-day capacity is the largest commercial recycling process for permanent magnets such as FeNdB and Co_mSm_n magnets outside China. Others are the RMF process for In, Ga, Ge and Sn and the PARFORCE process which provides phosphoric acid in whatever quality from primary and secondary sources. The COOL process digests both lithium ores and lithium accumulator black mass and is capable of converting primary lithium batteries, too. It delivers battery grade-Li_2CO_3 as a crude product and is at the verge of being commercialized as well [4, 9–12].

Future-directed approaches will increasingly incorporate this methodology. The time when we could afford disposing of by-components is over. Future-directed approaches of course will increasingly address recycling, but mining will get back into the focus, too, in the Western countries. The reason is clear: the aim is securing the raw material base. It is time to recognize mining and recycling as branches of the same tree.

In order to achieve this goal, a functionable chemical industry in Europe is needed, which not only is capable of processing the different raw material qualities, but also provides sufficient processing capacity. However, if the currently observed exodus of chemical industry to the Far East goes on, this will disable the continent in only some few years from providing itself even with recycling products.

Let us finalize the chapter with some thoughts on mining and recycling. Whether or not a mine will be established is the result of extensive preparatory work, such as sample drillings, environmental impact assessments and feasibility studies. Recycling is generally perceived as something that simply needs to be done. Both raw material streams require exact and deliberate preparation, though. No mine is being opened if the process chain is inefficient, no recycling process will be established if the composition of the recyclables is unclear. In both cases the intrinsic value of the processed product must exceed the operating costs. Otherwise, there will be no economically viable situation, and only external incentives may help to compensate for the costs. Something that is paid either by the taxpayer or through artificially overpriced products. In this context energy comes into play. In secondary raw materials, by-components are present in an entity in varying, sometimes minute amounts. In fact, the energy need to restore primary product quality from a recyclable product is the biggest cost item. The reason is obvious: removing impurities from a matrix is thermodynamically extremely unfavorable since being anti-entropic. Hence, the access to cheap energy is decisive. National economies which declare high-priced energy as reasonable and desirable tool to steer and control product streams will experience starvation from raw material supply.

It is important to stress that while a mine is something immobile that cannot be moved outside the borders of a national economy, recyclable products may very well. Politics will have to enforce laws that retain secondary raw materials in the recycling chain and prevent illegal exports. Last but not the least, the question that holds true for any raw material: What can/should/must be mined/recycled? As every time in chemical production, the question whether there is a business case or not will decide whether or not a process comes into action. The good news: origin-independent holistic approaches avoid costly deposition of by-products. They produce only marketable goods. As such they are not only far more environmentally friendly, they are even more economic.

References

[1]   Schmidt M. Rohstoffrisikobewertung – Lithium – DERA Rohstoffinformationen 33, Deutsche Rohstoffagentur (DERA) in der Bundesanstalt für Geowissenschaften und Rohstoffe (BGR), Berlin, 2017.

[2]   Rahimzei E, Sann K, Vogel M. Kompendium: Li-Ionen-Batterien. VDE Verband der Elektrotechnik Elektronik Informationstechnik e.V, Frankfurt/Main, 2020.

[3]   Nakamura T, Halada K. Urban Mining Systems. Briefs in Applied Sciences and Technology, Springer, Heidelberg, 2014.

[4]   Lorenz T, Bertau M. Recycling aspects – magnets. In: Pöttgen R, Jüstel T, Strassert CA. (Eds.) Rare Earth Chemistry – Basics for Master and PhD Students, De Gruyter, Berlin, 2020, 55–68. ISBN: 978-3-11-065360-1.

[5]   Lorenz T, Bertau M. Recycling of rare earth elements from FeNdB-magnets via solid-state chlorination. J Cleaner Prod, 2019, 215, 131–43. doi: 10.1016/j.jclepro.2019.01.051.

[6]   Lorenz T, Bertau M. Recycling of rare earth elements from SmCo$_5$-magnets via solid-state chlorination. J Cleaner Prod, 2020, 246, 118980. doi: 10.1016/j.jclepro.2019.118980.

[7]   Fröhlich P, Eschment J, Bertau M. Wertstoffchemie – Die Rohstoffbasis sichern. Nachr Chem, 2017, 65, 1206–09. doi: 10.1002/nadc.20174068276.

[8]   Bertau M. Future securing of the raw materials base. ChemBioEng Rev, 2019, 6, 1–11. doi: 10.1002/cben.201900004.

[9]   Pavón S, Kaiser D, Mende R, Bertau M. The COOL-process – a selective approach for recycling lithium batteries. Metals, 2021, 11, 259.

[10] Bertau M, Fröhlich P, Brett B, Lorenz T, Martin G. Valuable metals – recovery processes, current trends and recycling strategies. Angew Chem Int Ed, 2017, 56, 2544–80.

[11] Fröhlich P, Eschment J, Martin G, Lohmeier R, Bertau M. The PARFORCE-technology (Germany). In: Schaum C. (Ed.) Phosphorus: Polluter and Resource of the Future: Removal and Recovery from Wastewater, IWA Publishing, London, 2018, 417–24. ISBN 978-1780408354.

[12] Vostal R, Pavón S, Kaiser D, Bertau M. Separation of indium from acid sulfate-containing solutions by ion exchange with impregnated resins. Chem Ing Tech, 2021, 93, 1859–67.

2.2 Special steels and alloys for industrial use

Rainer Behrens

2.2.1 Introduction

Within the metallic materials, carbon steels, stainless steels, special steels and nickel alloys play a major role in a lot of domestic uses and for industrial uses. In daily life, steels are known from examples as cars, kitchen and cooking ware, concrete reinforcement, architectural uses, ships or railways.

In industrial uses, well-established application fields of special steels and nickel alloys typically are the chemical process industry (CPI) for chemical reactors, storage and transport of chemicals, conventional and nuclear power generating plants, environmental engineering including flue gas desulfurization, medical equipment or marine engineering. Special properties are required from materials to guarantee the functional demands in such applications as high corrosion resistance, heat resistance, creep resistance, high ductility and special physical properties like magnetism or expansion behavior. In many cases the used materials have to bear high temperatures and high pressures in the applied processes. In most cases wrought alloys are used, and for certain applications also cast alloys. The focus in the following descriptions is set intentionally to special steels and alloys with austenitic microstructure, knowing that these represent only a small cutout in the wide field of metallic materials [1].

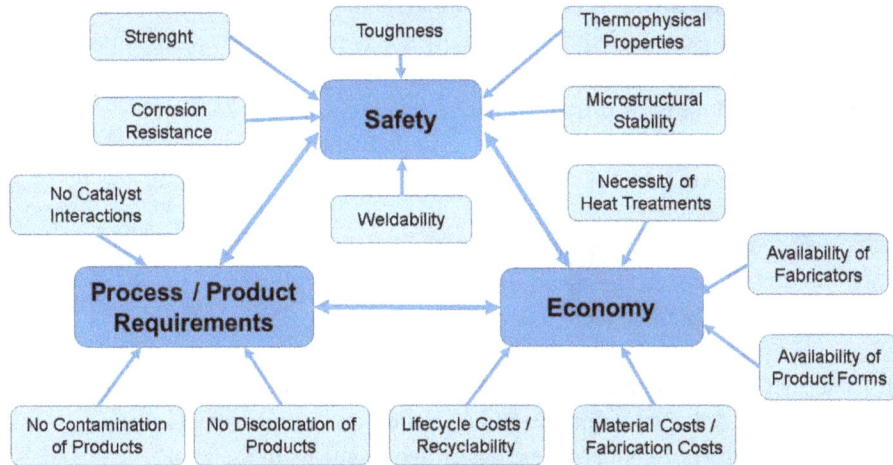

Criteria of materials selection in the Chemical Process Industry (based on J. Korkhaus BASF 2005 VDM Symposium)

Figure 2.2.1: Aspects of materials selection for the chemical process industry.

Dedication: In loving memory of my wife Brigitte Fröhlich – wish you were here.

https://doi.org/10.1515/9783110733143-009

The selection of a suitable material which has to be safe and reliable for equipment in the CPI and other industrial applications is influenced by several factors. The chemical substances themselves or resulting reaction products and their corrosive properties under the specific service conditions are some of these factors. To address the aspects to the different forms of corrosion resistance of the metallic material, the service conditions have to be known as temperature profiles, pressure, flow conditions, concentrations of starting substances, intermediate substances, final product, used catalysts or contaminants. As a matter of course, the mechanical material properties (strength, toughness and ductility, microstructural stability in long time service conditions) play an important role. They are directly connected to the aspects of safe use and reliable plant operation. The possibility of shaping the steels and alloys to plates, sheets, tubes, bars by mechanical work as hot and cold working, machining operations (drilling, milling, grinding) welding, heat treatment, etc. has also to be taken into consideration.

The aspect of safety indeed has a paramount importance as expressed in Figure 2.2.1 in the central position of the diagram.

The industrial process and product requirements bring in additional factors linked to the steel or alloys themselves. The steels and alloys have to be suitable to avoid product contaminations or discolorations by corrosion products. Other factors can be catalytic reactions of the chemical substances with the materials or interactions with catalysts.

From the point of an economic use this includes the view to chosen materials economy in terms of life cycle costs, materials costs, fabrication costs including processing steps as welding and/or heat treatments. The aims of the material selection are long use times, a safe use and cost efficiency in their applications. Availability of all needed shapes by fabricators and equipment producing engineering companies also have to be proved [1, 2].

After the end of equipment's lifetime, especially the more costly high-alloyed materials are recycled in terms of circular economy by melting the scraps to new alloys.

2.2.2 Selected industrial application fields for stainless steels, special steels and alloys

In Table 2.2.1 some examples of industries and application fields for the use of these materials are listed. Depending on the operation conditions – including corrosive load, temperature, pressure and mechanical requirements in the application fields – the complete range from stainless steels to high-alloyed materials can be found. Recent key figures and data on the distribution of steels in the European market and distribution between different flat products and long products can be found in the business reports from the European Steel Organizations in updated versions [3].

Some chosen practical examples for the uses of high-alloyed steels and high-performance alloys (HPA) will be shown in picture in a later subchapter.

Table 2.2.1: Different industrial application fields for stainless steels, special steels and alloys.

High-performance alloys (HPA) and special steels	Application fields
Chemical Process Industry (CPI)	Heat exchanger, high-temperature reaction vessels, furnaces, corrosion resistant tubes, storage tanks, road and rail transportation, pumps
Pharmaceutical industry/food processing and nutrition	Mixers, coating equipment, reactors, storage vessels, tubes, pumps
Conventional energy generation/ pollution control	Fossil fuel power plants, flue gas desulfurization, waste incineration, nuclear industry, wet scrubbers
Regenerative energy and environment	Solar power, geothermal power, marine scrubbers, biogas production
Oil and gas	Pump shafts, umbilicals, wirelines, pipelines, drilling equipment, LNG transportation (liquid natural gas)
Automotive	Valves, spark plugs, glow plugs, exhaust gas cleaning, catalytic converters, turbochargers
Aerospace	Gas turbines and engines: rotating and non-rotating parts, combustion chambers, landing gears
Electronics and electronic techniques	Solar energy, heating application, magnetic shielding
Metal processing	Furnace equipment, hot working dies, burners, exhaust fans, transport rolls and belts
Medical engineering	Surgery, medical equipments

2.2.3 Description of metallurgical aspects in special steels and alloys

For a better understanding how stainless steels, special steels and alloys work, some basic metallurgical aspects have to be given first [4–7]. The materials described in this chapter are mixtures of different metals that form the steel or an alloy. The metallic element that is present with the highest amount is called the base metal or matrix metal. In the examples in this chapter this is usually iron (structure type alpha/body cubic centered) or nickel (structure type gamma, face cubic centered). In this solid matrix, other elements like chromium or molybdenum can be dissolved at high temperatures. As comparable with liquid solutions, there are natural limits for solubility of such additions, depending on the given atomic structure and radius. The solubility of elements in the base matrix is temperature dependent and at high temperatures much higher as at low temperatures, for example, at 600 °C. This means, that it is possible to dissolve the alloyed elements in the base metal at a high temperature (solution annealing) and freeze this condition of solid solution by a rapid cooling or so-called quenching to a temperature where movements of atoms for diffusion in the microstructure is low. So, the alloyed elements are locked in place in an oversaturated or supersaturated solid solution [8].

This is the used condition for many materials of these steels and alloys for wet corrosion applications. Upon exposure and temperature excursions above 500–600 °C, where diffusion has a noticeable effect, stainless steels and alloys are prone to the precipitation of various classes of secondary intermetallic phases [9]. Such temperature excursions can occur in practical applications or during processing operations like welding. The precipitates have a preferred tendency for nucleation and precipitation at grain boundaries. Some of the precipitated phases are common for all of these materials, but other phases are not. The most prominent common phase is the $M_{23}C_6$ carbide.

This is not a general rule, but in most cases these precipitated phases have a negative influence on both the mechanical and the corrosion properties of the materials and have to be avoided for optimal materials performance in application. Due to the use of modern melting technologies, the carbon contents in these alloys can be kept to low levels (Table 2.2.9). This reduces the tendency for intercrystalline corrosion attack that can be caused by precipitated chromium carbides and chromium depletion at the grain boundaries.

For certain alloys such as high-temperature alloys or superalloys that are used in aerospace applications, precipitated secondary intermetallic phases can be used intentionally with advantage to strengthen the alloys against creep at high temperature. For high-temperature alloys, carbides can be used for strengthening as found in Alloy N08120 or 602 CA. A second mechanism is the precipitation or age-hardening that is often used for superalloys. The age-hardening is done through special heat treatments in the temperature range between 480 and 800 °C for the precipitation of strengthening phases throughout the microstructure. Representatives for these strengthening phases are γ'-Ni_3(Ti/Al) in Alloy 925 or γ"-Ni_3Nb in Alloy 718.

In Table 2.2.2, examples of typical intermetallic phases and the approximate temperature range of formation are given, which can be found in stainless steels and nickel alloys. It should be mentioned that some stable phases like titanium carbide, niobium carbides and carbo-nitrides or borides are found in the microstructure as primary precipitated phases from the solidified melt [8].

It should be noted that the temperature ranges and phase compositions given in Table 2.2.2 show only approximate values to give an impression and can differ remarkable not only for a certain alloy but also within this alloy. The chemical composition of a phase and resulting lattice parameters may vary remarkably as a function of the formation temperature and the local inhomogeneous distribution of elements segregation in the solidification of the melt. In Figure 2.2.2 an example in

Table 2.2.2: Structure type, major composition and temperature range of formation of intermetallic phases in special steels and alloys; modified from [9–11].

Intermetallic Phase	Crystal structure	Elements in the phase: major composition	Typical temperature range
alpha (α)	bcc	Fe	Up to 911 °C
alpha prime (α ʼ)	bcc	Fe-Cr	280–475–500 °C
gamma (γ)	fcc	Ni	Until liquidus
Gamma prime γʼ	fcc	$Ni_3(Ti/Al)$	600–950 °C
Gamma double prime γʼʼ	bct tetragonal	$Ni_3(Nb/Ta)$	600–1000 °C
Sigma (σ)	bct tetragonal	$(Fe,Ni)_x(Cr,Mo)_y$	600–1180 °C
Chi (χ)	bcc	$(Fe,Ni)_x(Cr,Mo)_y$	600–900 °C
mü (μ)	rhombohedral	$(FeCo)_7(Mo,W)_6$	Up to 900 °C
Eta (η)	hexagonal	Ni_3Ti	800–1000 °C
Laves	hexagonal	Fe_2Mo	550–1200 °C
Delta (δ)	orthorhombic	$Ni_3(Nb/Ta)$	550–1050 °C
$M_{23}C_6$	fcc	$(Cr,Fe,Mo)_{23}C_6$	600–950 °C
M_6C	fcc	$(Mo,W,Cr)_3(Ni,Co,Fe,)_3C$	700–1050 °C
MC	ordered fcc	$(Ti,Nb,V)C$	700 °C – liquidus

thermodynamic equilibrium for the calculated composition range of the main elements Cr, Mo, Fe, Ni (wt%) in the sigma-phase that can occur in Alloy 825 [12] is given.

In the example, the chromium content varies between 47 wt% at the lower end of the calculated temperatures and decreases to 40 wt% at the upper end (predicted sigma-solvus-temperature 905 °C). At the same time, the content of molybdenum increases from 14 to 19 wt% in the intermetallic phase.

The advantage of nickel with its gamma phase structure (austenite) as matrix element in comparison to iron will be described together with the examples of nickel-based alloys (HPA).

2.2.3.1 Influence of alloyed elements on corrosion properties

As described in the Introduction, special requirements exist for the application of special steel and alloys. Materials of construction for the CPI must resist uniform corrosion and have sufficient resistance to localized corrosion such as pitting, crevice corrosion and stress corrosion cracking. To achieve these properties most metals

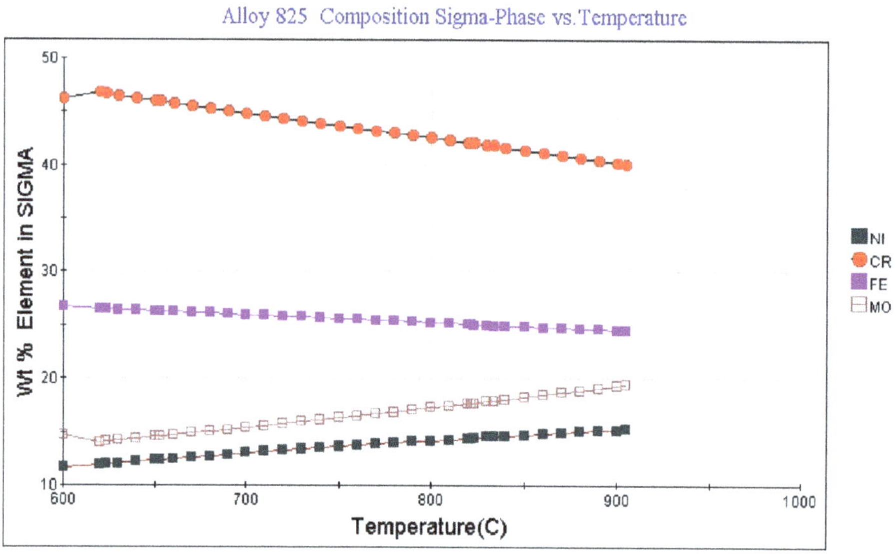

Alloy 825 Composition Sigma-Phase vs. Temperature

Figure 2.2.2: Example for composition change of intermetallic phase composition with temperature (Calculated with JMatPro-v9) in the range 600–1000 °C.

are not used in their pure form but will be alloyed with other elements to get the aimed application profiles. The presence of a second or third element like chromium and molybdenum added to iron or nickel will change the materials' properties significantly.

A brief description of the effects of common alloying elements and their function in the alloy composition with respect to wet corrosion properties can be found in Table 2.2.3.

Table 2.2.3: Contribution of alloying elements to wet corrosion properties [13].

Alloying element	Main feature	Benefit to wet corrosion properties
Nickel	Base gamma matrix for metallurgical compatibility for solving additional alloying elements, improvement of thermal stability	Facilitates repassivation in reducing acids, enhances corrosion resistance in alkaline media, improvement to stress corrosion cracking resistance
Chromium	Important element for passivation and resistance in oxidizing media	Decisive for resistance to localized corrosion, solid solution strengthening
Molybdenum	Provides resistance to reducing media, improves effect of Cr	Facilitate repassivation, solid solution strengthening

Table 2.2.3 (continued)

Alloying element	Main feature	Benefit to wet corrosion properties
Iron	Lower costs/alpha matrix	–
Tungsten	Similar behavior as molybdenum, but less effective	Solid solution strengthening
Nitrogen	Effective austenite microstructure stabilizer	Enhancement of localized corrosion resistance, thermal stability and mechanical properties, solid solution strengthening
Copper	Corrosion resistance	Improvement of resistance to reducing acid as low to medium concentrated sulfuric acid and hydrofluoric acid
Titanium, Niobium	Carbide formers, higher contents provide age-hardening properties in nickel matrix	Stabilization towards intercrystalline corrosion resistance

For forming a stable passive layer consisting of protective oxides, the most important element is chromium. By adding more than 10.5–11 wt% chromium to the base elements iron or nickel functioning as host, an adhering self-healing chromium oxide layer will form in the presence of oxygen, resulting in corrosion resistance to rusting, staining or corrosion under ambient atmospheric conditions. To cope with tougher conditions, higher chromium contents in the alloy as the minimum requirement are usual.

Another important alloying element is molybdenum, which increases the resistance to corrosion attack in reducing and chloride containing environments. In combination with chromium the corrosion resistance to pitting and crevice corrosion is increased. From the viewpoint of corrosion, nickel has a more noble electrochemical standard potential [12] as compared to iron and facilitates repassivation behavior in reducing media.

At the same time these elements work as solid solution hardeners due to the differences of the atomic size of the nickel atoms and the solute atoms [13]. This will be picked up again in the description of elements influence on the mechanical properties. Nitrogen as a non-metallic element can contribute both to the corrosion properties and solid solution strengthening by improvement of the microstructural stability of materials with austenitic (γ-matrix) structure.

2.2.3.2 Sigma-phase as an example for the influence of secondary phases to the properties of a super austenitic 6-Mo steel

One example from the list in Table 2.2.2 is the sigma-phase that is brittle at room temperature and can form in several different alloys like stainless steels, duplex steels, nickel-iron-chromium steels and some more. This phase enriched with chromium and molybdenum with a tetragonal (bct) structure can form in a wide temperature range between 600 and 1180 °C or even higher temperature, depending on the concentrations of phase forming elements in the alloy composition. This brittle phase is detrimental to the mechanical properties and harms the corrosion resistance against localized pitting corrosion [14–16].

Nowadays software tools like JMatPro and similar tools with associated databases for stainless steel, nickel-based superalloys, nickel-iron alloys and other systems allow thermodynamic calculations of alloy systems and an estimation of the phases to be expected in a certain alloy system. This includes calculations in thermodynamic equilibrium of alloys as well as time-temperature transition diagrams, phase compositions and much more. Details can be found in the technical documentations of these tools.

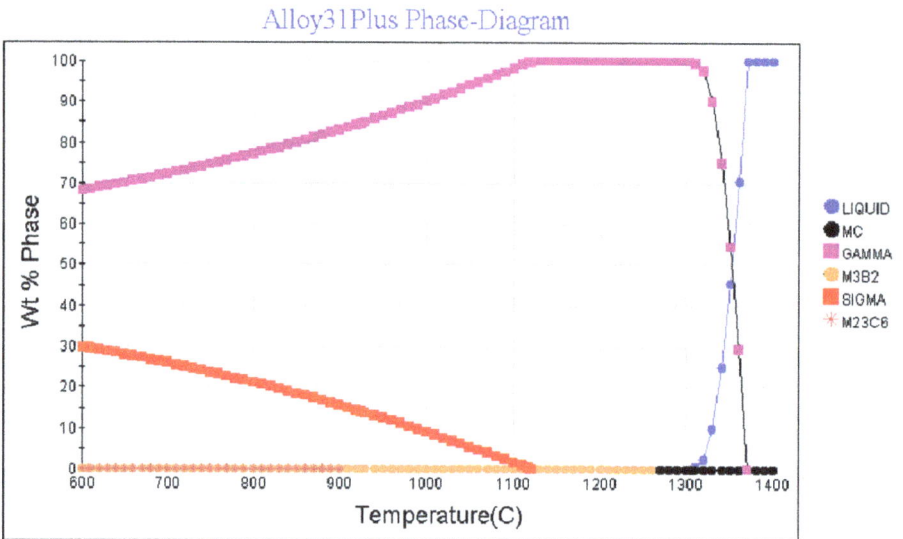

Figure 2.2.3: (a) Typical calculated phase diagram of Alloy 31 Plus and (b) time-temperature-transition diagram (calculated with JMatPro-v9) for the range 600–1400 °C.

Alloy 31Plus TTT-Diagram

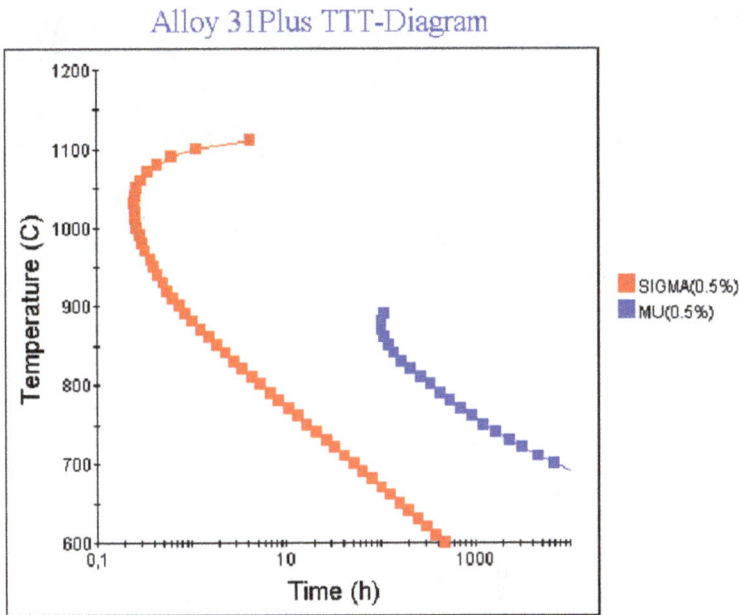

Figure 2.2.3 (continued)

The phase diagram allows an estimation of the phases to be expected for the given alloy composition, the amount of phases in thermodynamic balance and the solvus temperatures needed to dissolve a certain phase. In this example a temperature above 1120 °C is needed to dissolve the sigma phase. In the diagram (b) the kinetic of phase formation is shown, the line indicates the presence of a phase in amounts greater than a predefined value. In this case the calculation limit was set at 0.5 weight % for the secondary phases.

Figure 2.2.4 shows samples of the super austenitic 6-Mo Steel Alloy 31 Plus: (a) optical micrograph of the microstructure of the alloy prepared under solution annealed conditions at 1160 °C after fast cooling. This microstructure is free of secondary phases and in optimal condition for service application. Picture (b) shows a microstructure with a strong decoration of the grain boundaries with the detrimental brittle sigma-phase. The pictured phase was produced by annealing the sample at a temperature of 950 °C for 30 min.

2.2.3.3 Laboratory corrosion testing

Samples of the alloy material were immersed in a corrosive testing solution that is widespread used among material producers for measuring the Critical Pitting Temperature. This is an aggressive synthetic solution consisting of 11.5 wt% H_2SO_4 + 1.2 wt% HCl + 1 % $CuCl_2$ + 1 % $FeCl_3$ wt%, known as "Green Death" solution. The material is immersed at 40 °C in the testing solution and after 24 h inspected for

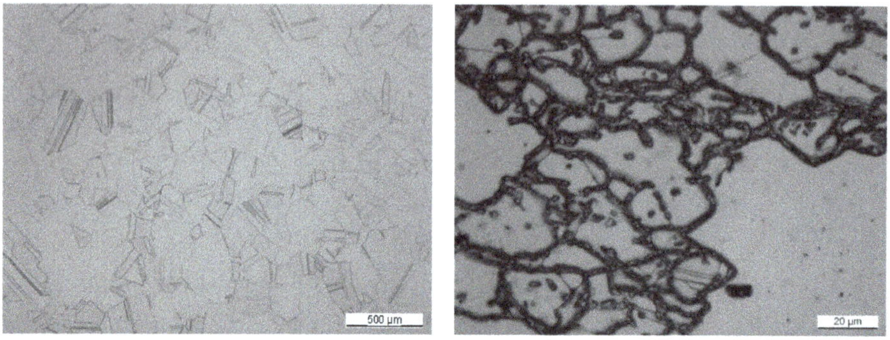

25:1; 1160 °C / 30 min; Alloy 31plus
heat #135755 HCl / HNO₃ 6:1

500:1; 950 °C / 30 min; Alloy 31plus
heat #135755 HCl / HNO₃ 6:1

Figure 2.2.4: Sigma phase in 6-Mo Steels/microstructure of Alloy 31 Plus: (a) Microstructure in solution annealed condition: free of intermetallic phases and (b) Microstructure in sensitized condition with intermetallic Sigma Phase.

corrosion attack. If no visible attack or mass loss of more than 1 mg is detected, the sample is set again in the testing solution at a temperature 5 K higher for further periods of 24 h and so on – until a corrosion attack can be detected. The solution annealed sample in correct metallurgical condition is able to reach 75 °C in this test, the samples with sigma phase showed the first signs of localized pitting corrosion attack already at temperatures between 45 and 50 °C and showed strong pitting at 75 °C. The influence of pitting temperature and solution annealing temperature is shown in Figure 2.2.5 as bar chart and the photographs of the samples from corrosion testing in Figure 2.2.6.

In laboratories of research & development or education, the use of glassware for chemical synthesis on small scale with fume hoods is common. This is no option for industrial productions where high temperatures or pressures are beyond the capability of glass. In industrial environments, safe and reliable materials are indispensable. The consequences by a corrosion leak or crack in a reactor vessel or tubing may result (in many cases) in severe consequences to workers, citizens nearby, environment and loss of production. The example above should emphasize the fact that materials have always to be carefully selected and used in proper condition.

2.2.3.4 Influence of alloyed elements to mechanical properties

In combination with the corrosion resistance and microstructural stability, the mechanical properties such as strength, ductility and creep resistance play a major role for selecting a suitable material for the planned application. The differences in yield strength for softer austenitic steels in solution annealed condition can be from about 200 to more than 2000 N/mm² for cold worked sheets or wires [17].

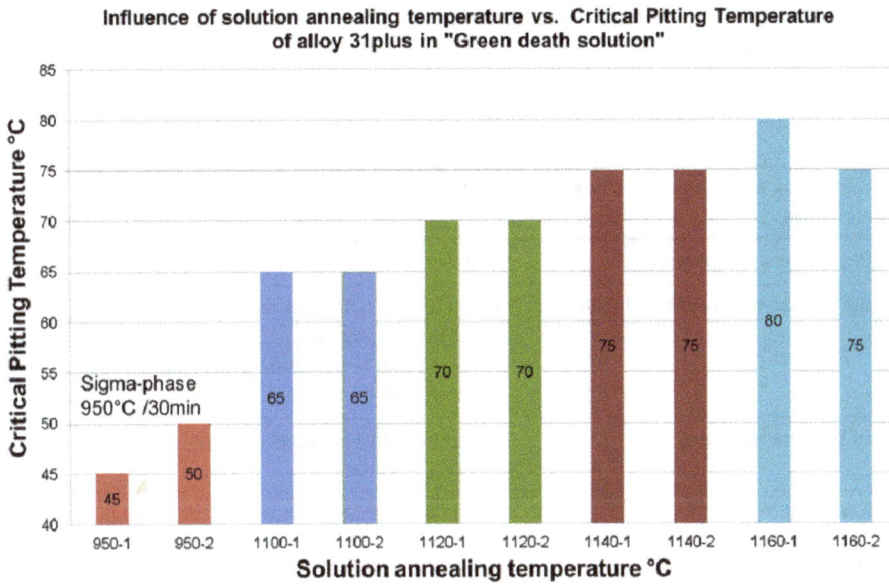

Figure 2.2.5: Influence of solution annealing temperature on critical pitting corrosion temperature.

Figure 2.2.6: (a) Correct solution annealed condition (left L); (b) improper sensitized condition (right S) (samples exposed to testing solution "Green Death" at 75 °C for 24 h).

The strength of an alloy can be influenced by different metallurgical principles as found in Table 2.2.4. The fact that metals can be strengthened is known since thousands of years to mankind when it was discovered that copper and tin alloyed together form bronze, which is stronger than the single metals. The most important strengthening principles are listed in Table 2.2.4 and the working principles will be described shortly.

Table 2.2.4: Basic mechanisms for the strengthening of austenitic steels and alloys.

Strengthening mechanisms of austenitic steels/alloys	Effect
Solid solution strengthening	Soluble elements: Cr, Mo, Co, W, N
Carbide/Carbo-nitride strengthening	Carbon 0.02 to 0.1% and carbide/nitride forming elements Cr, Ti, Nb, Mo; mainly for high-temperature alloys, not for wet corrosion alloys
Grain size strengthening	Finer grains provide higher strength (Hall-Petch equitation)
Cold working	Plastic deformation, increase of dislocation density
Precipitation hardening	Gamma prime (Ni_3/Ti/Al) gamma double prime (Ni_3/Nb) up to 6 wt% element addition

2.2.3.4.1 Solid solution strengthening

Dissolved alloying elements in the base matrix are not only effective to reach the corrosion resistance of the alloys but are also capable to increase the strength of the base material by an effect called solid solution hardening. If elements like chromium, molybdenum or tungsten are added and dissolved in the base γ-matrix, the atoms will substitute nickel atoms in the host structure. The elastic distortion of the base matrix metal structure (here meant for nickel as example) will result in an inhibition of dislocation movement due to the misfit in the atomic size. This effect increases with the increasing difference of atomic size. The size difference varies between + 1% in the case of cobalt and + 13% for tungsten [19]. In the case of atoms smaller than nickel (like nitrogen or carbon), these atoms can be located on interstitial lattice places.

Molybdenum, tungsten and also chromium are effective solid solution strengtheners and are found in many commercial alloys, as in nickel-based superalloys which find application at high temperatures, and need high creep strength [20–22]. In austenitic stainless steels or the super austenitic steels nitrogen not only delivers a contribution to the mechanical strength, but also to the resistance to pitting corrosion and stabilizes the austenitic microstructure and therefore is used in several commercial alloys of this type. Additional to the basic principle of solid solution strengthening by alloying, higher strength can be reached by using several other mechanisms.

2.2.3.4.2 Carbide/carbo-nitride strengthening

The strengthening of austenitic steels and alloys with carbides or carbo-nitrides is used for the high-temperature alloys. Carbon in combination with elements like Ti, Nb or W will form stable carbides during the solidification from the melt with a heterogeneous distribution in the alloys microstructure. Titanium is used for alloying up to 2 wt% and niobium up to 4 wt% to form MC carbides (fcc). This MC particles finely divided in the alloys matrix provide strength for application temperatures up to 1100 °C [19, 23].

As already mentioned in alloys for wet corrosion purposes, the carbon contents are kept at low levels to avoid detrimental influence of chromium depletion at the grain boundaries and sensitization towards intercrystalline corrosion caused by precipitated chromium carbides and nitrides if exposed to temperature ranges where these phases can form. By the addition of titanium or niobium in some steels, for example in the stainless steel Alloy 316Ti, excess carbon can bond to TiC carbides that are more stable as the chromium carbides and avoid the grain boundary depletion.

2.2.3.4.3 Grain size strengthening

A targeted adjustment of materials strength can be done by the control of grain size. By choosing the deformation temperature and the grade of deformation in the forming processes together with a recrystallization heat treatment, the grain size can be influenced [19]. A finer grain will result in a higher materials yield strength. For more details see the work of Rösler et al. [23] with the keyword to the Hall-Petch equitation.

2.2.3.4.4 Cold working

The strengthening of stainless steels by cold work is used when the intended application for a component lies in the lower temperature range, because at temperatures higher than the recrystallization temperature or solution annealing temperatures the increase of materials strength is lost.

2.2.3.4.5 Precipitation hardening

The precipitation hardening, also called age-hardening, is an effective method that is found in many commercial alloys for aerospace, gas turbines, oil and gas industry or valves in traction applications in cars or diesel engines. Two examples of age-hardenable alloys, Alloy 718 and Alloy 925, used in the oil and gas industry, be will described later in this chapter. The strength of precipitation age-hardened alloys is usually higher than the alloys strengthened only by the carbides or carbo-nitrides from room temperature up to 600 to 800 °C. Beyond this temperature the strength will drop due to the dissolution of the strengthening phases and the rapid coarsening of the precipitates. For more details to the age-hardening mechanism see the literature for superalloys [20, 24].

2.2.4 Composition and microstructures of stainless steels and special steels

Since the focus in this work is set on the austenitic structures of special stainless steels and alloys, the other common types are addressed as overview in Table 2.2.5 and will be not described in detail. Among the commercial iron-based stainless steels there are classes of different microstructure types as ferritic, martensitic or mixed ferritic-austenitic types that are used for numerous applications [5, 6, 18]. More detailed information to all types of these materials can be found in the book of Cardarelli [25]. The typical composition range for the most important alloying elements is listed in Table 2.2.5.

Table 2.2.5: Classification of stainless steels (general examples, therefore incomplete selection) [25].

Classification of stainless steels by type and microstructure	Typical composition range
Ferritic stainless steels	17–30 wt% Cr carbon < 0.2 wt% C
Martensitic stainless steels	12–18 wt% Cr carbon < 1.2 wt% C
Austenitic stainless steels	18–25 wt% Cr 8–20 wt% Ni
Super austenitic steels (6-Mo-steels)	18–23 wt% Cr 6–7% molybdenum
Duplex stainless steels	18–26 wt% Cr 4–7 wt% Ni 2–3 wt% Mo
Precipitation-hardening (PH) stainless	steels 12–30 wt% Cr (Al, Ti, Mo)

2.2.4.1 Nominal chemical composition of wrought stainless steels

In Table 2.2.6 chemical compositions of some common standardized stainless steels with uses in industrial applications are shown. Since the exact analytical ranges and elements spans according to DIN-EN or ASTM specifications, these documents should be referred. At the time of writing, the system for the designation of non-ferrous alloys is described in the standard EN 17007-4 (2012–12). The DIN-EN numbers start with a "1" for steels and a "2" for non-ferrous nickel alloys. The first two decimal places are giving the variety of the steel and the last two numbers have a count up to "99" for the specific steel grade.

The UNS (Unified Numbering System) is a system from North America. In this system "Heat and corrosion resistant steels" start with the letter S followed by a five-digit number and in the case of nickel alloys with the leading letter "N" instead. The element ranges can differ between DIN-EN and ASTM and as determined by customers' own specifications. This is not only true for the chemical compositions but also for the mechanical technology properties that are not listed here and can be found in the official standards or in the technical datasheets of the producers to the referred

steel or alloy. A recommended good survey can be found in [27–29] and the Bro-
chures of Euroinox [5, 6].

Table 2.2.6: Nominal compositions of some stainless steels [16].

Alloy designation	DIN-EN	UNS	Corrosion resistant alloys for plate, sheet, tubes					
			Nominal chemical composition wt%					
			Ni	Cr	Mo	max. C	Fe	others
304L	1.4307	S30403	8–12	18–20	–	0.03	Bal	
316L	1.4404	S31600	10–14	16–18	2.0–3.0	0.03	Bal	
316Ti	1.4571	S31635	10–14	16–18	2.0–3.0	0.08	Bal	Mn 2 Ti < 0.7
Duplex 2205	1.4462	S31803	4,0–6,5	21–23	2.5–3.5	0.03	Bal	0.08–0,20 N
Super-Duplex 2507	1.4501	S32750	6,0–8,0	24–26	3.0–4.0	0.03	Bal	0.25 N 0.8 Cu 0.8 W
904L	1.4539	N08904	23–28	19–23	4.0–5.0	0.02	Bal	
926	1.4529	N08926	24–26	19–21	6.0–7.0	0.02	Bal	0.9 Cu 0.20 N
28	1.4563	N08028	30–34	26–28	3.0–4.0	0.02	Bal	1.3 Cu
31	1.4562	N08031	30–32	26–28	6.0–7.0	0.02	Bal	0.20 N 1.3 Cu

2.2.4.2 Application areas of selected stainless steels
In Table 2.2.7 application areas of stainless steels are given as examples. The greatest share in the use field of stainless steels have flat products of the austenitic types
as plates or sheets. With the exception of the two duplex steels in Table 2.2.7 as representative of this ferritic-austenitic type, all other stainless steels listed in that
table are characterized by an austenitic microstructure.

2.2.4.3 High-performance alloys (HPA), composition and microstructures
For application fields with demands beyond the capability of stainless steels, in many
cases nickel alloys are the choice [13]. The austenitic (face cubic center) structure
is responsible for a tough ductile matrix and phase stability from cryogenic temperatures up to the melting point. Nickel as base metal for alloys provides several
advantages compared to iron [20]. The austenitic microstructure of nickel provides
wide range of solid solubility for alloying elements like chromium, molybdenum,
copper, iron and tungsten. Commercial alloys can alternatively include up to 35 wt%
chromium, 28 wt% molybdenum or 24% tungsten in solid solution. Many binary,
ternary and complex nickel alloys are possible by alloying additional elements,

Table 2.2.7: Description of application areas of some stainless steels.

Alloy designation	DIN-EN	UNS designation	Application areas
304L	1.4307	S30403	Chemical and food processing equipment, breweries, cryogenic storage, liquid fertilizer tanks, low carbon grade for improvement of carbide precipitation during welding.
316L	1.4404	S31600	CPI, pharmaceutical industry, food, medical equipment, pulp and paper, low carbon grade alloyed with molybdenum with better corrosion resistance as 304L in moderately corrosive environments with chlorides or other halides, alloyed with about 2.3 wt% Mo it is the "workhorse" for pharmaceutical industry and food processing.
316Ti	1.4571	S31635	Applications as before with 304L stabilized by titanium for welding applications, and minimizing sensitization at higher application temperatures, not suitable for polished surfaces.
Duplex 2205	1.4462	S31803	Desalination, tubes, pipes, heat exchangers, biofuels, ferritic-austenitic steels with high strength.
Super-Duplex 2507	1.4501	S32550	Resistance against stress corrosion cracking, higher corrosions resistance in chloride containing environments as duplex 2205, ferritic-austenitic steels with high mechanical strength.
904L	1.4539	N08904	Pipes, tanks, heat exchangers, valves, fertilizer production, to uniform corrosion better resistance as 316 steels.
926	1.4529	N08926	Firefighting systems, seawater filtration systems, hydraulic and reinjection piping, polished rods for pumps in the offshore industry, reverse osmosis desalination, evaporators, heat exchangers, filters and mixers used in the manufacture of phosphoric acid, containers for transportation of aggressive chemicals. The chemical composition is similar to that of Alloy 904L, but with nitrogen increased to about 0.2%, and molybdenum to about 6.5 wt% Mo, high conditions to chloride-ion stress-corrosion cracking, higher strength as 904 L.
28	1.4563	N08028	Cooling water heat exchanger with high chloride contents, tubing, molybdenum 3–4 wt%.
31	1.4562	N08031	6 wt% molybdenum steel applications in chemical and petrochemical industries, phosphate winning, hydrometallurgy, flue gas desulfurization and environmental engineering, heat exchanger, evaporators.

creating unique and special properties profiles [13, 16, 30, 31] to withstand a wide variety of severe operating conditions in industrial applications. These include highly corrosive environments ranging between alkaline conditions to high concentration of acids, very high temperatures, high mechanical stresses and combinations of these factors [32–34]. In Table 2.2.8 an overview of common nickel alloys [16, 25, 26] is listed and some examples from the list are described later in this chapter. More details to common use fields and mechanical properties of certain alloys can be found in the alloys data sheets in actual versions on the Internet pages in the download area of the producer [35].

Table 2.2.8: Application fields of uses for high-performance alloys (HPA).

Alloy type	Fields of Use
Ni	Caustic (alkaline) media, hydrofluoric acid, brine electrolysis
Ni-Cu	Mild reducing conditions, chloride brines, hydrofluoric acid non-aerated, dilute sulfuric acid, marine environments, desalination
Ni-Cr Ni-Fe	Resistance heating, steam generation magnetic shielding, controlled-expansion alloys, carbon fiber molds, liquid natural gas (LNG) transport
Ni-Mo	Reducing corrosive media, hydrochloric acid
Ni-W	Superconductor substrates
Ni-Cr-Fe	High-temperature applications, heat treatment, chemical processing, solar power, fuel cells
Ni-Cr-Fe-Mo-Cu	High-alloyed special stainless steels, phosphoric acid production, sulfuric acid, seawater, environmental technology, hydrometallurgy
Ni-Cr-Mo	Most versatile: oxidizing and reducing media, highly corrosion resistant to localized corrosion and stress corrosion cracking, flue gas desulfurization
Ni-Cr-Fe-Mo Super alloys age-hardenable	Aerospace, gas turbines, oil and gas extraction

2.2.4.4 Nominal chemical composition of selected wrought nickel alloys

In Table 2.2.9 with the nominal chemical compositions, high-alloyed steels and nickel alloys are listed which are produced and supplied in solution annealed condition. Exceptions in this table are the alloys K-500, Alloy 925 and Alloy 718. The designation and classification of nickel alloys is not as systematic as known for steels. Many of the alloys are designated by their trade names assigned by the inventing company. Very often the alloy trade names contain in their name the last numbers of the UNS (unified numbering system) designation. As an example Alloy 59 is designated in the UNS System as UNS N06059.

Many of these alloys (which are fully austenitic in nature) provide high corrosion resistance and could be additionally strengthened by cold work. If thick sections or complicated shapes like thick walled tubes, large forging parts, or flanges need higher strength, it is possible to reach this strength by the use of age-hardening. For this metallurgical mechanism, it is necessary to add elements to the nickel alloy that will form fine precipitates in the microstructure.

Examples of age-hardened alloys are found in Table 2.2.9 as Alloy 718, Alloy 925 and K-500. By the addition of aluminum, titanium and/or niobium as alloying elements and an appropriate heat treatment, these alloys are capable to form secondary phases in the base austenitic microstructure, like γ"-prime (Ni_3Nb) in the case of Alloy 718 or γ'-prime phase as in the case of Alloy 925. By this aging mechanism, it is possible to reach approximately the double yield strength values as compared to the solution annealed condition [12].

2.2.5 Industrial applications of alloys

2.2.5.1 Super austenitic 6-Mo steels

These types of commercial alloys possess higher resistance to local corrosion attack in comparison to the general stainless steel grades in acidic and chloride-containing media such as seawater. They are characterized by higher alloyed additions of molybdenum, chromium and nickel. By the addition of nitrogen, the resistance to pitting corrosion as well as the mechanical strength can be enhanced and at the same time the precipitation of intermetallic phases (in most cases the sigma phase) can be slowed down due to the stabilization of the austenitic microstructure. This makes the components fabrication with thicker sections safer [2, 14]. The name "super austenitic stainless steels" originates from the molybdenum content that usually lies between 6 and 7 wt%. This class of alloys is an interesting option for offshore installations or at coasts, where seawater is used for cooling applications in plate or tube heat exchangers and the corrosion resistance of standard stainless steels is not sufficient anymore. Other fields where these alloys have found wide applications are the production of phosphate in the wet process (see also Chapter 6.2), hydrometallurgy with leaching of metals from ores, or recycling leaching for metals recovery, water-desalination, flue gas desulfurization in coal fired power plants, on ships and in sulfite digestion in the production of pulp and paper. These steels fit perfectly in the gap between classical austenitic stainless steels and the more costly nickel alloys. These materials offer excellent corrosion resistance of nickel alloys and provide cost effective solutions in many of the named applications.

Table 2.2.9: Nominal composition of selected common wrought Ni-Alloys (typical values; standardized values are listed in DIN 17,470).

Alloy system			Nominal chemical composition, wt%							
alloy	DIN	UNS	Ni	Cr	Fe	Mo	Cu	Co	C	others
Ni										
200	2.4066	N02200	≥ 99.2		≤ 0.4		≤ 0.25		≤ 0.1	
201	2.4068	N02201	≥ 99.0		≤ 0.4		≤ 0.25		≤ 0.02	
Ni-Cu										
400	2.4360	N04400	≥ 63.0		1.5		30		≤ 0.15	
K-500	2.4375	N05500	≥ 63.0		1.0		30		≤ 00.2	2.5 Al
Ni-Fe										
Pernifer	1.3912	K93600	36	≤0.20	Rest	–		≤ 0.5		
Ni-Cr										
Cronix 70	2.4658	N06008	68	30	≤ 1				≤ 0.07	0.025 rare earth
Cronix 80	2.4869	N06003	78	20	≤ 1				≤ 0.08	0.025 rare earth
Ni-Cr-Fe										
600 H	2.4816	N06600	≥ 72	16	8				≤ 0.1	
690	2.4642	N06690	63	30	9				≤ 0.02	
602CA	2.4633	N06025	62	25	9.5				0.20	2.1 Al, 0.1 Y
120	2.4854	N08120	≤ 39	≤ 27	Rest	≤ 2.5	≤ 0.50	≤ 3	≤ 0.10	0.30 N
Ni-Mo										
N10675	2.4600	N10665	≥ 65	1–3	1–3	27–32		≤ 3	≤0.01	≤0.5 Al
Ni-Cr-Mo										

(continued)

Table 2.2.9 (continued)

Alloy system			Nominal chemical composition, wt%							
alloy	DIN	UNS	Ni	Cr	Fe	Mo	Cu	Co	C	others
625	2.4856	N06625	62	22	≤ 3	9			≤ 0.025	≤ 3.5 Nb
C-4	2.4610	N06455	66	16	≤ 3	16			≤ 0.009	
C-22	2.4602	N06022	56	22	≤ 3.5	13			≤ 0.010	3 W
C-276	2.4819	N10276	57	16	5.5	16			≤ 0.010	3.5 W, 0.2 V
59	2.4605	N06059	59	23	≤ 1	16			≤ 0.010	0.3 Al
2120	2.4700	N06058	58	20	≤ 1.5	18.5	≤ 0.3	≤ 0.3	≤ 0.010	N
617	2.4663	N06617	54	21	≤ 2	9		11.5	0.075	Ti, Al, Zr
Ni-Cr-Fe-Mo										
X	2.4665	N06002	48	22	18.5	9		1	0.07	0.6 W, 0.6 Si
333	2.4608	N06333	46	25	18	3		3	0.05	3 W, 1 Si
Ni-Cr-Fe-Mo-Cu										
825	2.4858	N08825	42	21.5	31	3	2.2		≤ 0.025	0.9 Ti
926	1.4529	N08926	25	21	45	6.5	0.9			0.20 N
31 Plus	2.4692	N08034	34	26.5	Rest	6.5	≤ 2	≤ 0.50		2 Mn, 0.20 N
Age-hardenable alloys (oil and gas)										
718	2.4668	N07718	53	19	18	3				5.1 Nb
925	2.4852	N09925	43	20	29	3				2.2 Ti

2.2.5.2 Alloy 926

Alloy 926 is a super austenitic stainless steel with about 6.5 wt% molybdenum, developed by increasing the molybdenum and nitrogen contents to its forerunner Alloy 904L with 4.5 wt% molybdenum.

This results in elevated mechanical properties and improved resistance to corrosion in halide containing media [35]. Therefore, common industrial applications are tubes for seawater, e.g. for fire fighting lines, seawater filtration system, pumps and reverse osmosis desalination. In comparison to copper-nickel alloys, the 6 wt%-Mo alloys (like the Alloy 926) do suffer less corrosion caused by hydrogen sulfide. The Alloy 926 has a significantly higher resistance to localized corrosion in halide media and improved mechanical properties as compared to the Alloy 904L. Products made of Alloy 926 include evaporators, heat exchangers, filters and mixers in the manufacture of phosphoric acid, polished rods for pump shafts, tubing, couplings and flowline systems. For the application of pump shafts, Alloy 926 is used in the cold drawn condition to increase the yield strength. Figure 2.2.7 shows welding wire of Alloy 926 on spool and a flash drum for thermal water desalination.

Figure 2.2.7: (left) Alloy 926 welding wire on spool and (right) flash drum for thermal water desalination. (Courtesy Welders N.V. Belgium).

2.2.5.3 Alloy 31 Plus

Alloy 31 Plus is a so-called super austenitic 6-Mo steel belonging to the group of nickel-iron-chromium-molybdenum alloys with nitrogen addition. It is alloyed with 34 wt% nickel, 26.5 wt% chromium and 6.5 wt% molybdenum [14]. This recent alloy is an optimized successor of the forerunner Alloy 31. The complete nominal chemical composition is given in Table 2.2.9. It was designed for applications in chemical and petrochemical industries, components in plants for phosphate production [36] via the wet digestion process, hydrometallurgy and chemical processes with sulfuric acid [37], flue gas desulfurization, sea water and brackish water applications, environmental engineering and the storage or transport of corrosive chemical substances [38, 39, 41]. Alloy 31 Plus has a higher corrosion resistance to chloride containing media (like seawater) than Alloy 926 or Alloy 28. This was achieved by careful adjustments of the chemical analysis for an improved austenite stability. One of the elements to be named is nitrogen, which stabilizes the austenitic structure and reduces the solvus temperature in solution annealing of the intermetallic phases, e.g. sigma-phase, to lower values as compared to its forerunner. This and the lower kinetic of sigma phase formation and good weldability has widened the application field also for the use as hot roll cladding on carbon steels. In addition, nitrogen increases the mechanical strength and improves the resistivity against chloride pitting corrosion. The low carbon contents prevent the formation of chromium carbides and the sensitization during heat treatments regarding to intercrystalline corrosion. As an example, Figure 2.2.8 shows the upper part of a salt crystallisator/evaporator before installation.

Figure 2.2.8: Upper part of a crystallisator/evaporator Alloy 31 Plus (Ebner GmbH & Co. KG).

2.2.6 Nickel alloys

Nickel alloys are an important group of metallic materials and nearly irreplaceable in modern industry applications due to their unique properties [16, 22, 30–32]. These alloys are used when stainless steels are not suitable or beyond their capabilities. Several reasons were described already in Figure 2.2.1, like the central aspect of safety considerations, product purity or life cycle costs. Nickel and nickel alloys are readily fabricated as wrought alloys by conventional hot and cold forming methods and also as cast or powder metallurgy (P/M) products and 3D-additive manufacturing. The alloys are available from several manufactures and in several product forms. At cryogenic temperatures they offer freedom from ductile-to-brittle transitions with loss of ductility that is found in other metals and alloys, including steels.

2.2.7 Nickel and nickel alloys for wet corrosion applications

2.2.7.1 Nickel (unalloyed)

Unalloyed nickel [16] is corrosion resistant in normal atmospheres, in most non-aerated organic acids, in natural freshwaters and in deaerated non-oxidizing acids, and it has excellent resistance to corrosion by caustic alkalis such as sodium hydroxide or potassium hydroxide at high concentrations and temperatures. For this reason, nickel is used for the production and processing of caustic soda and caustic potash. In more concentrated ammonium hydroxide solutions it is not suitable. Unalloyed nickel offers very useful corrosion resistance in certain media as named and at the same time nickel provides a proven metallurgical base for alloying with chromium, molybdenum, tungsten, copper and other elements. In Figure 2.2.9 a cathode is shown for the electrolysis of brine in a membrane cell to produce sodium hydroxide and chlorine.

Figure 2.2.9: Cathode for electrolysis: Hoechst/Krupp Uhde membrane cell.

2.2.7.2 Alloy 825

Alloy 825 is a nickel-chromium-iron alloy that finds widespread applications as material for cladded pipes and separators. It is a solid solution alloy and not age-hardenable by heat treatment. The strengthening can be done by cold work if necessary. At moderate temperatures, Alloy 825 has a good resistance to carbon dioxide and high hydrogen sulfide loads [16, 35]. It has a limited resistance to chloride-induced crevice and pitting corrosion due to its low molybdenum content of 3 wt%. An improved version of Alloy 825 with a content of 5.7 wt% molybdenum for a better corrosion resistance to pitting named Alloy 825CTP was developed recently. In environments with very high chloride content, high temperature and H_2S present in the media Alloy 625 has an advantage over Alloy 825.

2.2.7.3 Alloy 625

Alloy 625 is especially resistant to severe corrosive environments with high chloride contents. It is widely used in production systems in claddings, overlay welding and for cold drawn pump shafts because of its excellent resistance to localized corrosion, its high resistance to stress corrosion cracking and its good weldability. It has a high molybdenum content of 9 wt%, which imparts a higher resistance to chloride type pitting [16]. For that reason, in environments with high chloride content, high temperature and H_2S present in the media, Alloy 625 has an advantage over Alloy 825. A common application of this alloy is flowlines. In many cases the highly corrosion resistant nickel alloy is used for cost saving reasons as a thin cladding on a thicker substrate made from carbon steel. For smaller diameter very often seamless tubes are used. These can be additionally strengthened by cold drawing of the tube. Another example of using this alloy in cold drawn condition is the enhancement of the yield strength is pump shafts.

Some other main applications of Alloy 625 include corrosion resistant weld overlay on flange faces (Figure 2.2.10), piping components as heating tubes in fossil-fuel-fired plants.

In the CPI the Alloy 625 is usually used in stabilized annealed condition. The sluggish response of Alloy 625 to aging treatments and the long time for the precipitation of the nickel-niobium phase (Ni_3Nb) is of advantage in welding applications to avoids unwanted strain in the Heat Affected Zones (HAZ) of the welds.

Figure 2.2.10: Flange of Alloy 625.

2.2.7.4 Alloy 59

Alloy 59 is a ternary nickel-chromium-molybdenum alloy with high concentrations of molybdenum. The molybdenum content in Alloy 59 is 16 wt% molybdenum and 23 wt% chromium. It is a highly corrosion resistant alloy with numerous applications in the CPI for the production of specialty chemicals [16, 30, 35]. It is well proven for applications [41, 42] in handling acids even when contaminated with halides and handling waste waters from flue gas desulfurization or corrosion problems in plate type heat exchangers with chlorinated seawater on the cooling side. Alloy 59 is a nearly pure ternary alloy with a low iron content and without the addition of other elements like copper or tungsten, resulting in a good thermal microstructural stability. This is an important point for overlay weldings and multi-pass welding of thicker plates. The high corrosion resistance is accompanied by a good workability and weldability. The main use of this alloy is for chemical reactors, flue gas desulfurization components and heat exchangers. Figure 2.2.11 shows applications of Alloy 59 for a chemical reactor in the pharmaceutical industry and a tube heat exchanger from a power plant.

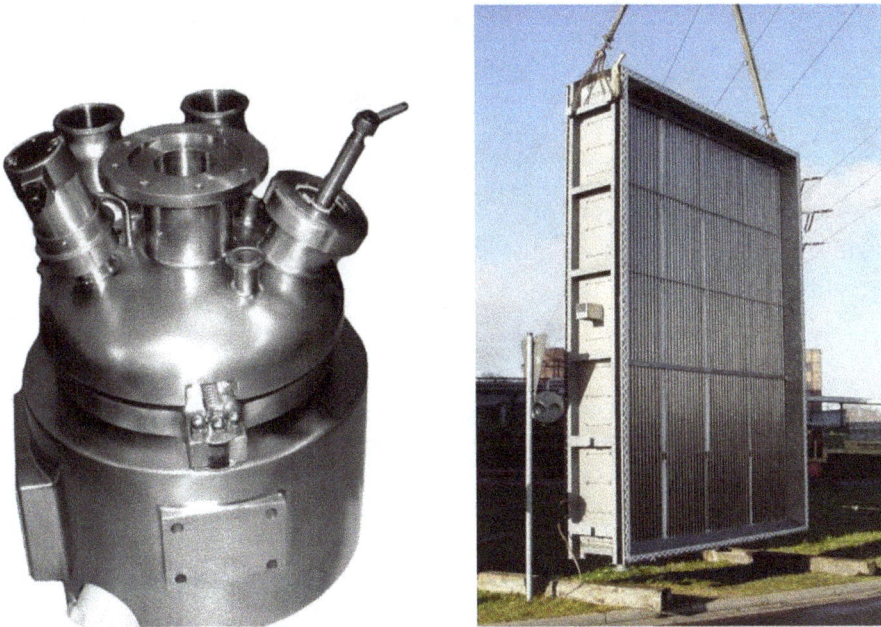

Figure 2.2.11: (left) Chemical reactor in pharmaceutical industry and (right) a heat exchanger from a power plant.

2.2.7.5 Alloy C-276

The alloy C-276 is a solid solution nickel-molybdenum-chromium alloy with the same high concentration of molybdenum as Alloy 59. The Mo content in Alloy C-276 is 16 wt%, the chromium content 16 wt% and the tungsten content 3.5 wt% [16, 30]. The main areas of use are corrosive acids such as sulfuric acid, phosphoric acid, hydrochloric acid and organic acids like formic, acetic, propionic or naphthenic acid. This alloy has good resistance to localized corrosion attack like pitting in seawater applications or brine solutions. In acetic acid it is used in a wide range of concentrations, temperatures and is used in places where acetic acid occurs in combination with inorganic acids, salts and halides, which restrict the use of standard stainless steels.

C-276 is resistant against chloride-induced stress corrosion cracking, even in the presence of elemental sulfur in sour gas environments. At temperatures above 100 °C there is sensitivity against hydrogen-induced stress corrosion cracking in such media. In the CPI, the alloy is used for components like reaction vessels, evaporators, heat exchangers and transfer piping. A typical application of Alloy C-276 tubulars is shown in Figure 2.2.12 for a high pressure gas cooler.

Figure 2.2.12: High pressure gas cooler of Alloy C-276 tubes (Courtesy Welders N.V Belgium).

2.2.7.6 Alloy 2120 MoN

Alloy 2120 MoN is a recent nickel-chromium-molybdenum in the C-alloy family that was developed with improved localized corrosion resistance and increased uniform corrosion resistance in mineral acids compared to the other family members [40–44]. It is the first nickel-chromium-molybdenum alloy to contain nitrogen as an alloying element. The chemical composition has the highest molybdenum content of this alloy family resulting in excellent corrosion resistance in reducing environments. The high chromium content of 20 wt% and molybdenum of 18.5 wt% together with nitrogen addition result in a pitting resistance surpassing all other commercial nickel-chromium-molybdenum alloys currently in use. Preferred applications are in flue gas desulfurization (Figure 2.2.13) with condensing sulfuric acid even when contaminations of chlorides or other halides are present, mixed mineral acid mixtures and hydrochloric acid. The material has been approved for use in pressure vessel applications for service temperatures from –196 to 450 °C (–320 to 842 °F) and shows a good fabrication characteristic and weldability. The filler metal of Alloy 2120 can also be used for overlay welding on carbon steels.

Figure 2.2.13: Heat exchanger tubes Alloy 2120 in a power plant (flue gas desulfurization).

2.2.7.7 Nickel-molybdenum

The nickel-molybdenum (28 wt% Mo) alloys like Alloy B-2 are the recommended choice for handling highly reducing media [30] at all concentrations as for example hydrochloric acid in the temperature range of 70–100 °C. For sulfuric acid these alloys provide corrosion resistance to pure sulfuric acid up to the boiling point for acid concentrations below 60 wt%. The lack of chromium in the chemical composition is the reason for the weak point of a poor corrosion resistance in the presence of oxidizing species like oxygen or iron(III) salts or other oxidizing metal salts. Alloy B-2 and its successor alloys (example UNS N10675) have been successfully used in the production of formic acid, acetic acid (Figure 2.2.14), pharmaceutical

Figure 2.2.14: Reactor for acetic acid production as example for the nickel-molybdenum Alloy B-2 (Courtesy Welders N.V. Belgium).

products, and alkylation of ethyl benzene, styrene, organic sulfonating reactions, melamine, herbicides and more.

2.2.7.8 High-temperature nickel alloys

High-temperature alloys are needed in industries such as in metallurgy, glass manufacture, waste incinerators, heat recovery, advanced energy conversion systems and more [23, 30]. The severity of the processes in today's industries in power generation and the technologies of thermal incineration of hazardous and municipal waste, biomass, the demands for higher plant efficiency and tougher environmental regulations and the use of "lower grade feedstock" requires the use of higher alloy systems. These can be iron-based, nickel-based and sometimes cobalt-based alloys. Today used alloy systems must have sufficient versatility to resist changing corrosive conditions as listed before. An overview of the high-temperature alloys and high-temperature corrosion mechanism is given in the book of Bürgel [23].

The target of high-temperature alloys are often applications in dry or gaseous corrosion systems. Depending on the composition of corrosive environments and temperatures, a variety of aggressive conditions and atmospheres are produced, which could be either reducing or oxidizing, carburizing, sulfidizing, halogenizing, nitriding in nature or a combination thereof. Additionally, in some cases, erosion by solid particles is a factor. Another required feature to their capacity to resist the aggressiveness of the high-temperature corrosive environments, these alloys also need to keep significant strength values in their operation temperature range. For an optimal material selection in high-temperature applications, a thorough understanding of the nature of corrosive attack and mechanical requirements at operation temperature are required.

Two examples for the use of nickel-based high-temperature alloys are described below.

2.2.7.8.1 Alloy N08120

Alloy N08120 is a nickel-chromium-iron alloy with about 37 wt% nickel, 25 wt% chromium and 0.7 wt% niobium used in industrial heating applications. It is an austenitic solid solution strengthened alloy with additions of carbon and nitrogen. The high creep strength is achieved by the formation of chromium-rich and niobium-rich nitrides and carbo-nitrides. At the same time the high chromium level provides a good oxidation resistance. Furthermore, nickel Alloy N08120 has good resistance to carburization and sulfidation and is providing high mechanical strength at operating temperatures up to 1100 °C.

Alloy N08120 has a wide range of applications in areas of elevated temperatures in furnace construction, like industrial annealing furnaces and waste incineration plants. Further it is used in the chemical industry, in environmental protection plants, in the automotive industry and in power plants. Alloy N08120 has a higher strength as compared to Alloy 330 or Alloy 800H.

Typical applications to be named are high-temperature chemical reactors (Figure 2.2.15) for the production of polycrystalline solar cell silicon, furnace muffles, heat treatment baskets, furnace conveyor belts, wire cloth, stators in gas turbines, super heaters for tubes and micro turbine recuperators.

Figure 2.2.15: UNS N08120 rolled slab in scaled condition (semi-finished product).

2.2.7.8.2 Alloy 602 CA

Alloy 602 CA is a high-temperature alloy composed with about 60 wt% nickel, 25 wt% chromium, 9.5 wt% iron and 2.2 wt% aluminum [35, 44]. The carbon content of approximately 0.18 to 0.2 wt% in combination with 25 wt% chromium leads to the precipitation of bulky homogeneously distributed carbides like $Cr_{23}C_6$. Micro alloying elements like titanium and zirconium allow the formation of fine MC carbides and carbo-nitrides that are stable at the application temperature. This alloy is especially characterized by very good oxidation resistance, which has an application range up to 1200 °C. Even under cyclic heating and cooling conditions, Alloy 602 CA retains this property due to its adhering and tight alumina Al_2O_3 layer, which is very resistant to chipping [30].

Typical examples of applications to be named for this alloy are bright annealing furnaces in the heat treatment industry, transport rollers for ceramic kilns, wire conveyor belts, anchor pins for refractories, tubes for bright annealing wires, calciners, rotary kilns for calcining and production of high-purity alumina, production of nickel and cobalt oxides and recycling of spent nickel catalysts from petrochemical industries. In the chemical/petrochemical industry Alloy 602CA is used for the production of hydrogen via steam reformer technology and pig tails in refinery reformer equipment.

Figure 2.2.16 shows furnace rolls for the heat treatment of plates in a continuous annealing furnace. Due to the mechanical properties at high-temperature and HT-corrosion resistance of this alloy it was possible to substitute rolls with inner water-cooling by uncooled all-metal furnace rolls to save energy.

Figure 2.2.16: Uncooled continuous annealing furnace rolls for temperatures up to 1200 °C (VDM Metals).

2.2.7.8.3 Age-hardenable alloys for offshore service

Among the high-strength materials, Alloy 718 (oil grade) and Alloy 925 are the most commonly used alloys in oil and gas production [45, 46] and will be taken as example for age-hardenable alloys.

Alloys for offshore service must be resistant to all the liquids and gases involved in the intended services. These include not only the liquids produced but also drilling muds and fluids, tensides, defoamers, solvents and other chemicals. Although the hydrocarbons from the oil itself are not corrosive, certain constituents in the produced medium are. Typical failure causes for piping systems exposed to seawater are pitting and crevice corrosion.

The liquid media pumped through the pipes often contain CO_2, H_2S, polythionic acids and elemental sulfur which induce stress corrosion cracking under tensile stress with the associated failure mechanisms. Corrosion attack in the produced media takes place at the interfaces of a multi-phase system consisting of the metallic material and gas/oil with salt water. These are conditions which can also give rise to crevice corrosion. The same corrosion mechanism applies to metal flanges while the integrity of welds is governed by their pitting resistance.

The mechanical strength levels desired in heavy-wall pipe, large forgings, flanges and rods from nickel alloys can no longer be attained through hardening by cold working. Here, alloying with aluminum, titanium and/or niobium in conjunction with a suitable heat treatment are the measures of choice to produce nickel alloys that are age-hardenable (like Alloy 718 oil grade, Alloy 925 and other alloys of this type). The increase in strength is achieved through the formation of fine distributed intermetallic phases in the austenitic microstructure.

2.2.7.8.4 Alloy 718 (oilgrade)
Alloy 718 is a niobium alloyed age-hardenable alloy [45, 47]. Apart from being widely used in gas turbine construction, it is the most preferred age-hardenable alloy for the oil and gas industry (Figure 2.2.17). In Table 2.2.9 the chemical composition of Alloy 718 is shown. Alloy 718 is also known from the aerospace industry for rotating parts in flying gas turbines. To meet the environmental requirements and toughness for the oil and gas industry, the properties of Alloy 718 have been modified through a slightly different chemical composition, thermo-mechanical processing and heat treatment (solution heat-treatment temperature and aging temperature). The result of that change is that the version of Alloy 718 does not reach the high strength of the aerospace version but the ductility is almost double. In the offshore industry, Alloy 718 is used in various oil field applications where high tensile and yield strength, high impact values at sub-zero temperatures, high hardness for mechanical design reasons and clean microstructure for superior corrosion resistance to sour gas environments are needed. The main applications are rods and tubes for downhole tubing and surface gas-well components, valves, hangers, landing nipples, tool joints and

Figure 2.2.17: Alloy 718 oilgrade, drill string equipment (rotary steerable system). (© 2021 image courtesy of Weatherford).

packers. Alloy 718 as well as Alloy 925 described below combine high strength with the required resistance to the typical forms of sour gas corrosion.

This alloy is mainly age-hardened by the formation of the intermetallic γ˝ (gamma double prime) phase consisting of $Ni_3(Nb/Ti/Al)$. This phase is the main strengthening mechanism for Alloy 718. The orthorhombic δ phase is detrimental for the corrosion properties and materials ductility and has to be avoided in the oil grade material. To get optimal performance of Alloy 718 thermo-mechanical processing conditions during hot working and heat-treatment have to be therefore carefully controlled [48].

2.2.7.8.5 Alloy 925

The field applications of Alloy 925 are very similar to those of Alloy 718. They are rods and tubes for downhole tubing, surface gas-well components, valves, hangers, landing nipples, tool joints and packers (Figure 2.2.18), where a high corrosion resistance to the hydrogen-induced forms of sour gas stress corrosion cracking is required [45, 47]. Alloy 925 is used where lower strength requirement for the components exist. The demands for yield strength in typical specifications for Alloy 925 are 110 ksi minimum (758 MPa) as compared for 120 ksi (827 MPa) for Alloy 718. Alloy 925 provides a little bit better resistance to pitting corrosion than Alloy 718. The austenitic nickel-based Alloy 925 contains molybdenum, copper, titanium and aluminum as alloying elements. The chemical composition of this alloy is a modification of Alloy 825 by the addition of titanium for the age-hardening mechanism. Alloy 925 owes its strength by the formation of γ′ phase $Ni_3(Ti/Al)$ during age-hardening treatment. The size and distribution of the γ′ phase depends on the heat treatment conditions. This phase is metastable and can transform by a wrong heat treatment to the stable η phase $Ni_3(Ti/Al)$ which is detrimental to the mechanical properties of that alloy. Also a brittle iron-rich sigma phase can form upon wrong heat treatment of that alloy. Hence, very

Figure 2.2.18: Alloy 925 forged bars with 215 mm diameter (semi-finished round).

precise controlling of the heat treatment parameters is critical in achieving the desired aging response and desired materials properties.

References

[1] Alves H. Material Selection and Recent Case Histories with Nickel Alloys, Corrosion 2018, Paper 11600, NACE International, Phoenix Arizona, 2018.

[2] Alves H, Behrens R, Paul L. Review of Corrosion Issues and Material Solutions in the CPI, Corrosion 2010, Paper 10338, NACE International, Houston, Texas, 2010.

[3] Eurofer. European Steel in Figures 2021, The European Steel Association, Brussels, 2021.

[4] Cunat P-J. Euro Inox, Alloying Elements in Stainless Steel, Brussels, 2004.

[5] Euro Inox. Stainless Steel: Tables of Technical Properties, 2nd edition, Brussels, 2007.

[6] Cunat P-J. Euro Inox, Working with Stainless Steels Brochure, 2nd edition, Brussels, 2008.

[7] Rebak RB, Crook P. Adv Materi Proc, 2000, 157, 37–42.

[8] Lo KH, Shek CH, Lai JKL. Mater Sci Eng, 2009, 65, 39–104.

[9] Plaut RL, Herrera C, Escriba D, Rios P, Padilha AF. Mater Res, 2007, 10, 453–60.

[10] Vander Voort GF. (Ed.). Metallography and Microstructures, ASM Handbook, Vol. 9, 9th edition, ASM International, Materials Park, OH, 1985.

[11] High Performance Stainless Steel, Nickel Development Institute, Brochure No.: 11021, Toronto, Ontario, 2000. www.nidi.org

[12] Brill U. Nickel und Kobalt, Nickel- und Kobalt Basislegierungen. In: Kunze E. (Ed.) Korrosion und Korrosionsschutz, Vol. 2, Chapter 3.1.2, Wiley-VCH, Weinheim, 2001.

[13] Agarwal DC. Nickel and Nickel Alloys. In: Wessel JK. (Ed.) Handbook of Advanced Materials, Chapter 7, John Wiley &Sons Inc, Hoboken, 2004. 10.1002/0471465186.

[14] Behrens R, Alves H, Stenner F. New Developed 6-Mo Super Austenitic Stainless Steel with Low Sigma Solvus Temperature and High Resistance to Localized Corrosion, Corrosion 2013, Paper 2228, NACE International, Orlando, Florida, 2013.

[15] Paulraj P, Garg R. Adv Sci Technol Res J, 2015, 9, 87–105.

[16] Heubner U, Kloewer J, Alves H, Behrens R, Schindler C, Wahl V, Wolf M. (Eds.). Nickel Alloys and High-Alloyed Special Stainless Steels Properties – Manufacturing Applications, 4th edition, Expert Verlag, Renningen, 2012. ISBN 978-3-8169-2751-8.

[17] McGuire MF. Austenitic Stainless Steels. In: Stainless Steels for Design Engineers, Chapter 6, ASM International, Materials Park, OH, 2008.

[18] Peckner D, Bernstein IM. Handbook of Stainless Steels, Mc GrawHill Book Company, New York, 1977.

[19] Stoloff NS. Fundamentals of strengthening. In: Sims T, Stoloff NS, Hagel WC. (Eds.) Super Alloys II, Chapter 3, John Wiley & Sons, Hoboken, 1987.

[20] Davis JR. ASM Specialty Handbook, Nickel, Cobalt, and Their Alloys, 2nd edition, ASM International, Materials Park, OH, 2007. ISBN-13: 978-0-87170-685-0.

[21] Rösler J, Harders H, Bäker M. Mechanisches Verhalten der Werkstoffe, 4th edition, Springer Vieweg, Wiesbaden, 2012, 198–220. ISBN 978-3-8348-1818-8.

[22] ASM. Handbook Volume 2, Nonferrous Alloys and Special-Purpose Materials, 10th edition, ASM International, Materials Park, OH, 1990.

[23] Bürgel R, Maier HJ, Niendorf T. Handbuch Hochtemperatur Werkstofftechnik, 5te Auflage, Springer Vieweg, Wiesbaden, 2015. ISBN 978-3-658-10590-7.

[24] Donachie M, Donachie S. Super Alloys: A Technical Guide, 2nd edition, ASM Technical Books, ASM International, Materials Park, OH, 2002. 10.31399/asm.tb.stg2.9781627082679.

[25] Cardarelli F. Materials Handbook, 2nd edition, Springer, London, 2008. ISBN 978-1-84628-668-1.

[26] Heubner U. (Ed.). Edelstahl Rostfrei – Eigenschaften, Merkblatt 821, 5te Auflage Informationsstelle Edelstahl Rostfrei, Düsseldorf, 2014. www.edelstahl-rostfrei.de.

[27] Fritz J. (Ed.). Practical Guidelines for the Fabrication of Austenitic Stainless Steels, International Molybdenum Organization (IMOA), 2nd edition, London, 2020. ISBN 978-1-907470-13-4.

[28] Design Guidelines for the selection and use of stainless steels, Publication No. 9014, Nickel Institute, Toronto, Ontario, republished 2020. www.nickelinstitute.org

[29] Boniardi M, Casaroli A. Stainless Steels, Gruppo Lucefin Polytecnico di Milano, Esine (Brescia) 2014. www.lucefin.com

[30] Alves H, Heubner U. Shreir's Corrosion, 2010, 3, 1879–915. DOI: 10.1016/B978-044452787-5.00092-5.

[31] Agarwal DC. Nickel and Nickel Alloys. In: Wessel JK. (Ed.) Handbook of Advanced Materials – Enabling New Designs, Chapter 7, John Wiley & Sons, Hoboken, 2004, pp. 217–68.

[32] Rebak RB. Pitting Characteristics of Nickel Alloys – A Review, Corrosion 2016, Paper 07450, NACE International, Vancouver, Canada, 2016.

[33] Heubner U. Nickel based alloys. In: Cahn RW, Haasen P, Kramer EJ. (Eds.) Materials Science and Technology, Structure and Properties of Nonferrous Alloys, VCH Verlagsgesellschaft, Weinheim, 1996.

[34] DuPont JN, Lippold JC, Kiser SD. Welding Metallurgy and Weldability of Nickel-base Alloys, John Wiley & Sons, Hoboken, 2009.

[35] https://www.vdm-metals.com/en/downloads, Material Datasheets: last retrieved December 27th 2021.

[36] Niespodziany D, Behrens R, Alves H, Wolf M. Characterization of High Performance Material UNS N08034, Corrosion 2019, Paper 13156, NACE International, Nashville, Tennessee, 2019.

[37] Alves H, Behrens R, Winter F. UNS N 08031 and UNS 08031plus – Multipurpose Alloys for the Chemical Process Industry and Related Applications, Corrosion 2016, Paper 7563, NACE International, Vancouver, British Columbia, 2016.

[38] Weltschev M, Bäßler R, Werner H, Behrens R. Suitability of more noble Materials for Tanks for Transport of Dangerous Goods, Corrosion 2004, Paper 04228, NACE International, New Orleans, Louisiana, 2004.

[39] Weltschev M, Baeßler R, Werner H, Behrens R, Alves H. Corrosion Resistant Higher Alloyed Metallic Materials for the Transport of Corrosive Dangerous Goods in Tanks, Corrosion 2006, Paper 06688, NACE International, San Diego, California, 2006.

[40] Alves H, Kurumlu D, Behrens R. A new developed Ni-Cr-Mo Alloy with improved Corrosion Resistance in Flue Gas Desulfurization and Chemical Process Applications, Corrosion 2013, Paper 2325, NACE International, Orlando, Florida, 2013.

[41] Alves H, Behrens R, Paul L. Evolution of Nickel base Alloys – Modification to traditional Alloys for Specific Applications, Corrosion 2014, Paper 4317, NACE International, San Antonio, Texas, 2014.

[42] Alves H, Behrens R, Paul L. Recent Experiences and Applications with a new Ni-Cr-Mo-N Alloy, Corrosion 2015, Paper 5683, NACE International, Dallas, Texas, 2015.

[43] Niespodziany D, Behrens R, Alves H. Alloy UNS N06058: A Solution for Demanding Applications Where Common Members of the Ni-Cr-Mo Alloys Experience Their Limits, Corrosion 2020, Paper 14557, NACE International, Online Conference.

[44] Agarwal DC, Brill U, Wilson J. Combatting High Temperature Corrosion with Alloy 602CA in Various Environments and Industries, Corrosion 2002, Paper 02372, NACE International, Denver, Colorado, 2002.

[45] Behrens R. Nickel alloys and high-alloyed special stainless steels for the Oil and Gas industry. In: Heubner U, Kloewer J, Alves H, Behrens R, Schindler C, Wahl V, Wolf M. (Eds.) Nickel Alloys and High-alloyed Special Stainless Steels Properties – Manufacturing Applications, Chapter 8, 4th edition, Expert Verlag, Renningen, 2012, pp. 238–54. ISBN 978-3-8169-2751-8.

[46] Lettner J, Klöwer J, Behrens R. Clad Plates and Pipes in Oil and Gas Production, Applications-Fabrication-Welding, Corrosion 2002, Paper 02062, NACE International, Denver, Colorado, 2002.

[47] Behrens R, Agarwal DC. Laboratory Testing of Age-Hardenable Alloys 925 and 718 in Sour Gas Environments, Corrosion 2005, Paper 05103, NACE International, Houston, Texas, 2005.

[48] Galakhova A, Prattes K, Mori G. Mater Corros, 2021, 72, 1831–42.

2.3 Metallic light-weight alloys: Al, Ti, Mg

Oliver Janka

Metallic light-weight alloys are usually based on Al, Ti, Mg and Be as main constituents. The applications of the Be-based alloys are addressed separately in Chapter 2.7. Due to their low density and high weight to strength ratio, light-weight alloys are widely used in numerous areas of everyday life. Additionally, all of these are usually non-ferrous alloys with a low toxicity (except for Be), enabling their use in a plethora of applications.

2.3.1 Aluminum-based alloys

2.3.1.1 Applications

Aluminum alloys are versatile materials that find application in almost every industrial and commercial sector. Especially the transport sector, e.g. automotive and commercial vehicle construction, rail vehicles (housing, brakes, etc.; Figure 2.3.1), as well as aerospace applications rely on these materials [1, 2]. In the automotive industry, the need for weight reduction to achieve CO_2 reduction goals along with increasing fuel prices have triggered a serious series of developments of these materials. For decades, the use of aluminum in the automotive industry was limited to molding applications in the drive system. However, the ability to fabricate motor blocks, heat exchangers or parts of the chassis from aluminum alloys have led to significant advances that have also transitioned to the field of commercial vehicles. The transportation sector has realized the advantages of aluminum alloys. Also, the fields of architecture, machine, apparatus and tool construction have employed various available aluminum-based alloys. One of the first attempts to use aluminum in architecture was in the construction of New York's Empire State building between 1930 and 1931 [3]. It was used in the basic structure of the building along with the lobby. During the decades more and more iconic landmarks, e.g. the Mercedes-Benz museum in Stuttgart, Germany, the train station Køge Nord in Copenhagen, Denmark or the Ravensbourne College of Design and Communication, London, England have been using aluminum alloys in their construction. Finally, significant amounts of aluminum alloys are used every year for the food industry. Starting from plain rolls of aluminum foil, different alloys are also used for coatings and lids of various food containers, bottle caps, packages for liquids etc. to the well-known and used beverage cans.

https://doi.org/10.1515/9783110733143-010

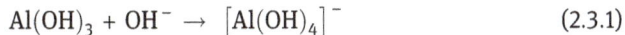

Figure 2.3.1: Aluminum brake discs used in the Munich subway made from Al-Si-Mg alloy. Picture supplied by Knorr-Bremse AG.

2.3.1.2 Advantages

The main advantages of aluminum alloys are the low specific weight (approx. 1/3 of the one of steel) and the manifold of fabrication methods, e.g. molding, rolling, forging or extrusion. This, combined with the shaping possibilities (machining, deep drawing, bending or punching), introduces a huge variety of possibilities to create all sorts of different parts. Finally, via anodic oxidation a good corrosion resistance can be achieved, which can be further improved by various surface coatings.

2.3.1.3 Aluminum production

The production of so-called primary aluminum is an extremely energy-consuming process since elemental aluminum must be produced by electrolytic reduction due to Al being an ignoble metal with a low standard potential (−1.67 V [4]). Bauxit is the main source for Al production, which is converted via the Bayer process [5]. In this process, the bauxit is first ground to a powder before being treated with a hot solution of sodium hydroxide under elevated pressure. During this process, the aluminum-containing part of the bauxit goes into solution (eq. (2.3.1)), while the other constituents (Fe_2O_3, $Fe(OH)_3$) remain solid.

$$Al(OH)_3 + OH^- \rightarrow \left[Al(OH)_4\right]^- \qquad (2.3.1)$$

After separation, the solution is seeded with $Al(OH)_3$ crystals to precipitate the dissolved aluminum. About 90% of the $Al(OH)_3$ is converted to Al_2O_3 via heat treatment, while the rest is used for the formation of e.g. $Al_2(SO_4)_3$ or aluminum chlorohydrate $Al_xCl_{(3x-y)}(OH)_y$.

In a subsequent step, the reduction of the Al_2O_3 to elemental metallic Al takes place. Here, aluminum oxide is dissolved in molten cryolite (Na_3AlF_6 [6]) at temperatures between 1213 and 1253 K. Cryolite exhibits a low melting point and vapor pressure and has a density lower than the one of molten aluminum (2.0 compared to 2.3 g cm^{-3}). This causes a separation of the molten aluminum from the salt melt, collecting the metal at the bottom of the electrochemical cell [7].

The so-called secondary aluminum is produced by recycling Al waste. The great advantage is the significantly lower energy consumption of only about 5% compared to primary aluminum. The scrap aluminum is mixed with CaF_2 and other chloride salts and heated to 923–973 K in a drum furnace under constant rotation. The oxidic impurities enrich in the slag. The recycling rates for e.g. aluminum-based beverage cans exceeded 76% in Europe in 2018, with Germany leading the chart with 99% [8]. In the USA, the U.S. Geological Survey published a recycling rate of 51% for the total aluminum recovery [9].

2.3.1.4 Mechanical properties

Elemental aluminum has a density of 2.68 g cm^{-3}, a melting point of 933 and a boiling point of 2740 K. It crystallizes in the cubic crystal system with space group $Fm\bar{3}m$ and a = 404.96 pm in the cubic closest packing, the so-called Cu-type structure. Its structure was first determined by Hull in 1917 [10]. Due to the face-centered unit cell, the closest contact of the Al atoms is along the face diagonal with $\sqrt{2} \times a/2$. In a cubic closest packed structure, the atoms form hexagonal close packed layers that get stacked according to ABC along the space diagonal of the cubic unit cell. In the case of plastic deformation, the dislocation movement is perpendicular to one of the four {111} slip planes. The motion occurs along one of the three the face diagonals <110>, since here the shortest distances are observed. This leads to a total of 12 possible dislocation motions.

During solidification and mechanical reforming, different types of defects and the formation of solid solutions as well as primary and secondary intermetallic phases can occur. In the following paragraph, these will be described and their consequences on the mechanical properties are evaluated.

Defects in solids are usually classified by their dimensionality [1, 11, 12]. The so-called point defects are zero-dimensional defects that can be further divided. The most common are vacancy and interstitial defects. Pure aluminum (99.5%) exhibits about 1.3×10^{-4} vacancies at 773 K, which decreases with increasing temperature [13]. These vacancies play a crucial role in the formation of precipitation hardening and different diffusion processes. An interstitial defect is observed when a position in a structure is occupied by an atom that is usually empty. Besides point defects, also dislocations (one-dimensional defects) can be observed. From a crystallographic point of view, edge and screw dislocations are distinguished. Dislocations can be moved under the influence of directional forces; the amount of force needed defines the strength of the material and its deformation resistance. Finally, stacking faults are two-dimensional defects that result in a different stacking pattern (with respect to the ideal ABC sequence) along the [111] direction.

Upon addition of alloying components, solid solutions can be formed. In this case, Al atoms are replaced by the atoms of the respective additive, leading to distortions of the structure. These distortions can increase the force needed for dislocations, therefore increasing the strength of the material. The different alloying components summarized in Table 2.3.1 have different influence on the respective mechanical properties.

Intermetallic phases are formed during (1) the casting process or (2) a subsequent annealing or homogenization step. While the first ones are called primary phases, the latter are characterized as secondary phases. Primary phases are stabilized due to the low solubility of the respective element in Al. During the solidification step primary phases with a length of several micrometers can be observed. The secondary phases can be further separated into dispersion and precipitation phases. Both phases are formed during thermal treatment of e.g. casted bars caused by a segregation of a supersaturated solid solution. However, dispersion phases, examples are the binaries $MnAl_6$ (*Cmcm*, [14]), $CrAl_7$ (*C2/m*, [15]), $ZrAl_3$ (*I4/mmm*, [16]) or the ternaries Mn_3SiAl_{12} (bulk structure unknown) or Mg_2CrAl_{12} (bulk structure unknown), are thermally highly stable leading to a very low solubility, especially at elevated temperatures. Precipitation phases in contrast are characterized by a high solubility at elevated temperatures. These intermetallics are formed during quenching and subsequent thermal aging. Prominent examples are Mg_2Si (*Fm$\bar{3}$m*, [17]), $MgCuAl_2$ (*Cmcm*, [18]), $MgZn_2$ (*P6$_3$/mmc*, [19]) or $CuAl_2$ (*I4/mcm*, [20]).

Although the addition of elements like Mg, Cu or Zn significantly improves the mechanical strength, especially the latter two lead to a significant increase in the density and weight of these materials. Therefore, especially for the use in aerospace applications, the addition of scandium (Sc) was developed. Its addition creates nanoscaled $ScAl_3$ (*Pm$\bar{3}$m*, [21]) precipitates with sizes typically below 10 nm [22], which hamper the grain growth especially in welded aluminum parts. Since the precipitated $ScAl_3$ forms smaller crystals, the size of the precipitate-free zone near the grain boundaries is reduced [22, 23].

The formation process of precipitates has been studied by various methods, especially, high-angle annular dark-field transmission electron microscopy (HAADF-TEM) along with theoretical calculations helped to understand the underlying mechanisms. The following examples are taken from the 2018 publication by Andersen and coworkers, which illustrates these processes in an impressive way [24] in the Al-Cu(-Mg) system. As shown in the sequences (2.3.2) and (2.3.3) below, the intermetallic phases $CuAl_2$ and $MgCuAl_2$ form by different routes, the nomenclature is taken from the literature [24]. The processes that form intermetallic precipitates like $CuAl_2$ or $MgCuAl_2$ originate from the decomposition of supersaturated solid solutions (SSSS). During this process, different stages have been identified, namely the formation of clusters, which continue to grow till so-called Guinier-Preston (GP) zones are formed. These zones are very small (on the nanometer scale) and can grow into different meta-stable phases until the final precipitate is formed. The formation of tetragonal $CuAl_2$ (*I4/mcm*) is illustrated in eq. (2.3.2).

$$SSSS \rightarrow solute\,clusters \rightarrow GP\text{-}I \rightarrow GP\text{-}II(\theta") \rightarrow \theta' \rightarrow CuAl_2(\theta) \qquad (2.3.2)$$

From the solute clusters, GP-I forms where the Cu atoms replace Al atoms in the *fcc*-type structure. This is followed by the GP-II type structure ($\theta"$) possessing in an ideal case a Cu to Al ratio of 1:3. The θ' phase exhibits already the composition $CuAl_2$, however, defects, e.g. vacancies and interstitials can be observed. This leads to a doubling of the unit cell. Further rearrangement of the atoms finally leads to the tetragonal θ-phase, $CuAl_2$ [20, 25]. Figure 2.3.2 illustrates this process by TEM images along the <001> projection.

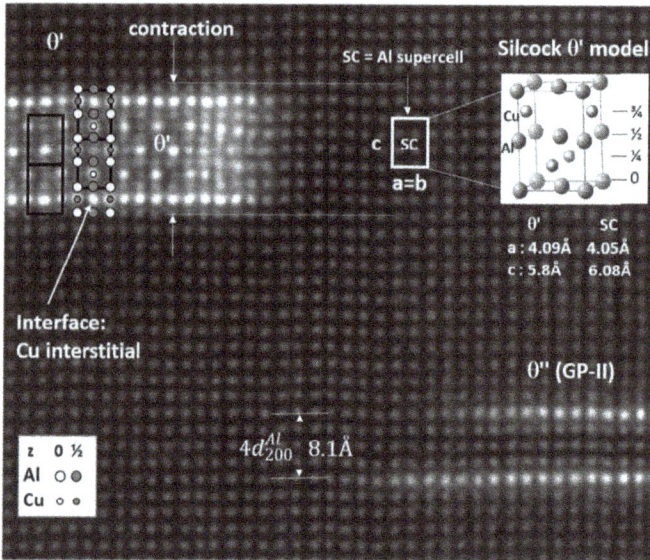

Figure 2.3.2: <001> Al projection of Al-5Cu alloy heat treated 2 h at 185 °C, showing $\theta"$ (GP-II) and θ' zones. The figure is taken from the literature [24] and the article is published under the Creative Commons Attribution License.

As a second example, the Al-Cu-Mg system was chosen (eq. 2.3.3). Here, after initial formation of clusters, so-called GPB zones (Guinier-Preston-Bagaryatsky) are formed, which were observed by Kovarik and coworkers [26, 27]. These lead to the precipitation of the orthorhombic S'-phase with the chemical composition $MgCuAl_2$.

$$SSSS \rightarrow solute\,clusters \rightarrow GPB \rightarrow S' \rightarrow MgCuAl_2(S) \qquad (2.3.3)$$

In contrast to $MgCuAl_2$ (S-phase, *Cmcm*, $a = 401.2$, $b = 926.5$, $c = 712.4$ pm [28]), the S'-phase exhibits a slightly different a lattice parameter of $a = 405$ pm (Figure 2.3.3). The structure of the S-phase was originally determined in 1943 by *Perlitz* and *Westgren* via Weissenberg film data [18]. In the following decades, this compound has been extensively investigated via powder X-ray diffraction experiments and electron microscopy due to its technological importance [29–31]. During these, the copper-magnesium ordering along with the space group symmetry were questioned numerous times [32, 33] before, in 2005, the structure was redetermined using state-of-the art diffraction techniques [28].

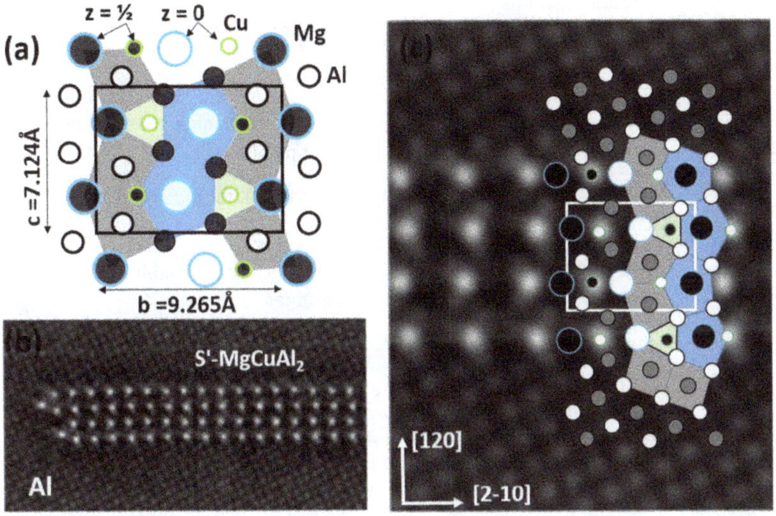

Figure 2.3.3: (a) Structure of the *S'*-phase (MgCuAl₂) phase along the *a*-axis. Mg, Al and Cu atoms are depicted by circles with blue, black and green borders. HAADF image of an <001> Al projection of an Al-3Cu-1Mg (wt%) alloy heat treated for 11 days at 170 °C, with Al-embedded S' plate along its coherent *a*-axis. Details of the particle of image (right) partly superimposed by the crystal structure. The figure is taken from the literature [24], the article is published under the Creative Commons Attribution License.

2.3.1.5 Alloys

For the formation of aluminum alloys, usually aluminum with a purity of 99.5% is used as starting material. The main alloying elements are Cu and Zn, followed by Mn, Mg and Si. The most common accompanying element is Fe. While Cu and Zn produce high-strength alloys, Mn, Mg and Si are used to obtain medium-strength alloys. Elements like Sc, Li, Sn, Pb, Bi, Ni and more are used for special alloys [2].

In the case of wrought alloys, the different materials are identified by a four-digit number and grouped with respect to their main alloying element (first digit). The different series are labeled as 1XXX to 8XXX (Table 2.3.1). The second digit, if not zero, indicates a variation of the alloy, while the third and fourth digits identify the specific alloy in the series. For the cast alloys (Table 2.3.2), a four-to-five-digit number with a decimal point is utilized. The potentially conducted heat treatments (Table 2.3.3) are added to the end of the number using a dash. Typical industrially used alloys are listed in Table 2.3.4, along with their compositions and trade names.

Table 2.3.1: Denomination of wrought aluminum alloys.

Series	Main alloying element	Description and use
1000	–	pure aluminum (min. 99 wt% Al), no hardening possible, high corrosion resistance, weldable – foil, chemical tanks, pipes
2000	Cu	precipitation hardening possible, high strength, low corrosion resistance, hard weldable – aerospace and space travel
3000	Mn	low strength, high corrosion resistance, no hardening possible, used as welding add – pots, car grille, power plants
4000	Si	alloys with the addition of Si can be hardened, for the others no hardening possible, welding add
5000	Mg (no Si)	medium to high strength, no hardening possible, weldable – shipbuilding, transportation sector, bridges, buildings
6000	Mg & Si	good weldable, addition for welding – used for extrusion forming
7000	Zn	high to very high strength, not all alloys are weldable – aerospace, space travel, casings of mobile devices, watches, sports equipment
8000	Other elements	strength, weldability and use depend on composition

Table 2.3.2: Denomination of cast aluminum alloys.

Series	Main alloying element
100.0	minimum 99% Al
200.0	Cu
300.0	Si with Cu and/or Mg
400.0	Si
500.0	Mg
700.0	Zn
800.0	Sn
900.0	other elements

Table 2.3.3: Overview over different temper designations, as well as wrought aluminum alloys.

Temper designation	Description
-F	as fabricated
-H	strain hardened (cold worked) with or without thermal treatment
-O	full soft (annealed)
-T	heat treated to produce stable tempers
-W	solution heat treated only

Table 2.3.4: Examples for industrially used alloys: trade name and compositions.

Trade name	Composition
A380	7.5–9.5% Si, 0.5% Mn, 0.1% Mg, 3–4% Cu, 1.3% Fe
Birmabright	<1% Mn, 1–7% Mg
Duralumin	0.2–1.0% Si, 0.5–1.2% Mn, 0.2–5% Mg, 2.5–5.5% Cu
Hydronalium	0.2–1.0% Si, 0.2–0.8% Mn, 3–12 Mg
Magnalium	5% Mg
Silumin	up to 14% Si

2.3.2 Titanium-based alloys

2.3.2.1 Applications

In contrast to aluminum-based alloys, the use of titanium alloys is rather limited to specific tasks, where their beneficiary properties are of crucial importance. Titanium alloys are characterized by a very high strength combined with an outstanding corrosion and extreme temperature resistance. The production costs of elemental titanium and its alloys limit their use in air- and spacecraft parts (e.g. engine parts), military applications (e.g. missiles), medical purposes (prostheses, orthopedic and dental implants), jewelry or premium sports equipment such as bicycles or golf clubs [34–36]. It has little importance in architecture, however, the Guggenheim museum in Bilbao, Spain, exhibits an extensive coating of 33,000 half-millimeter thick titanium plates over galvanized steel [37].

2.3.2.2 Advantages

The mechanical attributes are of high significance for their use. These properties are mostly based on the individual properties, the volume fractions and the local arrangement of the α and β phase [36]. The anisotropic crystal structure of the hexagonal α form (*vide infra*) has several advantages over the cubic β phase, most importantly anisotropic mechanical and physical properties, a reduced ductility and therefore a higher resistance towards plastic deformations. Further important properties are the high corrosion and erosion resistance, the excellent strength-to-weight ratios as well as their non-magnetic character and low thermal expansion co-efficient.

2.3.2.3 Titanium production

While primary aluminum is produced via electrochemical reduction, metallic Ti is produced by a chemical reduction from the available minerals. As starting materials, TiO_2 with its polymorphs anatase, brookite and rutile or ternary oxides, e.g. ilmenite ($FeTiO_3$) or perovskite ($CaTiO_3$), are available. Due to the high stability of titanium carbide (TiC), no direct conversion is possible. Therefore, the carbothermic reduction (2.3.4) must take place in the presence of elemental chlorine, leading to the formation of titanium tetrachloride $TiCl_4$, which subsequently can be purified via distillation (b.p. ($TiCl_4$) = 410 K).

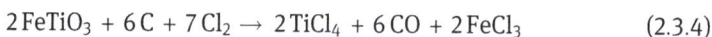

$$2\,FeTiO_3 + 6\,C + 7\,Cl_2 \rightarrow 2\,TiCl_4 + 6\,CO + 2\,FeCl_3 \qquad (2.3.4)$$

The purified $TiCl_4$ can now be reduced using elemental Mg (Kroll process, eq. 2.3.5) or Na (Hunter process, eq. 2.3.6) [36]. Further purification can be achieved via the van Arkel-de Boer process.

$$TiCl_4 + 2\,Mg \rightarrow Ti + 2\,MgCl_2 \qquad (2.3.5)$$

$$TiCl_4 + 4\,Na \rightarrow Ti + 4\,NaCl \qquad (2.3.6)$$

2.3.2.4 Mechanical properties

Elemental titanium exhibits two different crystal structures, depending on the temperature. Up to 1150 K, hexagonal primitive α-Ti (*hcp*, Mg-type, $P6_3/mmc$) with a = 297 and c = 472 pm (c/a = 1.589) is stable. Above this temperature, β-Ti, the body-centered high temperature (*bcc*, W-type, $Im\bar{3}m$, a = 332 pm) form is stable. The crystal structures of both α- and β-Ti were investigated by Hull [38, 39]. In contrast to the 12 possible dislocation motions in the *fcc* metals, the *hcp* structures exhibit only 3 dislocation glide opportunities, with the dislocation occurring along the a-axis. This leads to an unfavorable situation explaining the brittleness of many *hcp* metals. In the *bcc* structure, finally, 12 dislocations are possible that can occur along the space diagonal; therefore, the translation is $\sqrt{3} \times a/2$, favoring these deformations over the ones of the *hcp* structures. Finally, the density of α-Ti is higher than the one of β-Ti (74 vs. 68%), making atom diffusion in β-Ti significantly easier. The existence of two structurally different polymorphs is a key aspect for the properties of the respective alloys, since various mechanical properties are closely related to the structural arrangement of the atoms in the solid. Generally, one can summarize, α-Ti is stronger yet less ductile, while β-Ti is the more ductile. In comparison e.g. with other metals like aluminum, elemental Ti is twice as strong, however also 60% more dense. Therefore, alloying Ti with other elements leads to high-class materials.

The alloys of Ti can be grouped with respect to their alloying elements and the occurring stabilization of either the α- or β-Ti phase. Al, O, N or C stabilize the α-phase by increasing the phase transition temperature, while Mo, V, Ta or Nb reduce this temperature, stabilizing the β-phase. Finally, some elements, like Zr or Sn, do not influence the transition temperature, rendering them neutral [36]. Like the aluminum-based alloys, other elements also exhibit in elemental titanium a restricted solubility, which in turn leads to precipitation hardening upon annealing. In the field of light-weight alloys, especially the aluminum-containing materials are of great interest. Therefore, the binary Ti/Al phase diagram is probably one of the most investigated ones [40–44]. Several binary intermetallics have been identified in this system, of which $TiAl_2$ and $TiAl_3$ are too brittle to be of technical importance, however, α_2-Ti_3Al and γ-TiAl are of crucial importance to the field of titanium-based alloys [36, 45].

2.3.2.5 Alloys

Elemental titanium already has acceptable mechanical and physical properties (high strength-to-weight ratio, outstanding corrosion resistance, a density of $\rho = 4.51$ g cm^{-3} [36, 46, 47]) and is therefore used e.g. for dental and orthopedic implants [48, 49]. Titanium-based alloys have been developed to further increase the respective mechanical, physical and chemical properties, like their high thermal stability, their corrosion resistance or their high strength up to very high temperatures, often paired with a low density. The problem of brittleness has been resolved using additional alloying components and the optimization of the alloy microstructure, resulting in multiphase alloys [45, 50]. For most applications, titanium is alloyed with small amounts of vanadium and aluminum, usually around 4 and 6 wt%, respectively. This combination forms a solid solution; however, the solubility of the elements varies dramatically with temperature. Therefore, as already introduced for the Al-based alloys, precipitation of intermetallic phases can occur, leading to an increased strength of the alloy. Usually, the heat treatment process is conducted after the alloy has been worked into its final shape, allowing a much easier fabrication of a high-strength product.

Titanium alloys are generally classified into four main categories [36, 51]:

(1) α alloys: these alloys usually contain neutral elements (Zr or Sn) sometimes along with α-stabilizers (Al, O, N or C) and cannot be heat treated.

(2) near α alloys: these contain a small amount of the ductile β phase. To achieve this, they are alloyed with 1–2 wt% of β-phase stabilizers along with α-phase stabilizers.
(3) $\alpha + \beta$ alloys: include a combination of both α and β stabilizers; can be heat treated.
(4) β + near β alloys: β-Ti alloys offer the most versatility among all titanium-based alloys. They contain sufficient β stabilizers to maintain the beta phase even when quenched. Disadvantages, especially compared to $\alpha + \beta$ alloys are an increased density as well as higher costs.

An overview on different industrially used alloys is given in Table 2.3.5.

Table 2.3.5: Overview over different titanium-based alloys [36, 51].

Commercial name	Alloy composition	Category
Grade 1	Ti-0.2Fe-0.18O	α
TIMETAL 1100	Ti-6Al-2.7Sn-4Zr-0.4Mo-0.4Si	near-α
Ti-6-4	Ti-6Al-4 V	$\alpha + \beta$
Ti-6-6-2	Ti-6Al-6 V-2Sn	$\alpha + \beta$
Alloy C	Ti-35 V-15Cr	β
Ti-15-3	Ti-15 V-3Cr-3Sn-3Al	metastable
Ti-10-2-3	Ti-10 V-2Fe-3Al	metastable
TIMETAL 21S	Ti-15Mo-2.6Nb-3Al-0.2Si	metastable
Ti-17	Ti-5Al-2Sn-2Zr-4Mo-4Cr	β-rich

2.3.3 Magnesium-based alloys

2.3.3.1 Applications

The development of magnesium-based alloys was based on their potential for light-weight construction, especially in military applications. However, the importance of saving weight in the transport sector to achieve a more economic use of fuel along with a significant reduction of CO_2 emissions has dramatically increased [52]. Already in the 1960s, Volkswagen used 21,000 tons of Mg alloys for various parts, e.g. the gearbox housing, the crank case and various covers for the construction of the VW Bettle. This consumption increased up to 42,000 tons of magnesium alloys in 1972 [53]. And still today, the automotive industry is one of the large-scale users of Mg-based alloys [52].

2.3.3.2 Advantages

One clear advantage of Mg-based alloys is the low specific density of approximately 1.8 g cm^{-3}. They furthermore exhibit a high strength to density ratio, excellent casting and machining abilities (turning, milling, welding, etc.) as well as good damping properties. Besides their physical and mechanical advantages listed above, parts made of cast Mg alloys, in contrast to wrought alloys, have furthermore the advantage of shorter processing times and lower assembly costs [54, 55]. However, all the advantages and good properties face some disadvantages that must be noted [52]. The cold workability and the corrosion resistance of Mg-based alloys are poor and molten Mg alloys are furthermore very reactive. Finally, the shrinkage during solidification and cooling and the thermal expansion are further aspects that must be considered when choosing Mg-based alloys.

2.3.3.3 Magnesium production

Elemental magnesium can be obtained via the Dow process, which is an electrolytic reduction of molten $MgCl_2$. To lower the melting point, NaCl and $CaCl_2$ are used as additives. The $MgCl_2$ is obtained from seawater and brines, which are treated with CaO. Due to the alkaline reaction with water, $Mg(OH)_2$ precipitates (eq. 2.3.7).

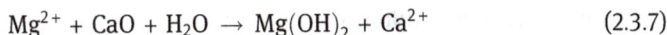

$$Mg^{2+} + CaO + H_2O \rightarrow Mg(OH)_2 + Ca^{2+} \tag{2.3.7}$$

The precipitate is removed by filtration and subsequently treated with HCl to generate $MgCl_2$, which is then reduced in the Dow cells.

In China, the main producer of Mg metal, however, the Pidgeon process is utilized. In this silicothermic reduction, the raw material, mostly magnesite, $MgCO_3$ or dolomite, $(Ca,Mg)CO_3$, is heated along with elemental Si. In the initial step, the carbonates decompose (eq. 2.3.8), forming the respective oxides. These react with the silicon forming elemental Mg along with SiO_2 (eq. 2.3.9) and calcium silicates during subsequent reactions (eq. 2.3.10).

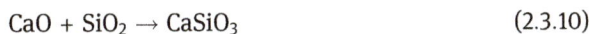

$$MgCO_3 \xrightarrow{\Delta T} MgO + CO_2 \tag{2.3.8}$$

$$2\,MgO + Si \rightarrow Mg + SiO_2 \tag{2.3.9}$$

$$CaO + SiO_2 \rightarrow CaSiO_3 \tag{2.3.10}$$

A carbothermic reduction is not possible, although MgO can get reduced by C at high temperatures forming CO (eq. 2.3.11). The formed Mg, however, readily reacts with the present CO forming MgO again (eq. 2.3.12).

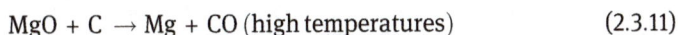

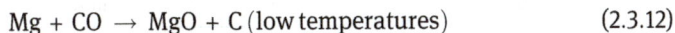

$$MgO + C \rightarrow Mg + CO \text{ (high temperatures)} \tag{2.3.11}$$

$$Mg + CO \rightarrow MgO + C \text{ (low temperatures)} \tag{2.3.12}$$

2.3.3.4 Alloys

Mg-based alloys are, like the titanium-based ones, characterized by a combination of the alloying element and a number, indicating the weight percentage of said element. In contrast to the Ti alloys, however, the constituent elements are coded by a letter (Table 2.3.6). The denomination is finalized by a letter (A, B, C, . . .), indicating the stage of development of the alloy [52, 56].

As for the aluminum alloys, also a distinction between cast and wrought alloys is made. The castable alloys usually contain 3–9 wt% Al with increasing castability for higher aluminum amounts. The most commonly used casting alloys are the AZ (Al + Zn; AZ63, AZ81, AZ91), AM (Al + Mn; AM50) and AS (Al + Si; AS61) ones. AZ alloys exhibit a low heat and creep resistance and good room temperature properties. AM alloys exhibit an improved ductility compared to the AZ by exchanging manganese for zinc. AS finally can only be processed by casting and exhibits an increased heat and creep resistance by the combination with rare-earth (*RE*) elements forming *RE*-Mg precipitates. It is important to note, that an increased aluminum content results in the formation of cubic $Mg_{17}Al_{12}$ (γ-phase, $I\bar{4}3m$, $a = 1054$ pm [57]) precipitates on the grain boundaries, leading to a reduction in strength and limited ductility [52]. In contrast to the cast alloys, wrought magnesium materials exhibit a more limited use due to the hexagonal structure of elemental Mg and the possibility of twin formation. The most important aluminum-based alloys are AZ31, AZ61, AZ80, while WE43 and WE54 represent typical commercial *RE*-based cast alloys [54]. Commercially available alloys are e.g. the *Elektron* alloys, which are available both as cast and wrought variants (Table 2.3.7) used for example in aerospace applications (Elektron 21) or the chassis and wheels in motorsports (Elektron 675).

Table 2.3.6: Codes for elements found in magnesium alloys.

Letter	Alloying element	Letter	Alloying element
A	aluminum	B	bismuth
C	copper	D	cadmium
E	rare-earth elements	F	iron
H	thorium	K	zirconium
L	lithium	M	manganese
N	nickel	P	lead
Q	silver	R	chromium
S	silicon	T	tin
W	yttrium	Y	antimony
Z	zinc		

Finally, the wrought alloy Magnox should be mentioned. It usually consists of 99% Mg along with 1% additives, usually Al, Be, Fe, Mn or Zn. Its name stands for "magnesium non-oxidizing" and is used in fuel-rod housings of gas-cooled "Magnox" nuclear reactors, which were mostly used in the 1950s to 1970s in the UK [58].

Table 2.3.7: Commercially available magnesium-based alloys.

Name	Alloying components	Processing
AZ63	6% Al, 3% Zn	cast
AZ81	7.5% Al, 0.7% Zn	cast
AM50	4.4–5.4% Al, 0.26–0.6% Mn	cast
WE43	4% Y, 2.3% Nd	cast
WE54	4.75–5.5% Y, 1–2% RE	cast
Elektron 21	3% Y, 1% Gd, 0.5% Zn	cast
AZ31	2.5–3.5% Al, 0.6–1.4% Zn	wrought
AZ61	5.8–7.2% Al, 0.4–1.5% Zn	wrought
AZ80	7.8–9.2% Al, 0.2–0.8% Zn	wrought
AE41	4% Al, 1% RE	wrought
Elektron 675	6% Y, 7% Gd, 0.5% Zr	wrought

2.3.3.5 Mechanical properties

Magnesium is (besides beryllium) the lightest metal that can be utilized as construction material. The element itself belongs to the alkaline-earth metals, exhibits a density of 1.74 g cm^{-3} and crystallizes in the hexagonal closest packing (hcp), the Mg-type structure ($P6_3/mmc$, a = 322, c = 523 pm, c/a = 1.624 [10]). However, to be a challenging competitor for Al-based alloys, several criteria must be met, since the mechanical properties of magnesium-based alloys are below their competitors. As for α-Ti-based alloys, the hexagonal structure of Mg causes a strong anisotropy leading to poor deformation properties at room temperature. The strength of Mg-based alloys on the other hand can already significantly decrease at temperatures as low as 373 K [59], leading to a limited temperature range in which these materials can be used. As for the aluminum alloys, also numerous intermetallic compounds have been identified and characterized in Mg-based alloys. Especially the Mg-Al-Zn system has gotten a lot of attention. Here, the τ_1 [60], τ_2 [61], q [62] and ϕ [63] phases have been structurally investigated. The τ_1, τ_2 and ϕ phases are so-called complex metallic alloys (CMA) and τ_1 and τ_2 are furthermore 1/1 and 1/2 approximants of the quasicrystalline q phase which are of the Bergman type.

References

[1]   Ostermann F. Anwendungstechnologie Aluminium, Springer Vieweg, Berlin, Heidelberg, 2014.

[2]   Kammer C. Aluminium-Taschenbuch 1: Grundlagen und Werkstoffe, Aluminium-Zentrale, Düsseldorf, 1995.

[3]   Empire State Realty Trust, Empire State Building Fact Sheet, https://www.esbnyc.com/sites/default/files/esb_fact_sheet_4_9_14_4.pdf.

[4]   Emsley J. The Elements, Clarendon Press, Oxford University Press, Oxford, New York, 1998.

[5]   Mortimer CE, Müller U. Chemie: Das Basiswissen der Chemie, Thieme, 2019.

[6]   von Náray-Szabó S, Sasvári K. Z Kristallogr, 1938, 99, 27.

[7]   Binnewies M, Finze M, Jäckel M, Schmidt P, Willner H, Rayner-Canham G. Allgemeine und Anorganische Chemie, Springer Spektrum, Berlin, Heidelberg, 2016.

[8]   European Aluminium, Metal Packaging Europe, Aluminium beverage can recycling in Europe hits record 76.1% in 2018, press release, 2020, https://european-aluminium.eu/media/3013/2020-12-16-european-aluminium-mpe-aluminium-beverage-ca+n-2018-recycling-rate_press-release.pdf.

[9]   Survey USG. Mineral Commodity Summaries 2021, U.S Geological Surve, 2021.

[10] Hull AW. Phys Rev, 1917, 10, 661.

[11] Siegel RW. Atomic defects and diffusion in metals, Argonne National Lab., IL (USA), Technical Report, 1981.

[12] Ehrhart P, Robrock KH, Schober HR. In: Johnson RA, Orlov AN. (Eds.) Basic Defects in Metals in Modern Problems in Condensed Matter Sciences, Vol. 13, Elsevier, 1986.

[13] Altenpohl D. Aluminium und Aluminiumlegierungen, Springer, Berlin, Heidelberg, 1965.

[14] Phragmén G. J Inst Met, 1950, 77, 489.

[15] Cooper M. Acta Crystallogr, 1960, 13, 257.

[16] Brauer G. Naturwissenschaften, 1938, 26, 710.

[17] Bol'shakov KA, Bul'onkov NA, Rastorguev LN, Tsirlin MS. Russ J Inorg Chem, 1963, 8, 1418.

[18] Perlitz H, Westgren A. Ark Kemi Mineral Geol, 1943, 16B, 1.

[19] Tarschisch L, Titow AT, Garjanow FK. Phys Z Sowjetunion, 1934, 5, 503.

[20] Friauf JB. J Am Chem Soc, 1927, 49, 3107.

[21] Rechkin VN, Lamikhov LK, Samsonova TI. Sov Phys Crystallogr, 1964, 9, 325.

[22] Dorin T, Ramajayam M, Vahid A, Langan T. In: Lumley RN. (Ed.) Chapter 12 – Aluminium Scandium Alloys, Woodhead Publishing, 2018.

[23] Ahmad Z. JOM, 2003, 55, 35.

[24] Andersen SJ, Marioara CD, Friis J, Wenner S, Holmestad R. Adv Phys X, 2018, 3, 1479984.

[25] Meetsma A, De Boer JL, Van Smaalen S. J Solid State Chem, 1989, 83, 370.

[26] Kovarik L, Court SA, Fraser HL, Mills MJ. Acta Mater, 2008, 56, 4804.

[27] Kovarik L, Mills MJ. Scr Mater, 2011, 64, 999.

[28] Heying B, Hoffmann R-D, Pöttgen R. Z Naturforsch, 2005, 60b, 491.

[29] Kilaas R, Radmilovic V, Dahmen U. Mater Res Soc Symp Proc, 2001, 589, 273.

[30] Laves F, Witte H. Metallwirtschaft, 1936, 15, 15.

[31] Mondolfo LF. Al-Cu-Mg Aluminum-Copper-Magnesium System, Butterworths, London, Boston, 1976.

[32] Cuisiat F, Duval P, Graf R. Scr Metall, 1984, 18, 1051.

[33] Yan J, Chunzhi L, Minggao Y. J Mater Sci Lett, 1990, 9, 421.

[34] Veiga C, Davim JP, Loureiro AJR. Rev Adv Mater Sci, 2012, 32, 14.

[35] Ribeiro MV, Moreira MRV, Ferreira JR. J Mater Process Technol, 2003, 143–144, 458.

[36] Peters M, Hemptenmacher J, Kumpfert J, Leyens C. Structure and Properties of Titanium and Titanium Alloys in Titanium and Titanium Alloys: Fundamentals and Applications, Wiley-VCH Verlag GmbH, Weinheim, 2003.
[37] FMGB Guggenheim Bilbao Museoa, 2021, https://www.guggenheim-bilbao.eus/de/, 08.06.2021.
[38] Hull AW. Science, 1920, 52, 227.
[39] Hull AW. Phys Rev, 1921, 18, 88.
[40] Massalski TB, Okamoto H, Subramanian PR, Kacprzak L. Binary Alloy Phase Diagrams, 2nd edition, ASM International, Ohio, U.S.A., 1990.
[41] Okamoto H. J Phase Equilibria, 1993, 14, 120.
[42] Raghavan V. J Phase Equilibria Diffus, 2005, 26, 276.
[43] Batalu D, Coşmeleaţă G, Aloman A. UPB Sci Bull, Series B, 2006, 68, 77.
[44] Schuster JC, Palm M. J Phase Equilibria Diffus, 2006, 27, 255.
[45] Sauthoff G Intermetallics. In: Ullmann's Encyclopedia of Industrial Chemistry, 2012.
[46] Barksdale J. In: Hampel CA. (Ed.) Titanium in the Encyclopedia of the Chemical Elements, Reinhold Book Corp, New York, 1968.
[47] Enghag P. In: Enghag P. (Ed.) Titanium, in Encyclopedia of the Elements: Technical Data – History – Processing – Applications, Wiley VCH Verlag GmbH, Weinheim, 2004.
[48] Özcan M, Hämmerle C. Materials, 2012, 5, 1528.
[49] Van Noort R. J Mater Sci, 1987, 22, 3801.
[50] Sauthoff G. Intermetallics, 2000, 8, 1101.
[51] Terlinde G, Fischer G. Beta Titanium Alloys in Titanium and Titanium Alloys, Wiley-VCH Verlag GmbH, Weinheim, 2003.
[52] Kainer KU, von Buch F. The Current State of Technology and Potential for Further Development of Magnesium Applications in Magnesium – Alloys and Technology, Wiley-VCH Verlag GmbH, Weinheim, 2003.
[53] Höllrigl-Rosta F, Just E, Köhler J, Melzer H-J. Metall, 1980, 34, 12.
[54] Pan F, Yang M, Chen X. J Mater Sci Technol, 2016, 32, 1211.
[55] Kulekci MK. Int J Adv Manuf Technol, 2008, 39, 851.
[56] Moosbrugger C Chapter 1 – Introduction to Magnesium Alloys in Engineering Properties of Magnesium Alloys, ASM International, Materials Park, Ohio, USA, 2017.
[57] Riederer K. Z Metallkd, 1936, 28, 312.
[58] Jensen SE, Nonboel E. Description of the Magnox Type of Gas Cooled Reactor (MAGNOX), Denmark, 1999.
[59] Sankaran KK, Mishra RS. In: Sankaran KK, Mishra RS. (Eds.) Chapter 7 – Magnesium Alloys in Metallurgy and Design of Alloys with Hierarchical Microstructures, Elsevier, 2017.
[60] Bergman G, Waugh JLT, Pauling L. Acta Crystallog, 1957, 10, 254.
[61] Berthold R, Mihalkovic M, Burkhardt U, Prots Y, Amarsanaa A, Kreiner G. Intermetallics, 2014, 53, 67.
[62] Takeuchi T, Mizutani U. Phys Rev B, 1995, 52, 9300.
[63] Berthold R, Kreiner G, Burkhardt U, Hoffmann S, Auffermann G, Prots Y, Dashjav E, Amarsanaa A, Mihalkovic M. Intermetallics, 2013, 32, 259.

2.4 Copper and copper alloys [1]

Bernd E. Langner

The red-orange element copper is a relatively soft metal, which is easy to shape and tough. Its use is based on its outstanding properties: high electrical and thermal conductivity, good corrosion resistance, good workability and antimicrobial abilities. With its recyclability without quality loss and low power losses in the production, transformation and distribution of electrical and thermal energy, it is also a very important element for the sustainable development. Because of the production of regenerative energy and electrification in the car industry there will be an increasing demand in the future. The reserves of copper may last for at least 40 years, with the identified resources even for more than 100 years [2].

The production of copper and copper alloys amounts to about 30 Mio. tons/year (2018), from which about 35% is from recycling. The main uses are electrical applications like copper cables, busbars, transformer strips or special copper alloys for the electronic industry.

The primary production of copper is based on two sources: sulfidic copper ores – mainly chalcopyrite ($CuFeS_2$) – which can be concentrated from 0.2–1% to 20–50% copper content by flotation, and this concentrate is processed by pyrometallurgy and refined by electrolysis and from oxidic copper ores which are processed by hydrometallurgy.

In the pyrometallurgical process (Figure 2.4.1), the copper concentrate is first partly oxidized to sulfur dioxide which is further processed to sulfuric acid, fayalite as the slag and copper matte with a copper content of about 50–70% and then further in a converter to so-called blister copper by the internal reaction:

$$Cu_2S + 2\,Cu_2O \rightarrow 6\,Cu + SO_2$$

After removal of excess oxide by reduction, the melt is cast to copper anodes which are refined by electrolysis.

In the solvent extraction process, the original oxidic ore is piled up and extracted by diluted sulfuric acid over weeks. The copper sulfate solution is collected in huge tanks and purified by solvent extraction. Afterwards, copper cathodes are produced by electrowinning (Figure 2.4.2). Sulfuric acid is recirculated.

The third important source of copper is recycling of copper scrap and copper containing residues, which are processed mainly by pyrometallurgy with a subsequent electrolysis in a copper tankhouse. If the scrap is very pure, it can be used directly by remelting it to products, especially when highest electrical conductivity is not so important like for roofing, plumbing tubes or certain copper alloys.

https://doi.org/10.1515/9783110733143-011

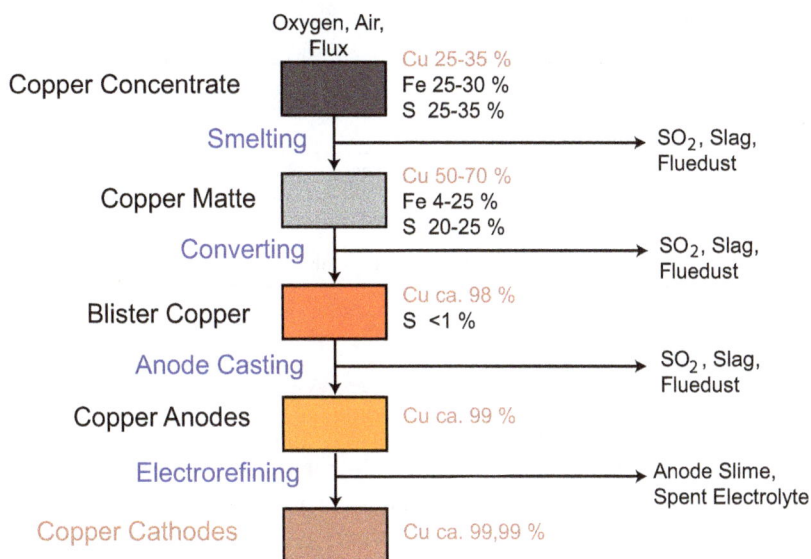

Figure 2.4.1: Flowsheet – copper production from sulfidic ores.

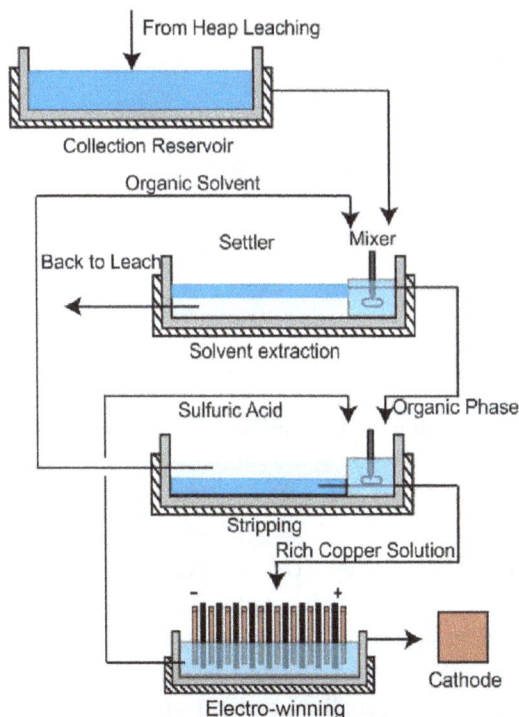

Figure 2.4.2: Flowsheet – copper production from oxidic ores.

Table 2.4.1: Properties of pure copper.

Density	8.96 g/cm^3
Melting point	1358 K
Electrical conductivity	58–59 × 10^6 Siemens/m = 100% IACS
Thermal conductivity	394 W/mm K
Tensile strength	220 (after recrystallization) – 385 (after cold working)

2.4.1 Copper cathode

The base of all high-quality copper products is the copper cathode (Figure 2.4.3) with a weight of about 100 kg and a size of some less than 1 sqm. The copper cathode has to be melted and cast to produce copper products.

Figure 2.4.3: Copper cathodes.

The electrolytic process for refining copper in a copper tankhouse is the reason for the very low level of impurities and the reason why copper cathodes from recycling and from primary sources has the same quality. Copper produced by solvent extraction and electrowinning has the same good quality as from pyrometallurgy.

Low levels of impurities are very important for the electrical conductivity, as even levels less than 60 ppm of impurities like phosphorus or iron reduce the electrical conductivity by 30–40%. Impurities like bismuth, tellurium or antimony have (even less than 5 ppm) influence on the workability of copper as they have a big impact on the recrystallization temperature. Therefore, highest quality standards are essential for the use of copper in most of the applications, especially for electrical applications.

2.4.2 Species of pure copper

Although copper cathode is the initial material, after melting, different species are produced according to applications (Table 2.4.2). In international standards, copper with other elements less than about 1% belongs to pure copper.

Table 2.4.2: Different pure copper species [3].

Name	Characteristics	Electrical conductivity (S/m)	Recrystallization temperature (°C)
ETP-Copper, Cu-ETP	100–350 ppm oxygen	58–>59 = 100% IACS	170–180
OF-Copper	0–5 ppm oxygen	58–>59	200–210
SE-Copper, Cu-HCP, BE-Copper	<70 ppm phosphorus mainly as oxide	57–>58	230–260
DHP-Copper, DLP-Copper	120–600 ppm phosphorus	40–50	280–300
Copper-silver	0.1% silver	≥58	320

2.4.2.1 ETP-Copper

ETP-Copper (ETP = electrolytical tough pitch – historical name) is the most important species for electrical applications. It has a market share for the world copper demand of about 70–80%. The main use is for copper wires and cables and for busbars. It is also used in transformers for windings and components.

During melting of copper, some oxygen from air is absorbed in the melt and forms very small particles of cuprous oxide in the copper matrix. To avoid inclusions of bigger particles of cuprous oxide in copper, the oxygen content has to be controlled by adjustments of the atmosphere during melting and casting to about 100–350 ppm oxygen. The small amount of oxygen has the advantage, that it reduces metallic impurities by oxidation and refines the grain structure, so recrystallization may be improved. In contrast to alloying of metallic compounds, small amounts of oxides do not influence the electrical conductivity.

It has the highest electrical conductivity of 58–59 S/m, which corresponds to 100% IACS (International Annealed Copper Standard) and has a very good cold workability with low recrystallization temperatures of only 180–190 °C. It can be rolled, drawn, soldered and brazed. The only disadvantage is that it cannot be welded in a hydrogen containing atmosphere as the oxide reacts to water forming pores, leading to the so-called "hydrogen embrittlement".

2.4.2.2 Oxygen-free copper (OF-Copper)

Oxygen-free copper is a specialty and is used only for applications where hydrogen resistance is needed or for vacuum applications. Oxygen-free copper is also used, if a joint between glass and copper has to be made for vacuum tubes.

For the high quality of OF-Copper, the oxygen content has to be reduced to less than 5 ppm (second quality up to 10 ppm oxygen) and the hydrogen content also has to be very low. Some companies produce it by special proprietary processes, where the melt has to be treated by the exclusion of air or oxygen.

As there are no oxides in oxygen-free copper, the impurities are not tied up by forming oxides. Therefore, only the highest quality of cathodes can be used for OF-Copper and the recrystallization temperature is a little bit higher than for ETP-Copper.

Measuring this low content of oxygen and hydrogen in copper for quality control is difficult.

2.4.2.3 Phosphorus deoxidized copper (Cu-OF-XLP, Cu-HCP, SE-Copper)

In Europe and in the United States, there exists an alternative to OF-Copper, when only hydrogen resistance and high electrical conductivity is necessary, which is sold by different brand names (like BE-Copper at Aurubis AG). Normally, phosphorus would decrease electrical conductivity drastically. But in contrast to higher phosphorus containing copper, the phosphorus content in SE-Copper is carefully adjusted in a way that it only neutralizes the oxygen content – like a titration – without forming Cu-P compounds in a copper matrix. So, the electrical conductivity remains at a high level of 58 S/m and more. Another advantage compared to OF-Copper is a fine-grain structure because of the incorporation of finely divided oxides which leads to an excellent workability, although it has a higher recrystallization temperature of about 230 °C.

SE-Copper can be used as a substitute for OF-Copper with the exception of vacuum applications.

2.4.2.4 Phosphorus containing copper (DHP-Copper, DLP-P Copper)

As phosphorus decreases the electrical conductivity, phosphorus containing copper with 120–600 ppm phosphorus is not used for electrical applications, but for applications where corrosion resistance and thermal conductivity are important. Furthermore, phosphorus containing copper has an excellent workability and is resistant against hydrogen embrittlement. It has also a distinctly higher recrystallization temperature of 280–290 °C, which allows the usage of copper at higher temperatures before it becomes soft.

DHP-Copper made from cathodes is used for industrial tubes e.g. for air conditioning. DHP-Copper – made mainly from directly melted pure copper scrap or off-grade cathodes – is used for roofing and for plumbing tubes.

The production of DHP-Copper is made by adding a copper-phosphorus pre-alloy to a copper melt.

2.4.2.5 Low alloyed copper

There is no sharp limit between pure copper species and copper alloys. So, copper alloys with less than 1% alloying elements are counted often for pure copper species.

One reason to add small amounts other metals to copper is to increase the recrystallization temperature so that it can be used at higher temperature without becoming softer. So, for special applications – like catenary wires for railway overhead lines – 400–1000 ppm silver is added as it does not influence electrical conductivity.

But silver is very expensive and so other species are used, which are often a compromise between the recrystallization temperature, tensile strength, electrical and thermal conductivity and price. Some examples are copper-iron with 0.15% iron, copper-tin with up to 0.25% tin or copper-nickel or in former times and in China even today copper-cadmium wires with about 1% cadmium.

To increase recrystallization even more, dispersion-hardened alloys (see mechanism of this effects in Chapter 1.4.3) like copper-chromium or copper-zirconium are used – especially for heat exchangers at high temperatures – e.g. for water-containing tubes cooling a metal melt.

2.4.3 Copper alloys

Copper is very flexible in forming alloys as molten copper dissolves a lot of other metals and metalloids. There exist more than 400 copper alloys, which are commercially used. The production of copper alloys accounts for about 25% of the total copper production. The objective of alloying is to adjust properties of the metal like electrical and thermal conductivity, tensile strength at normal and higher temperatures, recrystallization temperature, color and corrosion resistance. Especially with the development of electronic devices, the variety of alloys has increased a lot. In many applications compromises have to be made e.g. between tensile strength and conductivity. The properties of copper alloys are not only dependent on the chemical composition but also on the treatment of the alloy during processing. Whereas pure copper has a fixed melting point, copper alloys melt over a range of temperatures.

There are two types of alloys, wrought alloys and direct-cast alloys. Wrought alloys are alloys which are first cast to billets, cakes and wire rod and get their final geometry by rolling, drawing or forging, whereas direct-cast alloys are cast to their final geometry only processed further by machining or sawing.

The most important alloy of copper is brass, which accounts to about 70% to the total copper alloy market. But it is not only one copper-zinc alloy but also a family of different alloys with 5–45% Zn. The next is bronze, a copper-tin alloy with about 5–15% tin. Nickel-silver (which does not contain silver but zinc) and copper-nickel are other important alloys. The rest divides in hundreds of alloys.

The production of alloys is made for a major account from copper and copper alloy scrap, which is directly melted to alloys. Only for adjustments of the composition or a shortage of scrap in the market, copper cathodes are used for the production. As during the production, up to 40% scrap is produced, this scrap is directly remelted. Furthermore, it is very important to sort the scrap according to alloys and to avoid impurities in the scrap so that it can be processed directly to the same or similar alloy.

2.4.3.1 Brass

Brass represents the biggest group of alloys within all copper alloys. The group comprises alloys with 5–45% zinc. The generic terms brass covers a wide range of alloys with different properties like strength, machinability, ductility, wear resistance, conductivity and corrosion resistance. Alloys with more than 85% copper are called tombac. For a lot of applications even a third or fourth element is added like lead, aluminum or nickel. Brass is a substitutional alloy that means that zinc atoms substitute copper in the crystal structure, as zinc and copper form both a solution in a melt and in a solid material.

Brass crystallizes in different structures depending on composition, casting conditions and temperature treatment. In the range up to 37% zinc brass crystallizes in the same structure as copper by substitution of copper by zinc atoms. This is called α-brass. In the range 37% to 46% zinc, the structure changes as the copper structure cannot include more zinc atoms. In this range we have a mixture of alpha and beta phase. With even more zinc we have only the β-phase. The differences in crystal structures are shown in Figure 2.4.4.

The α-phase has a very good cold formability but bad shape-cutting properties – like copper. So, it can be extensively deformed at room temperature and is used for complex components. In contrast to that, β-brass can be extensively hot worked and it has a good shape cutting property.

α-Brass has a wide range of applications: in mechanical and apparatus engineering for tubes, for oil lines, radiators, valves and many others. In electrical engineering, it is used for sockets, springs, connection terminals and others. Furthermore, it is used for spectacle frames, in watch cases or musical instruments and many others. In alloys with copper contents >80%, it is used for decorative applications in the jewelry industry. Also, for coins the brass alloys are used in many countries.

β-Brass is used for free machining and cutting applications, e.g. where the metal has to be milled to its final shape. To get better turnings in milling and increase the productivity of milling often up to 2% lead is added – more and more substituted by other metals like bismuth or silicon. One important application is screws made

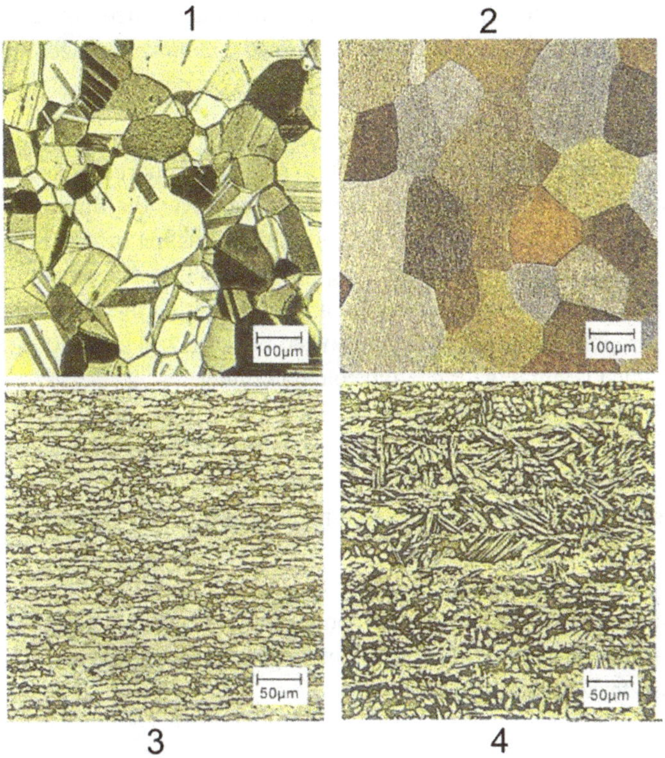

Figure 2.4.4: Different grain structures of brass after etching [4]. 1: α-brass, 2: β-brass, 3: (α + β) brass granule, 4: needle-like (α + β) brass.

from α/β brass. As in milling there is a lot of production scrap, direct recycling of the lead containing brass is an important issue in the brass industry.

For special applications, other metals are added e.g. aluminum for increasing tensile strength and hardness without influencing ductility very much.

Tin is added to increase the corrosion resistance e.g. in mining and marine applications.

Silicon is alloyed together with manganese to produce very hard intermetallic manganese silicide in the basic matrix which increases the wear resistant properties to brass.

2.4.3.2 Bronze

Bronze is an alloy of copper and tin, including red brass – an alloy of copper tin, lead and zinc – it is the second biggest group of copper alloys. The standardized copper-tin alloys are classified in wrought alloys with a content of max. 10% tin and cast alloys with a tin content of max. 20%. Bronze is a substitution alloy as up to 16% tin is soluble both in the liquid and solid phase.

Sometimes alloys which do not contain any tin are named as bronze, like aluminum bronze – containing aluminum or manganese bronze and nickel bronze.

2.4.3.2.1 Phosphorus bronze

Normal tin bronze also contains in nearly all application some phosphorus as deoxidizing agent ("phosphorus bronze"). The phosphorus also improves the fluidity of the molten metal which is important in casting. The problem in casting bronze is its tendency to segregation, as it exhibits a wide melting interval, so that differences of the tin content of up to 10% tin between the center and the outer surface of a cast strip may occur. Therefore, bronze has to be homogenized between 700 and 1000 °C for getting a structure which is suitable for hot and cold deforming.

Wrought phosphor bronze is used for springs, bolts and various other items, where resistance to fatigue, wear and chemical corrosion are required. So, it finds applications in electronic devices, connectors and switches. Components of phosphorus bronze retain springiness even at elevated temperatures, which is often the case with electrical switches and connectors.

Cast alloys of bronze with normally 9–12% tin have a variety of applications because of the corrosion resistance, low wear and good gliding properties. So, they are used in mechanical engineering, for propellers in ships and for casting church bells. They are also used for valves, fittings and rotors.

2.4.3.2.2 Red bronze

Red bronze also named red brass or gunmetal, as it was used in former times for the production of cannons, is an alloy of copper, tin, zinc and in many cases also lead. The composition varies between 3–7% tin, 2–8% zinc and 3–7% lead.

Red bronze is normally used as a cast alloy (e.g. sand casting), not as a wrought alloy. It is used for valves and because of its excellent gliding properties as a bearing metal. Because of the lead content red bronze has good cutting and milling properties and it has a high-corrosion resistance.

2.4.3.2.3 Other bronzes

Aluminum bronze is an alloy of copper with 5–12% aluminum often with the addition of other elements like iron, silicon or nickel. They are stronger than brass and phosphorus bronze with a better corrosion resistance. They also have an attractive golden color. The major use of aluminum bronzes are seawater applications like pipe fittings or heat exchangers.

Manganese bronze (in USA) is not really a tin bronze as it contains 40% zinc, 1% lead and 1% manganese. Outside USA, it is an alloy of copper with 12% manganese – used for electrical resistors.

Silicon bronze is an alloy of copper with 3% silicon. Some applications are door fittings and railings. They are also used for welding wires in the car production with welding robots.

2.4.3.3 Copper nickel alloys

Copper nickel alloys contain normally 10–30% nickel. This alloy improves strength and corrosion residence without losing ductility. They have an excellent resistance to marine corrosion and biofouling. As the alloys are expensive because of the price of nickel, they are used only for special applications, e.g. for seawater pipework, heat exchangers, in seawater desalination plants and for sheathing of legs and risers of offshore platforms. They are also used for the production of coins (1 euro, 2 euro).

Nickel silver is another alloy, which does not contain any silver but has a silvery appearance. A typical composition is 60% copper, 20% nickel and 20% zinc (therefore also sometimes named as brass). It is used in zippers, better-quality keys, for making musical instruments and sometimes for connectors (but only with an electrical conductivity of less than 10% of pure copper).

2.4.3.4 High-performance alloys (HPA)

The term "high-performance alloy" is not clearly defined. In this context, HPAs are defined as copper alloys which have a high strength, even at elevated temperature and at the same time a high electrical conductivity.

In normal alloys, which are substitutional alloys in a solid solution, the electrical conductivity is reduced by alloying elements but higher strength and hardness can be achieved like in phosphorus bronze. To achieve both high conductivity and high-tensile strength the so-called precipitation hardening or aged hardening is employed for certain alloys (Figure 2.4.5). It is a special heat treatment and is based on the different solubility of foreign particles at different temperatures.

The mechanism can be explained as follows:

Assuming an alloying element is totally soluble in the liquid phase, but in the solid phase only at high temperatures, during solidification, two phases will be formed – the matrix and the crystals of the mixture of the alloying elements. But this only happens if the melt is cooled very slowly, so that the crystals have enough time to agglomerate. If the melt is cooled very rapidly, the alloying element will not have the time to grow to agglomerates, so it will remain as so-called supersaturated solution. But by a heat treatment of the alloy, only few of 10–100 nm crystals of the alloying element or intermetallic phase by diffusion are formed. The particles in the matrix disturb the free gliding of the matrix leading to a high-tensile strength. On the other hand, the second phase of precipitates does not reduce the electrical conductivity so much as they are not alloyed but only incorporated.

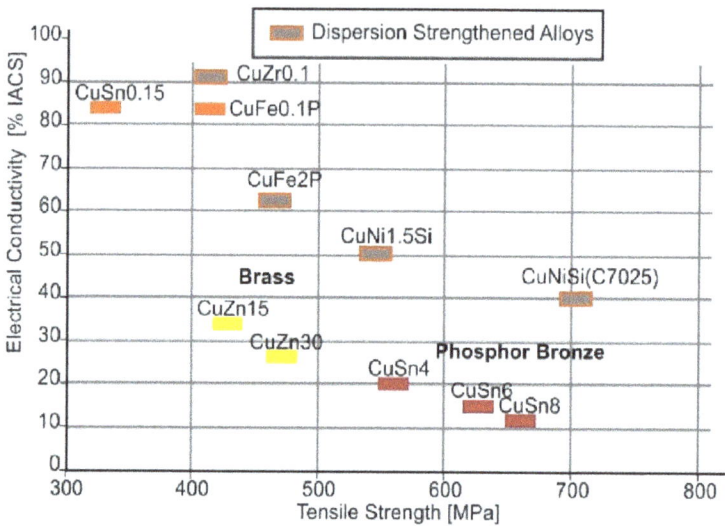

Figure 2.4.5: Tensile strength and electrical conductivity of copper alloys.

Some examples for such alloys are:

Copper iron: Even 0.1% iron and phosphorus can increase the tensile strength by 30% but reduces the electrical conductivity only by 15%.

Copper zirconium: 0.1% zirconium can increase tensile strength by 30% while reducing electrical conductivity only by 10%.

Copper nickel silicon: This is a very important group of HPAs. With about 2–3% nickel and silicon as this special alloy – often modified by addition of other elements like magnesium – one can achieve a tensile strength which is 2.5 times higher than steel but has still an electrical conductivity of 40% of pure copper.

Copper beryllium: This alloy with 0.25% to 2% beryllium has the highest tensile strength but is restricted because of the toxicity of beryllium.

The main field of applications of HPAs is the electronic industry needing high electrical conductivity and high-tensile strength and a good relaxation behavior, especially for connectors or springs which are switched several thousand times.

Often, these alloys are tin plated by hot dip tinning or electrolytic plating – to increase corrosion resistance and reduce the force which is needed for making contacts between connectors. Because of the electronic industry with its trend to miniaturization, the demand for these alloys has increased in the last decades.

2.4.4 Shapes and products of copper and copper alloys

As copper and many of the alloys are very ductile, they can be formed easily by hot rolling, cold rolling, drawing and forging to their final shape. Some copper alloys – often lead containing brass – can be easily milled and cut.

There are three big groups of intermediates, from which most of the copper products are made by further processing: copper wire rod, cakes and billets.

2.4.4.1 Copper wire rod

Copper wire (see Figure 2.4.6) rod is the most important copper semis product normally made only of pure copper cathodes and sometimes by very high-quality copper scrap. About 70% of all copper cathodes are used for the production of wire rod.

Figure 2.4.6: Coils of copper wire rod (Source: Aurubis AG).

Producing high-quality wire rod is also the requisite for the registration of a cathode on the London Metal Exchange. Wire rod is the intermediate from which all wire and cable applications are made.

Furthermore, wire rod is a real commodity, as it is standardized all over the world and it is traded all over the world. So, wire rod is defined as:
– Diameter of 8 mm – so all machines for further processing are constructed for 8 mm wire rod
– ETP-Copper with an oxygen content of 100–400 ppm
– Delivering in coils with a weight of 2–10 tons

The production of wire rod is a continuous process: Cathodes are melted – very energy efficiently in a shaft furnace and then cast in a water-cooled casting wheel with a diameter of about 3 m or a moving mold caster continuously to a trapezoid strand of about 10 × 5 cm. Afterwards, it is continuously rolled down in 9–14 steps in the same line to a 8 mm wire rod. At the end, the wire rod is coiled with a speed of about 100 km/h to the final wire rod coil of 2–10 tons. Capacities of big wire rod plants may be up to 300,000 t/year.

The invention of the wire rod production by three manufactures in the 1960s with slightly different systems was the base for the mass production of wires and cables with a high and constant quality.

2.4.4.2 Billets (round bars)

Billets are round bars with a weight of 80 to more than 500 kg and diameters of 100 to 500 mm. They are the base for extrusion products e.g. for copper and copper alloy tubes, for profiles, forging products and for certain brass alloys also for the production of alloy wires. In most cases, they are cast continuously (pure copper and some alloys) with a flying saw or semi-continuously (alloys) in hollow water-cooled molds.

2.4.4.3 Cakes (slabs)

Cakes of copper or copper alloys are the base for the production of flat products like strips or foils of copper. In most of the production lines they are cast semi-continuously with a width of up to 1200 mm, a thickness of up to more than 300 mm and a weight of up to 30 tons. The further processing to strips and foils is made by hot rolling to about 2–5 mm and cold rolling in several steps to the final thickness to foils of less than 50 µm thickness.

2.4.4.4 Important products of copper and copper alloys

2.4.4.4.1 Wires and cables

The most important copper products are wires of ETP-Copper used as the basis for all cables, for motors and power generators. They are produced by drawing wire rod in several steps down to a few µm, as copper becomes harder by mechanical treatment. Therefore, between the drawing steps – often after about 90% reduction area – the wire has to recrystallized by heating. Annealing is done with a continuous resistance annealer, using the thin copper wire as a resistance itself, producing heat when connected to a voltage.

Cables are copper wires, which are insulated by a plastic shield. According to the voltages, there are different cables:

- Housing cables with three connectors for plus, minus and grounding with a cross-section of about 1.5 mm².

–   Low voltage cables (up to 1000 V) consist of single conductors or multi-
    conductor cables. In this range there are a lot of special cables like instrumenta-
    tion, audio, coaxial cables or even airport lightning cables as some examples.
–   Medium voltage cables for 5 to 45 kV and high-voltage cables with voltages be-
    tween 50 and 750 kV. They are used as underground cables over long distances,
    whereas aluminum is used for long distance aerial cable because of the weight.

Magnet wire, winding wire or enameled copper wire is a copper wire which is insulated by small layers of different polymer films according to voltage and temperature ranges. Smaller diameter magnet wires have a round cross-section, whereas
thicker magnet wire is often square or rectangular to permit more efficient use of
available winding space. Magnet wires are used for electrical motors and transformer windings.

Trolley wires or overhead wires have a special cross-section with a groove to ensure a good fastening to the overhead line. Because high-tensile strength is needed
and also sparking should be avoided often not only pure copper but also low alloyed
copper like copper magnesium, copper silver or copper cadmium is used.

EDM wires a made mainly from high-quality brass, which are used as electrode
for spark erosion cutting. They are used especially for cutting in the manufacturing
of molds and tools. During the cutting process, the wire is consumed.

2.4.4.4.2 Tubes

Tubes are made by extrusion and drawing of billets. Plumbing tubes which do not
need a high electrical conductivity are made mainly of DHP copper produced directly from copper scrap if available. Tubes for ACR (air conditioning and refrigerat-
ing) are made from higher quality copper sources like cathodes and have lower
phosphorus content. In many applications they have an inner groove to improve
heat transfer from the liquid to the tube. In Asia, ACR tubes are often produced by a
"cast and roll", in which at the first step hollow billets are cast horizontally which
are rolled down with on planetary rolls within one line.

Brass tubes are used for interior architectural designs and fabrication. Brass
tubes are also used for lamps and other light fixtures. Other applications are stair
carpet rods, handrails and railings.

2.4.4.4.3 Profiles, bars and rod

Profiles, bars and rod are produced mainly in the same way as tubes by extrusion
of billets and subsequent drawing. If soft material is wanted, the profiles have to be
annealed after drawing.

Copper profiles are used mainly for high current applications to ensure a low
voltage drop as they have bigger cross-sections than cables. Therefore, the main
use are busbars often made from ETP-Copper of OF-Copper. Copper busbars are

also produced by rolling of cakes. As an intermediate, from busbars also electrical components are produced by bending, folding, punching and cutting.

Profiles and rods made from brass are the most important products for brass. They are used in a variety of industries as gears, shaft, lock bodies, pinions, nuts and screws. One key industry for brass profiles is the building industry for brass hinges and surface bolts. They are used also for parts in the automotive industry and for bicycles, in electrical and mechanical engineering and plant construction. Leaded brass is used where milling and cutting is necessary like in the production of screws. During the manufacturing of brass profiles, up to 70% production scrap is formed which has to be recycled directly to the cast plant.

A new application of copper and copper alloy bars has been discovered to prevent the increasing danger of killing pathogenic microbes especially in hospitals. Copper and copper alloys are anti-microbial and can reduce the risk for infections. So, they are used increasingly for doorknobs or shower fittings.

2.4.4.4.4 Strips and sheets

Flat-rolled products like strips and sheets are another big class of copper products. They are produced by casting cakes, then hot rolling down to about 15 mm and then cold rolling down to foils of minimum 50 µm. As during cold forming hardening increases, the strip is mostly annealed after the rolling process. At the end the strip with a width of 300–1000 mm is split to the final width and coiled. In contrast to the commodity copper wire rod, strips and sheets are produced according to customer specifications.

One major application of copper strip is roofing and as well as being used as gutters. Roofing strip typically has a thickness of about 0.5–0.7 mm and is made mainly from DHP copper as the electrical conductivity does not play a role for this application, but good welding and soldering properties are essential for roofing. In the atmosphere copper changes its color within decades from orange to brown to green because of the formation of different basic copper salts like basic copper carbonate from the air (see Chapter 13.1.4).

Cable strip is used for the shielding of copper cables – especially for high-frequency cables, e.g. for the transmitters of mobile phone stations, to avoid the influence of disturbing electromagnetic interference. Even aluminum cables are shielded by copper strip. Because of good welding properties cable strip often is produced from OF-Copper or SE-Copper.

Connector strip can consist of a lot of different copper species and copper alloys. Besides pure copper, copper alloys like brass, bronze, nickel, silver and HPAs are also used depending on the application. Especially, the combination of spring properties, tensile strength, relaxation behavior and thermal and electrical conductivity are important for the choice of the copper alloy for certain applications. Thickness of the strip can vary in a range between 0.1 and 3 mm. In many applications, the connector strip is tin coated. Connector strip is produced according to detailed specifications of the customer, who is often an "original equipment manufacturer" (OEM). Connectors and lead frames (Figure 2.4.7) are produced from strips by stamping – producing single parts from a coil. Sometimes more than 50% of the strip is stamped out leading to a scrap rate of more than 50%.

Figure 2.4.7: Lead frames for microprocessors (Source: Wieland Werke AG).

Copper foils are strips with a thickness of less than about 100 μm. They are produced by rolling or by electrolytic deposition. To manufacture foils by rolling, special rolling mills with 20 rolls are used to get this small thickness and a flat surface with low tolerances. A totally different process for coil production is the electrolytic deposition of copper on a rotating drum.

One main application of copper foils are printed circuit boards. A printed circuit board (PCB) is the base for mounting electronic components using conductive copper pathways which are etched from copper foils. In lithium batteries copper foils are used for the anode. Lithium batteries contain about 5–10% copper.

The production of coins starts with copper alloy strip. A coil of strip is hoisted onto a wheel that feeds into a blanking press. There is a huge variety for coin material like solid metal blanks from copper-nickel, electroplated blanks or cladded blanks where copper alloys are cladded with steel or bi-color blanks with different cores and rings. After the blanks have been produced, they are minted in another stamping press.

Commutators are the core of nearly all electrical motors. Commutators are used in window lifters in cars, washing machines, drilling machines and nearly all other application in which electrical motors are used. They consist of strips of copper about 1–4 mm thickness and a width of some millimeters to some centimeters according to the power of the motor. They are produced not only from strips but also from extruded material like profiles.

Brass strips are used for decorative applications and also for connectors.

2.4.4.5 Copper powder

Copper powder is a specialty. There are different types of copper powders which differ mainly by their surface properties. Reduction of copper oxides is the oldest process for making copper powders, but it is still of an important commercial impact. Because of the fineness and the reactive surface, this copper powder takes up oxygen from the open air. So, normally, the oxygen content is more than 0.1%, which is much higher than in solid ETP-Copper. As it is sensitive against moist atmospheres, it should be stored in sealed containers or under an inert gas.

By reducing copper oxides at elevated temperature with a reducing gas, a sintered porous copper cake is achieved which is milled to powder afterwards. In another process, copper powder is made from an aqueous suspension by a reduction agent leading to very fine copper powder of less than 3 μm.

Another method is atomizing molten copper. This procedure is used if a high purity is required or if powders of copper alloys like bronze powders are produced. In this process, copper cathodes are molten in an electric furnace – if necessary, alloying elements are added – and then the melt is atomized through nozzles at high pressure. Both air and water as atomizing media are used for this process. After cooling in a water bath, the powder is classified according to sizes.

The third method for producing copper powder is electrolysis, where as an anode a copper cathode is used. In contrast to a copper electrolysis a very high current density of 700–1100 A/m^2 is applied which leads to dendritic copper crystals instead of solid copper cathodes.

Another subsequent method used ball milling of copper or copper alloy powder to flat copper particles. This leads in pigments to a luster appearance.

Copper powder metallurgy is used for producing single parts by sintering with the advantage of producing minimum amounts of scrap and for combinations of copper with materials, which normally cannot be achieved by alloying, if the second material does not dissolve in the molten copper. For this purpose, the powders of the components are mixed and sintered or pressed below the melting point of copper. By combinations of copper with carbon, copper with tungsten or copper with finely divided ceramics are possible. Examples of applications are carbon containing brake lines and carbon brushes. Other applications are conductive pastes for joints in electronic devices – partly substituting silver pastes – metallic pigments or catalysts. Bronze powder is used for self-lubricating friction bearings as the sintered bronze parts can take up some oil in the pores of the sintered part.

References

[1]   Langner BE. Understanding Copper – Technology, Markets, Business, 2011. ISBN 78-3-00-036273-6.
[2]   International Copper Study Group the World Copper Factbook, www.icsg.org, 2019.
[3]   Kupfer in Bestform, Aurubis AG, https://www.aurubis.com/binaries/content/assets/aurubis relaunch/files/produkt–und-imagebroschuren/shapes/aurubis-shapes_de.pdf
[4]   Kupferwerkstoffe, Wieland Werke, Ulm, 1999.

2.5 Solder materials in electronics

Jörg Trodler

2.5.1 Introduction

In modern electronics manufacturing, both for printed circuit boards and power electronic modules, solder pastes are essentially used for a substance-to-substance connection. The development came from the change of the components to be soldered from the so-called through hole technology (THT) to the surface mount technology (SMT) with the use of surface mount devices (SMD). In THT and in the early days of SMT, solders were used that consisted of eutectic tin-lead (Sn63) or with a 2% silver content (Sn62). Due to a European law to avoid the use of lead, the so-called lead-free solders (LF) were developed, which consist mainly of tin, silver and/or copper (SAC) with various modifications or additions of alloying elements. Since SMT is now the main technology in electronics, the chapter refers to this technology.

2.5.2 SMD Assembly Technology (SMT Technology)

SMT uses solder paste, which consists of about 50% flux by volume and 50% metal powder by volume as shown in Figure 2.5.1.

Figure 2.5.1: Volume ratio of flux (left) and solder powder (center) to solder paste (right) (Source: Stannol).

The volume ratio is important to wet the powder sufficiently with the flux and thus also to protect it. However, if the mass ratio is taken as a basis, then the density of the flux and the metal powder results in a ratio of approximately 10% by weight flux and 90% by weight metal powder.

https://doi.org/10.1515/9783110733143-012

The assembly manufacturing steps include:

I. Applying the paste to the substrate by printing, dispensing or pin transfer. The amount of solder powder between printing and dispensing can be varied.

   a. Printing has the most importance because it is a bulk production. Here, the solder paste is printed through a metal stencil onto the landing areas called pads using by a squeegee. The paste itself must roll well according to the direction of the squeegee (Figure 2.5.2).

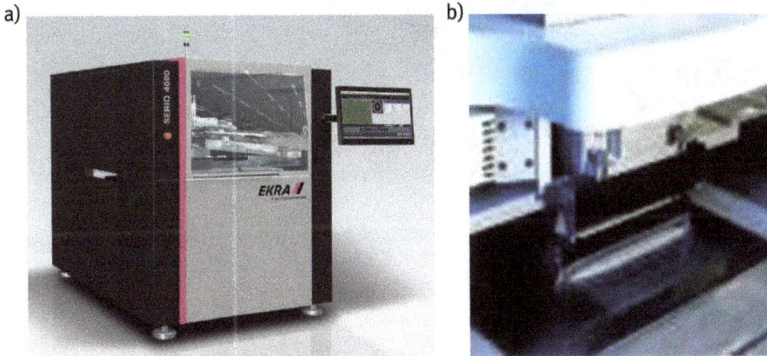

Figure 2.5.2: a) Typical automatic printer (Source: Ekra [1]) and b) detailed view into the printer with stencil and squeegee.

   b. Dispensing (Figure 2.5.3) is suitable for flexible changeover. In this process, the paste is pressed through a dispensing needle by means of a pressure or a screw valve, recently also by means of a jet valve and a typical solder paste depot is created.

II. Solder paste inspection (SPI), Figure 2.5.4. The added value in assembly production increases with the vertical range of manufacture. This means that if quality problems lead to insufficient quality of the manufactured assembly, the costs for scrapping also increase. In addition, some end users, such as automotive manufacturers, do not allow repair for their electronics. This means that manufacturers of these electronics need to identify a potential failure as early as possible. For this reason, inspection of solder paste prints/deposits by means of 3D inspection has become widely accepted in modern electronics manufacturing. Here, the deviations are determined based on the theoretical volume and corresponding substrates with defective prints being sorted out.

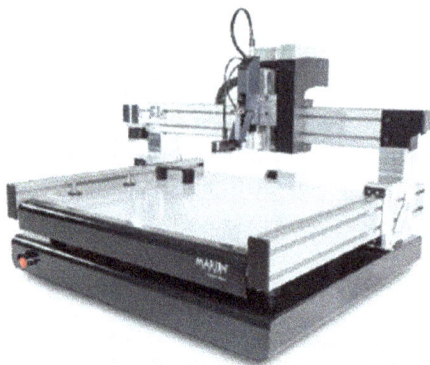

Figure 2.5.3: Dispensing machine from Martin [2].

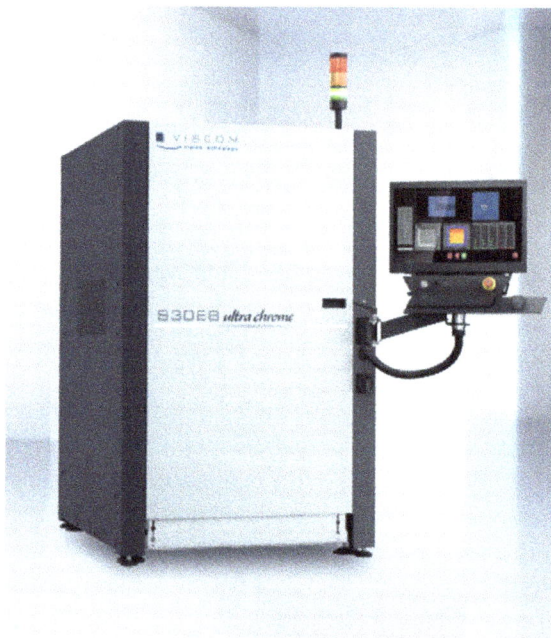

Figure 2.5.4: 3D-SPI, Viscom [3].

III. Component placement, which is carried out by so-called pick-and-place machines, Figure 2.5.5, now achieves placement rates of several thousand components per hour and a placement accuracy of a few micrometers.

IV. Soldering by using in so-called reflow systems, Figure 2.5.6. For this purpose, special temperature curves are set up, so-called temperature profiles (Figure 2.5.7),

Figure 2.5.5: Siplace pick-and-place machine from ASM [4].

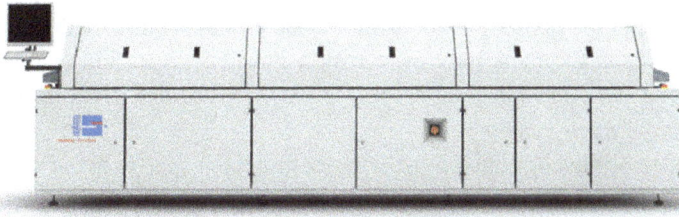

Figure 2.5.6: Vision XC reflow system (convection) from Rehm Thermal Systems [5].

which on the one hand activates the flux in the preheating so that this reduced the oxides from the metal powder and then, in a further temperature increase (above liquidus), causes the solder powder to melt so that it joins with the metals to be soldered. Depending on the assembly or substrate package, it must also be taken to ensure that the temperature is distributed as homogeneously as possible, so that overheating of the substrate (particularly important here with printed circuit boards) and the components is also avoided.

V. After soldering, quality control is carried out by AOI (automatic optical inspection, in which the assembly is optically inspected according to appropriate criteria, e.g. J-STD610), electrical functionality testing and in some cases the quality of the solder joints is also determined by X-ray equipment.

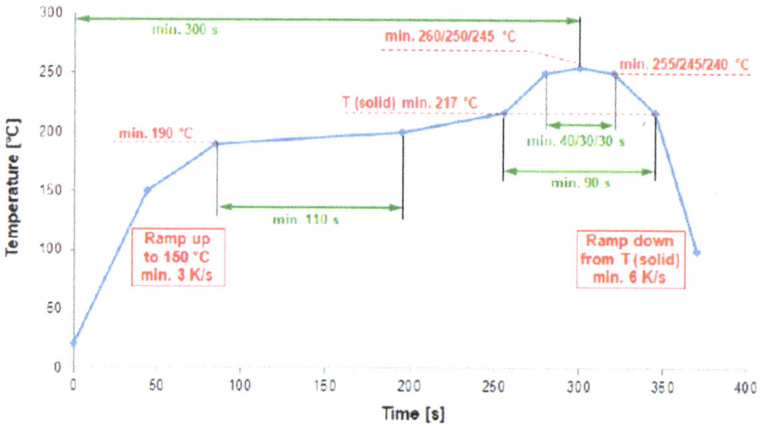

Figure 2.5.7: Typical reflow profile (Source: IEC TR 60068-3-12 CD-Draft).

2.5.3 Solder powders

Nowadays, solder powder is produced in a mass process since the demand should be several thousand tons by now. As a result, powder production (schematic diagram, Figure 2.5.8) takes place on several levels and includes (from top to bottom) the following production steps:

1) Production of the alloy/melt mostly under protective gas (to avoid oxide formation of the melt = dross)
2) Atomization
3) Sieving (powder size and fractionation)
4) Packaging

The most complex manufacturing step is atomization. Three types are distinguished here, e.g.:

– Gas atomization
   – A molten stream of tin or alloy through an annular nozzle and a gas stream (e.g. with nitrogen) splits the alloy into droplets. These droplets then solidify and are fractionated through several stages of screening.
– Spinning disk
   – As the molten tin stream hits a disk rotating at high speed, it is split. The resulting particles form particles/balls which are then spun off the disk. These particles then solidify and are fractionated through several stages of screening.

- Ultrasonic
  - The molten tin stream is directed onto a sonotrode and broken down into particles according to the drive frequency. These particles then solidify and are fractionated through several stages of screening.
- Dispersion in a liquid (dispersion medium), e.g. Welco process
  - In this process, the alloy is melted in a temperature-stable medium and rotating tools break down this melt into particles, which in turn are enclosed by the dispersion medium and, in combination with the surface tension, form high-quality fine powder particles. However, this process is a batch process, thus quantitatively limited and only suitable for ultra-fine particles.

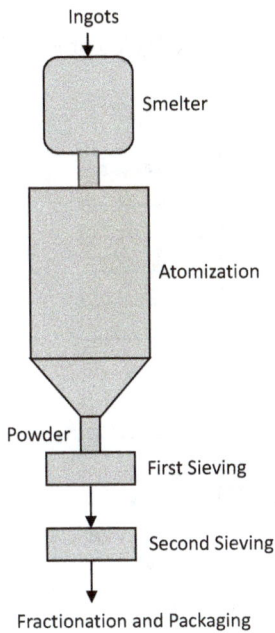

Figure 2.5.8: Schematic representation of a powder production plant.

The size of the powder particles is fractionated according to international and national standards, such as J-STD-005 (IPC-TM-650) and DIN 32513 (Table 2.5.1). The size value shown in Table 2.5.1 is the range of the particles of at least 90%. However, the ranges above and below this range are also clearly standardized. Important is the measuring method to determine the size distribution. Here, both transmitted light (thus measuring the individual diameters) and determination of the weight are possible. Types 7 and 8 are currently not standardized. They are still in the industrial development stage.

Table 2.5.1: Solder powder fractionation according to J-STD-005 and DIN 32513.

Type	Powder size
1	150 to 75 µm
2	75 to 45 µm
3	45 to 25 µm
4	38 to 20 µm
5	25 to 10 µm
6	15 to 5 µm
7	11 to 2 µm
8	8 to 2 µm

A comparison between the standard procedures and the Welco process is shown in Figure 2.5.9.

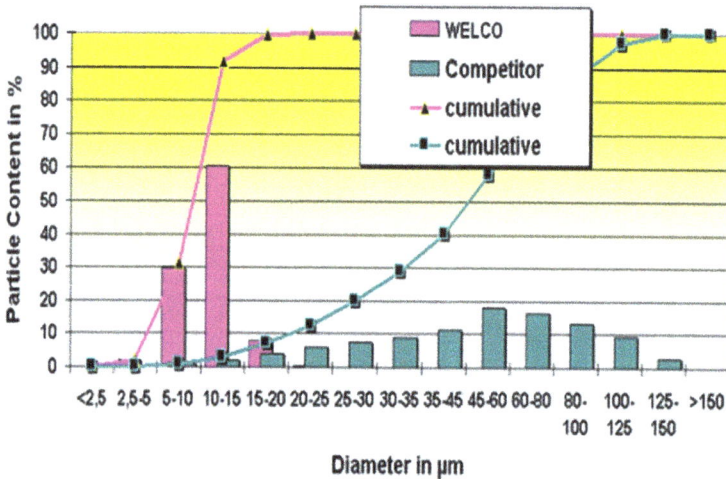

Figure 2.5.9: Yield during powder production (Source: Heraeus, Symposium: Löten in der Elektronik, October 13 and 14, 2021).

The diagram shows that the Welco process gives a very high yield, whereas screening is essential for the standard processes.

The powder is the most important component in the soft solder pastes. The quality of the powder determines:

- Formation of the intermetallic phases
- Alloy composition
- Liquidus and solidus temperature
- Surface tension
- Oxides and O_2 uptake

Defects or insufficient qualities due to roughness in the powder surface, satellite formation, partial remelting, contamination (Figures 2.5.10 and 2.5.11) and the oxide layer leads to quality variations in the paste in terms of rheology, which can have a negative effect during application (e.g. printing), in the soldering properties, e.g. too much oxide reduces the solderability, and in the stability of the pastes, e.g. during storage, i.e. the long-term stability.

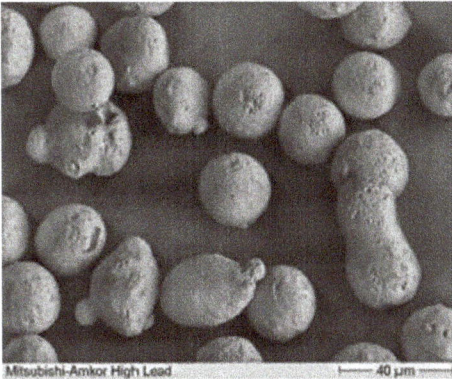

Figure 2.5.10: Defects in powder production due to conglomerates (dog bones) (Source: Heraeus, Symposium: Löten in der Elektronik, October 13 and 14, 2021).

Figure 2.5.11: Satellites and contamination with residual melt (Source: Heraeus, Symposium: Löten in der Elektronik, October 13 and 14, 2021).

A decisive factor with the powder is the oxide formation as the surface or the metallically bound oxide on the powder surface. An increase in the tin content has resulted from the conversion of lead-containing alloys to lead-free alloys and was already described in 2004 [6] (Figure 2.5.12).

The present situation in Fujitsu

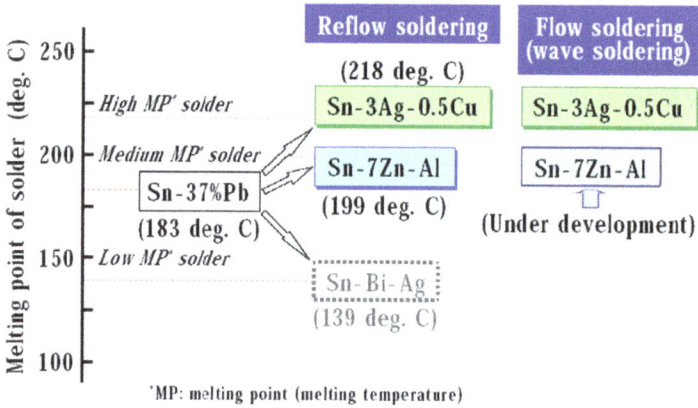

'MP: melting point (melting temperature)

Figure 2.5.12: Change in paste viscosity due to a reaction of tin (Sn) and the flux by intensive tin salt formation [6].

The tin salt formation shows up in pitting of the powder and is shown in Figure 2.5.13.

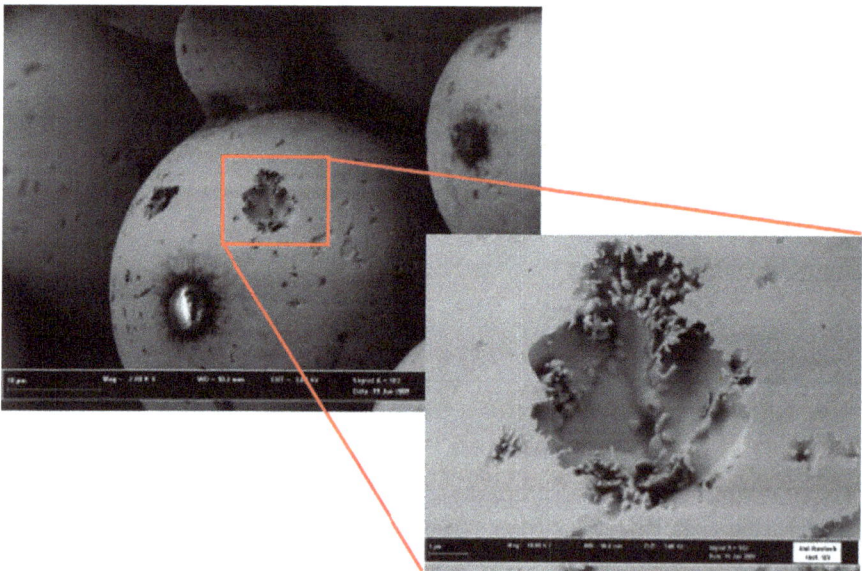

Figure 2.5.13: Pitting of SAC powder due to reaction of the flux with the powder (Source: Dr. Beck University Rostock, Institut für Gerätesysteme und Schaltungstechnik).

This results in a minimum for the metal-bonded oxide, which must protect the powders for interaction with the flux (like the anodizing of aluminum), but this oxide must also not be too large, so that no wetting inhibition builds up during subsequent soldering. The flux can only chemically reduce these oxides, since thermal reduction is not possible due to the required temperatures of more than 400 °C in soft soldering (soft soldering < 450 °C, brazing > 450 °C).

In summary, the oxide influences the storage stability, the so-called voiding, i.e. gas inclusions and thus three-dimensional defects in the solder joint, the viscosity of the paste and the wetting properties.

2.5.4 Fluxes

The flux consists essentially of a resin, solvent, activators and auxiliary materials. Together with the solder powder, this produces the solder paste. After soldering, residues of the flux are the resins, small proportions of the activators and auxiliaries, but these must then be bound in the flux residue to prevent any harmful interactions with the substrate and the components, such as surface resistance (SIR = surface insulation resistor), which leads to unwanted leakage currents and/or corrosion.

The resin, solvent, activators and auxiliary materials determine the behavior of the flux:

a) Resin
  - Wetting/residues after soldering
  - Rheology/printing behavior
  - Slumping, i.e. the quality of the printed solder paste deposit (no bridging)
  - Tackiness (after component placement, the paste must be able to fix the component by adhesion before soldering)
  - Solder balling (no secondary solder balls, which can also lead to short circuits)
  - Surface resistance

Figure 2.5.14 shows good and poor wetting. In the picture with insufficient wetting, the cut edge of the component connections is visible.

b) Activator
  - Solderability/wetting, i.e. the activators remove the oxides and impurities from the metal powder
  - Surface resistance and thus the classification of the solder pastes into so-called no-clean solder pastes [7]
  - Storage stability

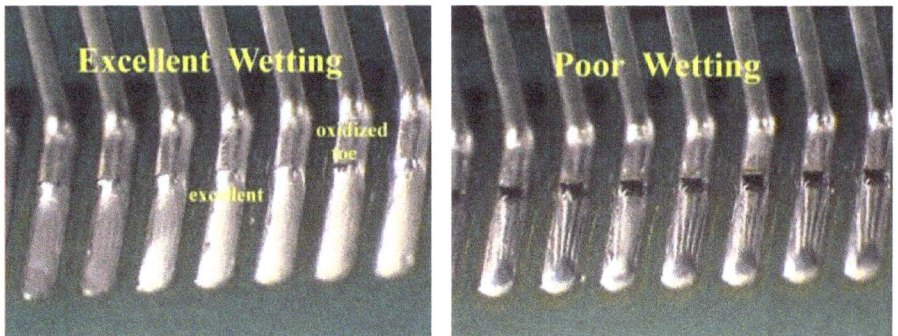

Figure 2.5.14: Comparison of good wetting (left) with insufficient wetting (right) (Source: Heraeus, Symposium: Löten in der Elektronik, October 13 and 14, 2021).

Figure 2.5.15 shows a comparison between two pastes with the same no-clean properties. According to the standard test [7], they can also be classified as suitable in terms of wetting, but show significant differences in the wetting quality on a nickel surface.

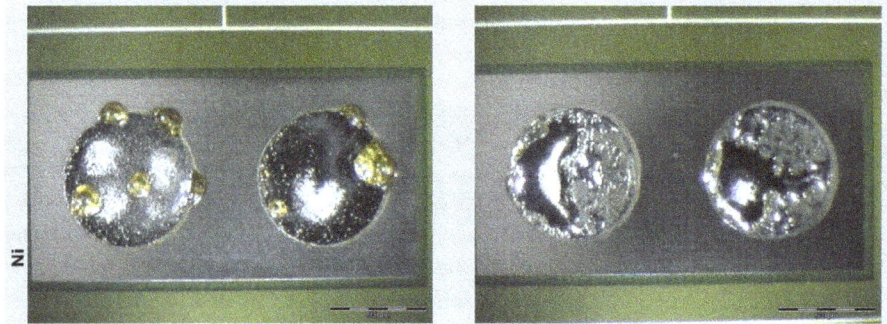

Figure 2.5.15: Results of solder pastes with respect to a wetting test on nickel surfaces (Source: J. Albrecht, R. Knofe, Siemens AG).

c) Solvent
   The solvent evaporates during the reflow process and affect:
   – Application behavior
   – Stencil life and placement time
   – Print after wait (quality of first print after wait)
   – Slumping of the paste

There are differences, for example, in application by printing and dispensing. Dispensing usually involves working with two solvents, a low-boiling solvent and a

high-boiling solvent. The low-boiling solvent is used to make the material as thin as possible so that it can be dispensed through the needle, but it must evaporate as much as possible after application to enable an optimum dot shape, i.e. no slumping.
d) Additives
   – Adjustment of the viscosity
   – Settings of the thixotropic properties
   – Very strongly influence the quality of the printed deposit

Another new criterion is the formation of voids. Here, a minimum void behavior is required. However, this behavior is strongly dependent on the wetting properties. With so-called arey-arey components such as BGAs, where solder balls are located on the bottom side of the components, a wetting reduction has a positive effect, while the exact opposite occurs with flat solder contacts like passive components or heat sink areas (Figures 2.5.16 and 2.5.17).

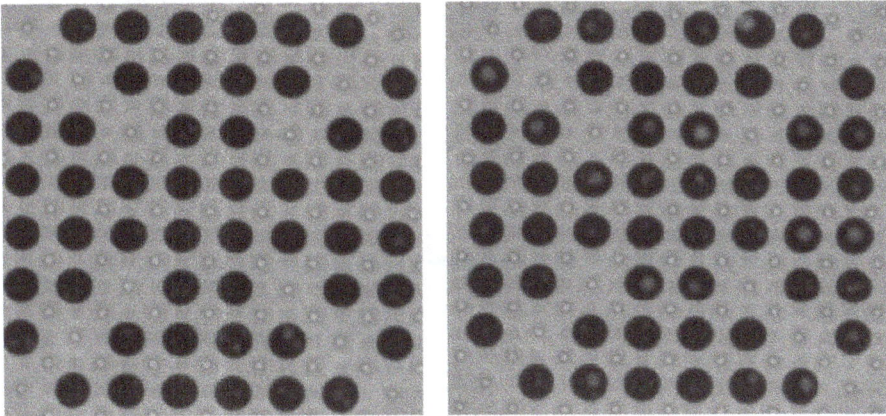

Figure 2.5.16: Void formation, soldered with nitrogen atmosphere, alloy SAC405 (Source: Heraeus, Symposium: Löten in der Elektronik, October 13 and 14, 2021).

This means that such quality criteria must be considered in the process control and the soldering parameters must be set accordingly.

2.5.5 Alloys

As already mentioned, the use of soft solder alloys is mostly covered by lead-free alloys. These differ due to the product requirements. For example, in standard applications (printed circuit boards), the alloy SAC305 (with 3% silver and 0.5% copper) is the most widely used. In power electronics, where high currents and voltages are

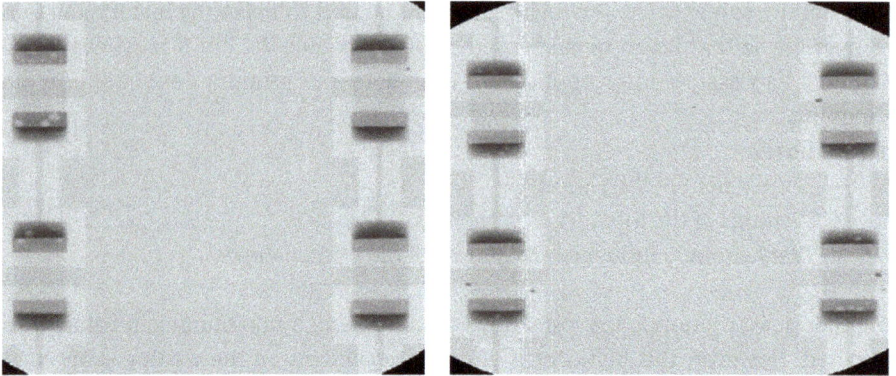

Figure 2.5.17: Void formation passive components (capacitor), soldered under nitrogen atmosphere, alloy SAC405 (Source: Heraeus, Symposium: Löten in der Elektronik, October 13 and 14, 2021).

expected, the eutectic SnAg3.5 (3.5% silver). SnBi58 is also used as a so-called low-temperature alloy. For a higher temperature range, the so-called Innolot has become widespread, especially in the automotive sector. Table 2.5.2 shows an overview of the most important alloys.

Table 2.5.2: Summary of the most important lead-free solder alloys.

Alloy	Liquidus temperature (°C)	Remarks
Sn96.5Ag3.5	221	SnAg eutectic
Sn96.5Ag3Cu0.5	217–219	SAC standard
Innolot SnAg3.8Cu0.7Ni0.125Sb1Bi3	206–218	Higher reliability on pcb, field temperature up e.g. up to 150 °C
HT1 SnAg2.5Cu0.5In2CM (CM = crystal modifier)	210–217	Higher reliability on ceramic substrates, field temperature up to e.g. up to 150 °C

In the automotive sector in particular, the requirements for assembly and joining technology (eAVT) are increasing, which includes the expected temperature range and the number of cycles that can lead to early failure. In this context, it is precisely the shear strength that has proven to be an indicator. After a temperature test, which can consist of a shock or alternating test, the assemblies to be tested are cycled, e.g. in a range from −40 °C to +125 °C with a certain test frequency. By regularly removing the test subjects, shearing off the components, the shear force is determined and the difference to the initial shear force value is determined. The

50% value has become accepted as the error here, i.e. the shear force drop should not be more than 50% of the initial value. A comparison of different alloys [8] is shown in Figure 2.5.18.

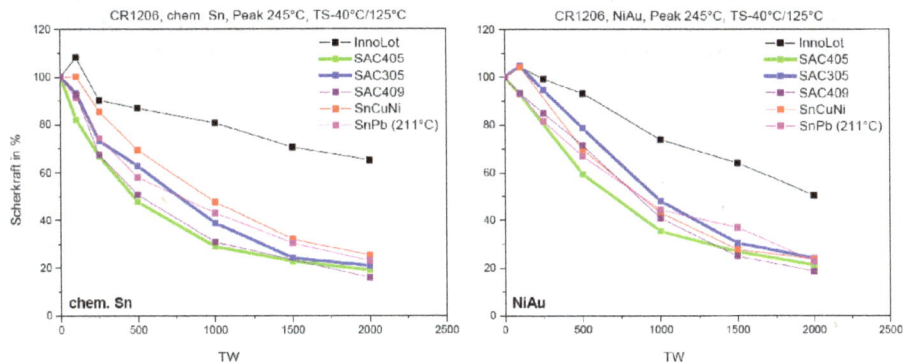

Figure 2.5.18: Investigation of the shear strength of different alloys [8].

2.5.6 Summary

Nowadays, solders are mainly used in paste form. By the specific selection and development of the fluxes, the properties for the application and the soldering quality can be specifically adjusted. However, these must always align to the requirements of the assembly and no undesirable effects, such as leakage currents. This also results in special requirements that make it necessary to clean the assembly with appropriate media.

The choice of alloy depends on the reliability requirements. In this respect, products for an automobile differ significantly from those for a television set or desk/laptop computer.

References

[1] Webpage: https://www.asys-group.com/de-de/produkte/electronics/drucken/, downloaded 29th November 2021, Data Sheet Fa. Ekra, Serio 4000.
[2] Webpage: https://www.martin-smt.de, downloaded 29th November 2021, product description Dotliner 08 with dosing head 5 ccm, Company Martin.
[3] Webpage: https://www.viscom.com/de/, downloaded 29th November 2021, Product brochure S3088 ultracrome of the company Viscom.
[4] Webpage: https://www.asm-smt.com/de/produkte/placement-solutions/, downloaded 29th November 2021, Product brochure Siplace SX of the company ASM.
[5] Webpage: https://www.rehm-group.com, downloaded 29th November 2021 Product brochure Vision Series from Rehm Thermal Systems.

[6]   International Microelectronics and Packaging Society (IMAPS), IMAPS Seoul, September 2nd and 3rd, 2004, K. Hasihimoto (Fujitsu Laboratories Ltd./Japan) Assembly Technology Using Pb-free Solders: The State of the Art and Issues.

[7]   IPC Standards, Homepage: https://www.ipc.org/,J-STD-004,J-STD–005

[8]   Scheel W, Wittke K, Nowottnick M. Materialmodifikationen für geometrisch und stofflich limitierte Verbindungsstrukturen hochintegrierter Elektronikbaugruppen „LiVe", 1st edition, Verlag Dr. Markus A, Detert, Templin, Germany, 2009. ISBN-13: 978-3-934142-57-2.

2.6 Metallic coatings

Ralf Feser

Metallic coatings fulfill a wide range of functions. These range from purely decorative aspects to mechanical properties such as hardness or wear resistance to increasing corrosion resistance. Metallic coatings are usually deposited on metal substrates, but they can also be applied to other substrates, such as plastics.

In principle, metallic coatings can be applied to components as a piece process or to sheet metal as a strip process.

Most processes are already used in the industrial manufacturing process of the components or semi-finished products. However, some processes, such as thermal spraying, can also be applied to the component on site.

A key requirement for the coating is adhesion to the substrate. This can be improved, for example, by post-treatment of the surface. This often involves thermal processes in which stresses in the coating metal are relieved or interdiffusion between the coating and the substrate can take place.

Metallic coatings for corrosion protection must be applied without pores to prevent local corrosion processes. Different metals have different corrosion behavior. The presence of pores in the coating can lead to local dissolution of the substrate as a result of galvanic corrosion, which is undesirable. Zinc-based coatings have proved successful for the corrosion protection of steel, since zinc coatings have a more negative rest potential than steel and therefore a relatively small defect in the zinc layer results in cathodic corrosion protection of the steel. The deposition of metallic coatings must therefore always be considered from the point of view of corrosion, since coatings can never be applied without defects or, at the latest, damage to the coating can occur again during practical use. To avoid such effects, multi-layer coatings are applied in various ways, either of the same type, so that defects from the coating process do not propagate, or of different coating metals in order to gradually create potential differences between the outer surface and the substrate.

In general, for the application of coatings and of course for metallic overlays, the surface of the part to be coated must be clean. The degree of cleanliness can be defined in different ways, but certainly the surface must be free of greases, oxides and other covering layers. Otherwise, the adhesion of the coating metal to the substrate is disturbed and gaps in coverage or spalling will occur.

In this chapter, only a very brief introduction to metallic coatings can be given; more detailed explanations can be found in the literature [1–3]. In particular, there are a large number of variants for each process, which differ in terms of modifications and additions to the basic process. Therefore, only the basic processes will be explained in the following; for a more in-depth consideration, please refer to the special literature.

https://doi.org/10.1515/9783110733143-013

2.6.1 Vapor deposition

In recent years, processes have gained in importance in which the substrate is coated with a metal from the reaction with a gas phase. These are characterized by the fact that the amount of waste or process media to be treated is relatively low, but their throughput is also limited.

In principle, two processes are available, each of which is available in different modifications in order to optimize the layer formation in terms of speed or properties. In this chapter, these processes will be considered in summary. These are CVD (chemical vapor deposition) and PVD (physical vapor deposition) processes.

In general terms, this means that metals are vaporized in a vacuum or a defined atmosphere and condense on the substrate. In the CVD process, chemical deposition takes place from a liquid medium, e.g. carbonylene or organometallic compounds, which is usually evaporated by heating the liquid. A gas, used as a carrier gas or itself as a component of the reaction, transfers the reactive substance to the surface of a heated substrate, and a reaction then takes place at the surface. There is a wide temperature range in which CVD processes are used, ranging from 150 to 1300 °C. Figure 2.6.1 shows a schematic diagram of the process.

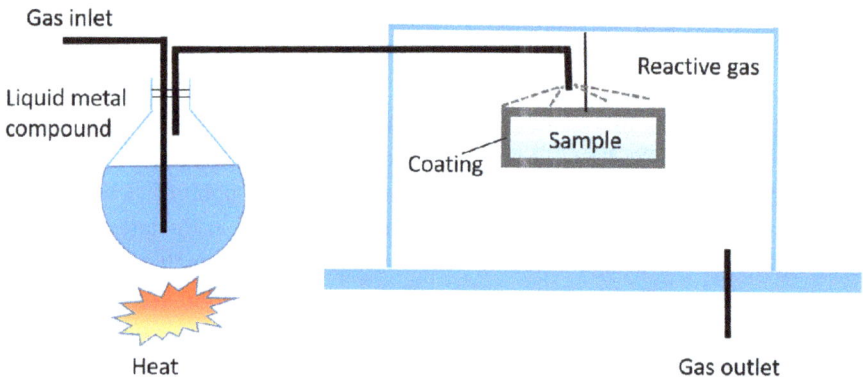

Figure 2.6.1: Schematic representation of the CVD process.

The PVD process is characterized by the fact that the starting material is converted into the gas phase using physical processes, such as heating the metal or ionization as a result of applying a high voltage. Here, too, condensation of the metal on the substrate occurs, and the temperatures are usually in the range of room temperature. Figure 2.6.2 shows the basic process.

The low temperatures of the PVD processes allow the application of coatings without thermally affecting the substrate.

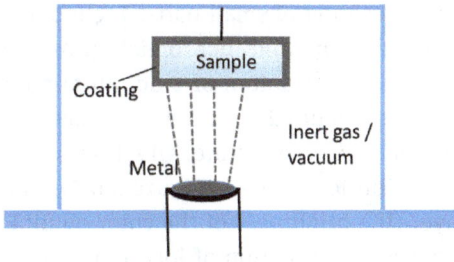

Figure 2.6.2: Schematic diagram of the PVD process for the application of metallic coatings.

2.6.2 Diffusion coatings

Diffusion processes produce an adhesive coating that is formed by solid-state diffusion of the coating metal into the substrate. Since diffusion is a thermally activated process, the rate of reaction increases with temperature. However, the temperature remains below the melting temperature (T_m) of the coating metal, so that the reaction takes place in a powder of the coating metal and even small parts can receive a metallic coating without sticking together during the process.

Diffusion coatings are metal coatings that are typically formed in a powder of the metal coating at temperatures below the melting point of the coating metal. This results in solid-state diffusion of the coating metal into the substrate. A coating with a solid solution structure is formed which is characterized by high adhesion. Alloy phases are therefore produced on the substrate surface, which differ in their properties from those of the base material. Figure 2.6.3 shows the simplified coating process.

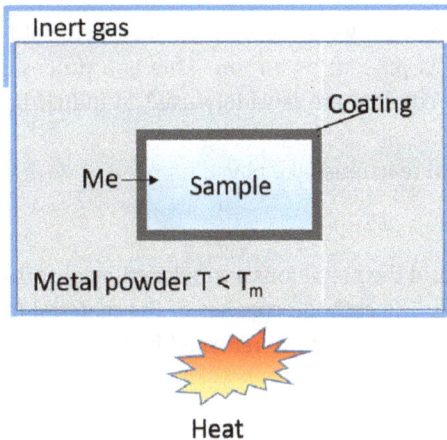

Figure 2.6.3: Schematic representation of the diffusion processes for generating alloy phases on a substrate.

This technology is preferably used for the coating of small parts. Technologically important processes are sherardizing, alitizing and inchromizing. As the process names already indicate, this involves the reaction of zinc, aluminum and chromium powders with metal substrates. Figure 2.6.4 shows the metallographic cross-section through a zinc layer on a steel substrate, which was produced by sherardizing. In the element distribution images (Figure 2.6.5), one can see a low iron content in the zinc layer (the brightness of the color qualitatively indicates the concentration differences as a function of location) and at the same time the formation of an intermetallic phase in the area of the junction zone between zinc and steel.

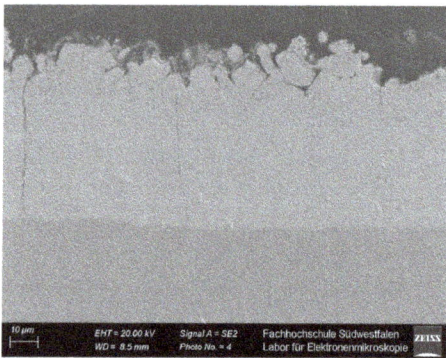

Figure 2.6.4: Zinc coating (top) on steel (bottom) after the sherardizing process.

2.6.3 Electroplated coatings

Electroplated coatings are created by polarizing a work piece in an electrolyte solution in which the metal to be deposited is present as an ion. This is a process, which is suitable for both strip and piece coating and is widely used in industrial surface technology.

Basically, the following electrochemical reactions take place:

$$Me^{z+} + ze^- \rightarrow Me.$$

The work piece is connected as a cathode and the metal ions are reduced by absorbing electrons and deposited on the surface. The cathode consists of the material to be deposited. The ions thus removed from the solution are returned by the dissolution of the anode.

Dissolution reaction at the anode:

$$Me \rightarrow Me^{z+} + ze^-.$$

This procedure is shown schematically in Figure 2.6.6.

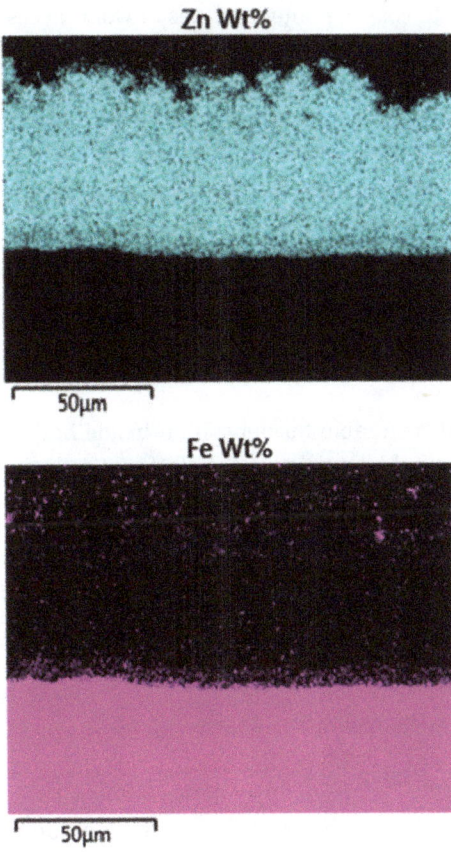

Figure 2.6.5: Elemental distribution images of zinc and iron in layers produced by sherardizing.

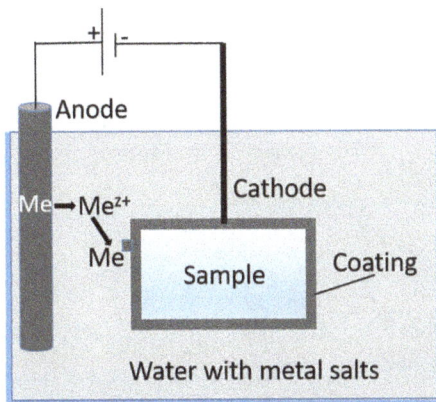

Figure 2.6.6: Schematic representation of galvanic metal deposition.

In principle, this process can be used to coat the outer surfaces of work pieces. Due to the Faraday effect, there is no polarization of the inner surface of pipes or containers. There is indeed a small amount of grip in the edge area and thus a partial coating. However, the circumference depends on the diameter of the opening and the conductivity of the electrolyte.

Such processes are often used in the sanitary industry, where high-gloss and wear-resistant coatings are also deposited on individual components. In addition, barrel plating is used to provide small parts with metallic coatings.

2.6.4 Electroless plating

Electroless plating of metal coatings is also suitable for deposition inside hollow bodies, i.e. also in places where current cannot be applied to the electrolytically deposited coatings. The deposition of metallic layers from aqueous electrolytes can also be carried out without external current. In this case, it is necessary to add a reducing agent to the bath in addition to the metal ions in order to be able to deposit a layer. Over time, the bath becomes depleted in metal ions, which then have to be added again as a metal salt.

In general, this electrochemical reaction can be described as follows:

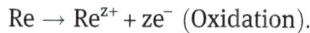

$$Me^{z+} + ze^- \rightarrow Me \text{ (Reduction)}$$

$$Re \rightarrow Re^{z+} + ze^- \text{ (Oxidation)}.$$

Re is the reducing agent which is itself oxidized while the dissolved metal ion is reduced to a metal atom.

Figure 2.6.7 shows the scheme of a coating without external current.

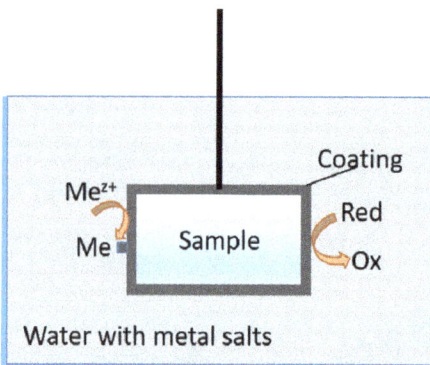

Figure 2.6.7: Schematic of the electroless deposition of metal coatings from baths.

The following reactions take place for the tin coatings deposited without external current shown in Figure 2.6.8 as an example:

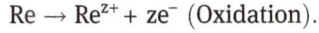

$$Sn^{2+} + 2e^- \rightarrow Sn \ (\text{Reduction})$$

$$Re \rightarrow Re^{z+} + ze^- \ (\text{Oxidation}).$$

While a relatively uniform layer can be seen at low magnification, at higher magnification individual punctual thickenings can be seen which are due to the nucleation on the surface and the locally different nucleation growth rate. Basically, the industrially used electrolytes are characterized by the fact that the deposition takes place evenly on the surface due to additives.

Figure 2.6.8: (top) Tin coating deposited without external current and (bottom) enlargement of the top figure.

2.6.5 Hot-dip coatings

Hot-dip coatings are produced on metals by dipping the component or semi-finished product into a liquid metal melt. Starting from the surface, the part to be coated is heated up to the temperature of the melt. As the temperature of the component increases, solid-state diffusion processes take place.

Figure 2.6.9 shows a schematic diagram of the hot-dip metallization process.

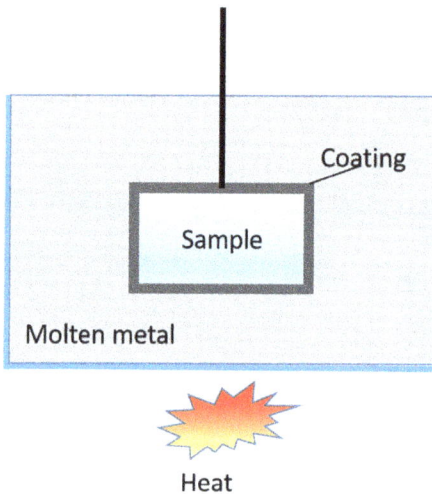

Figure 2.6.9: Schematic representation of the production of hot-dip coatings.

As a rule, the components of the melt diffuse into the work piece and form intermetallic phases. On the one hand, this leads to layers that adhere well; on the other hand, the intermetallic phases are brittle and can also flake off when subjected to mechanical stress. An example of such layers is the zinc-iron alloy layer formed during piece galvanizing, which is characterized by a sequence of η, ζ, δ_1 and Γ phases. The layer sequence can be taken from the binary Zn-Fe phase diagram. With increasing distance to the steel/zinc layer phase boundary, the iron content decreases until one has an almost iron-free pure zinc layer on the outside. Figure 2.6.10 shows a zinc layer as it is formed during piece galvanizing. The structure of the layer with the different intermetallic phases is shown.

Hot-dip coatings are widely used for galvanizing steel in corrosion protection. In addition to the process of piece galvanizing, in which the components are individually dipped into the melt on racks, strip galvanizing is widely used. In this process, the sheets are drawn through the molten metal and a metallurgical reaction takes place. Besides the use of pure zinc alloys, zinc-aluminum or zinc-aluminum-magnesium alloys are also used.

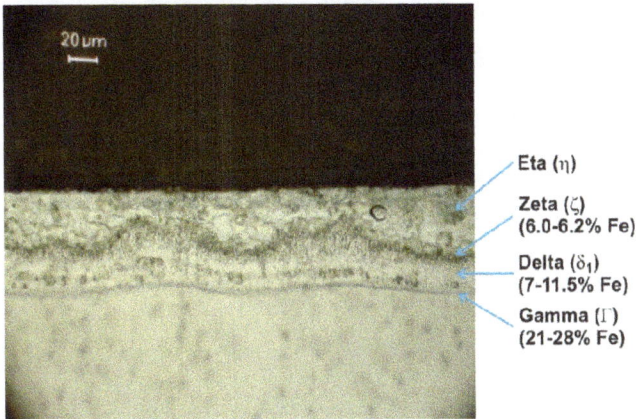

Eta (η)

Zeta (ζ)
(6.0-6.2% Fe)

Delta (δ_1)
(7-11.5% Fe)

Gamma (Γ)
(21-28% Fe)

Figure 2.6.10: Zinc coating on steel with indication of the intermetallic phases formed during galvanizing.

2.6.6 Metal spray coatings

Metal spray coatings can be applied to virtually any substrate. By liquefying the metal and atomizing it with the aid of a carrier gas or air, the metal droplets are transported onto the substrate, where they solidify and mechanically bond with the roughened substrate during cooling. Figure 2.6.11 shows the process schematically.

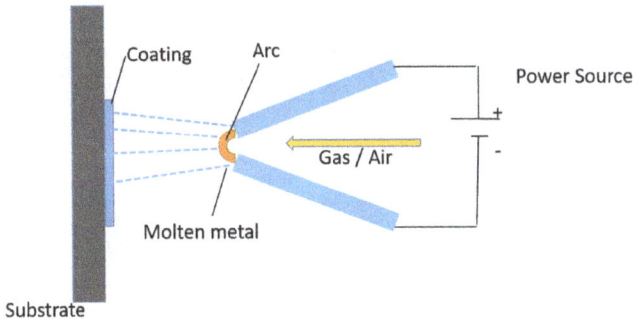

Figure 2.6.11: Schematic representation of the production of metal coatings by thermal spraying.

Interesting applications for metal spray coatings are zinc or zinc-aluminum coatings. These are mainly used for corrosion protection. In addition to the pure protective effect of the zinc in the atmosphere due to the protective layer formation, these coatings also have a cathodic protective function. This protective effect can also be applied to concrete structures, where a thermally sprayed Zn layer on the concrete surface protects the steel in the cracks of the structure. In principle, all metals can be melted and atomized. Depending on the metal, however, the melting process must be selected. Different processes can be used, ranging from melting in an electric arc to plasma processes.

Basically, it must be taken into account that metal spray coatings cannot be applied without pores due to the process [4]. An example of the porosity of the coatings is shown in Figure 2.6.12 and is caused by the application process. Droplets already solidified in flight, oxide skins or abrasives from roughening the surface provide porosity that is usually <10%. With the help of computer tomographic images, one can make the pore network visible, see Figure 2.6.13, whereby an internal oxidation of the layers can take place.

Figure 2.6.12: Spray coated ZnAl15, porosity 6–8% [4].

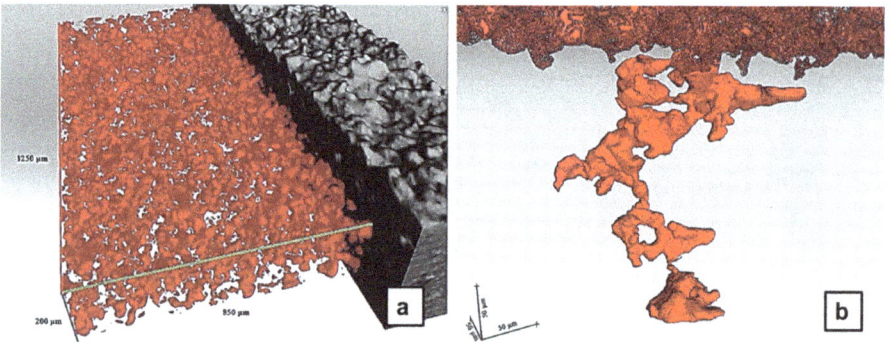

Figure 2.6.13: Three-dimensional depiction of pores inside thermally sprayed coatings, measurement resolution: 1 µm per voxel size, porosity 9%, sprayed coating ZnAl15. a) Overview of the pore structure, b) detailed view of a pore [4].

2.6.7 Plating

In cladding, there are two processes that lead to a metallurgical bond between the substrate and the cladding material. One is roll cladding and the other is explosive cladding. Through a forming process such as rolling, new, fresh surface is created, whereby the atoms from the substrate and the coating come closer to each other on an atomic level and form a bond. The principle of the process is shown in Figure 2.6.14.

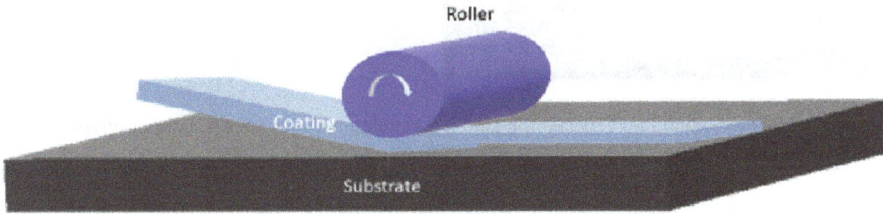

Figure 2.6.14: Roll cladding is shown schematically.

In explosive cladding, two metal plates are placed on top of each other. Through the controlled ignition of explosives on the upper side of the overlay material, the two metals are pressed together so strongly that the interface melts and a metallurgical bond is formed.

In principle, metals can also be joined together in this way where a joint would otherwise not have been possible, e.g. by hot dipping, due to the formation of brittle intermetallic phases. In roll cladding, a clad sheet is produced by rolling the substrate and the cladding material together. Clad sheets are characterized by a good adhesive strength of the cladding material and combine the properties of the substrate, which can give the strength of a low-alloyed steel at the outside of a vessel, with a good corrosion resistance of the cladding material on the inside of the same. In the following Figures 2.6.15 and 2.6.16, the phase boundary of an aluminum sheet with a copper layer on top is shown. It can be seen, that the phase boundary has a micro roughness which enhances adhesion. The elemental distribution shows a slight interdiffusion of aluminum and copper at the interface of the metals which gives the adhesion.

2.6.8 Overlay welding/brazing

Deposition welding or brazing is a way of providing metal surfaces with an additional metal layer, whereby a bond is created with the substrate by melting the filler metal or brazing alloy. In soldering, only the solder is melted and the substrate is not; in welding, both the filler metal and the base material are melted. In both cases, bonded joints result. The introduction of heat into the substrate creates heat-

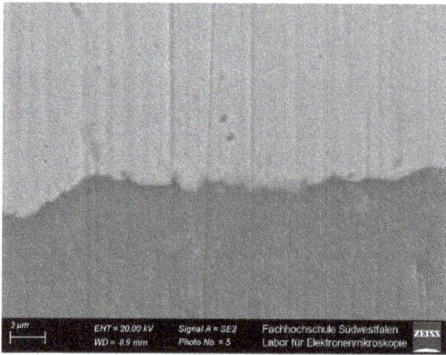

Figure 2.6.15: Phase boundary between roll-cladded aluminum/copper sheet. Copper on top, aluminum below.

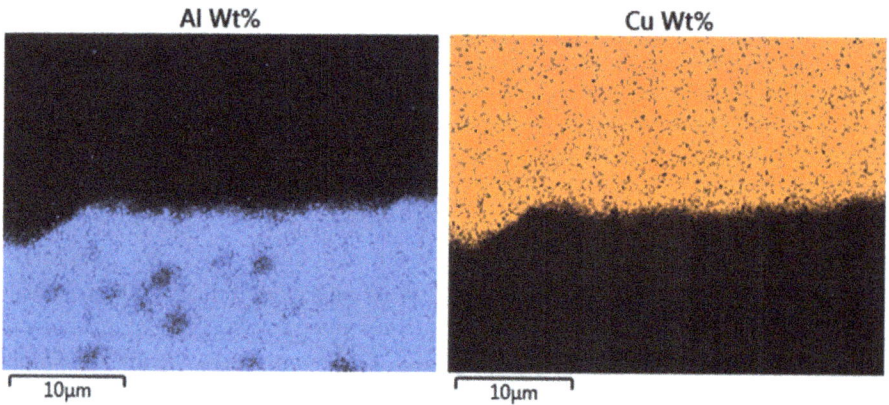

Figure 2.6.16: Element distribution at the phase boundary roll-cladded aluminum/copper sheet. Aluminum left, copper right.

affected zones (HAZ) in the substrate with the formation of different phases, which then also influence the mechanical behavior and thus also the adhesive strength. The process is shown schematically in Figure 2.6.17.

Hard facing is important for the production of wear-resistant coatings on steel surfaces. In particular, nickel-based material or hard coatings containing, for example, tungsten carbide are used. In this way, only those areas can be specifically protected against wear that are subject to particular stress.

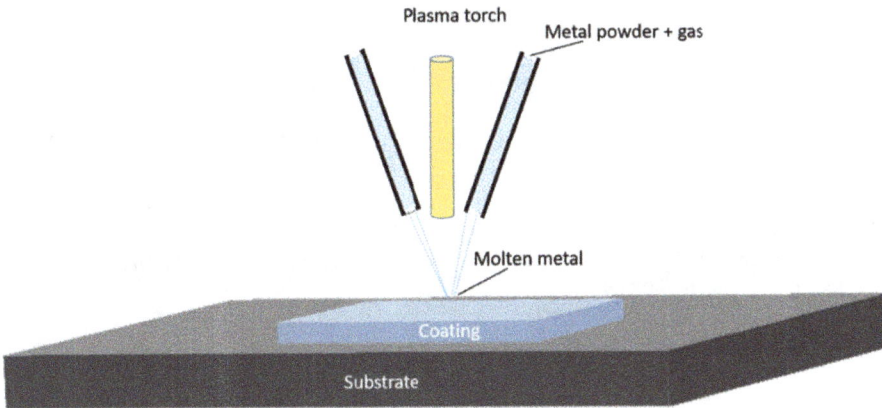

Figure 2.6.17: Schematic representation of build-up welding.

2.6.9 Summary

The application of metal coatings (Table 2.6.1) allows the production of composite materials in which the combination of different metals provides cost-effective and functional solutions for complex stresses. The development of processes for the production of these material composites is far from complete and the available alloys allow a wide variety of solutions. The functionality and durability of the layer system must be tested by means of suitable tests before industrial use in order to avoid costly repair measures.

Table 2.6.1: Selected metal coating along with their application.

Coating metal	Coating technique	Application
Ti, Cr, Mo	Vapor deposition	Corrosion protection
Zn, Cr	Diffusion coating	Sheradizing of steel for corrosion protection, wear protection
Zn, Cr, Ni	Electroplated coating	Decorative function, corrosion protection
Zn, Ni		Electroless coating, wear and corrosion protection
Zn, Al	Hot dip coating	Corrosion and oxidation protection
Zn, Mo, Ni	Metal spray coating	Corrosion protection, metal sliding layer, increase of adhesion
Ni-alloys	Plating	Corrosion protection
Nickel base alloys, stellite	Overlay welding	Wear and corrosion protection

References

[1]   Kunze E. Korrosion und Korrosionsschutz, Band 3: Korrosionsschutzverfahren, Wiley-VCH, Weinheim, 2001.

[2]   Hoffmann H, Spindler J. Verfahren in der Beschichtungs- und Oberflächentechnik, 3. Aufl., Fachbuchverlag Leipzig, 2015.

[3]   Bobzin K. Oberflächentechnik für den Maschinenbau, Wiley-VCH, Weinheim, 2013.

[4]   Mertke A, Feser R. Transport Processes in Thermally Sprayed Zinc and Zinc Aluminium Coatings considering Structural and Corrosive Influences, Proc EUROCORR 2017, Prague, Czech Republic, 02.–07.09.2017.

2.7 Be and Be alloys

Magnus Buchner, Oliver Janka

2.7.1 Metallic Be

2.7.1.1 Physical and mechanical properties

Beryllium (Figure 2.7.1), the fourth element of the periodic table, belongs to the alkaline earth elements and exhibits an electron configuration of $1s^2\ 2s^2$. With only four electrons but a filled s-orbital, Be is, especially in comparison with its neighbor Li, relatively unreactive since it forms a passivating oxide layer on the surface, hampering its reactivity. At the same time, the low electron count makes elemental Be nearly transparent for X-rays (*vide infra*). At room temperature, metallic beryllium has a density of 1.85 g cm^{-3}, a melting point of 1560 and a boiling point of 3243 K [1]. Be is dimorphic and its α-form crystallizes in the hexagonal Mg-type structure with space group $P6_3/mmc$ and lattice parameters of $a = 228.6$ and $c = 358.4$ pm ($c/a = 1.568$) [2]. β-Be exhibits a cubic body centered structure (W type, $Im\bar{3}m$, $a = 255.2$ pm) and is only stable between 1523 K and the melting point [3]. α-Be exhibits furthermore excellent mechanical properties. For example, the elastic modulus is about 1.5 times larger compared to the one of steel. The Poisson's ratio μ of α-Be is also very low (0.024–0.03), indicating that the cross section remains nearly constant under tensile strength. Additionally, high values for the specific heat (16.4 J mol^{-1} K^{-1}) and the thermal conductivity (200 W m^{-1} K^{-1}), while at the same time a low thermal expansion (11.5×10^{-6} K^{-1}), are observed [1, 4]. These properties result in an extreme stability of metallic Be at elevated temperatures.

Figure 2.7.1: Piece of polycrystalline beryllium metal.

https://doi.org/10.1515/9783110733143-014

2.7.1.2 Beryllium production

The production of elemental metallic Be was noted to be 260 metric tons in 2011 [5]. Due to its high affinity to oxygen, metallic Be does not occur in nature, and in air it immediately forms an oxide film on the surface (passivation), similar to e.g. aluminum. Therefore, naturally occurring Be sources are oxide-based minerals for example beryl ($Al_2Be_3[Si_6O_{18}]$), chrysoberyl (Al_2BeO_4) or bertrandite ($Be_4Si_2O_7(OH)_2$). Commercially, beryllium is usually extracted from beryl either via alkaline fluxes or fluoride-based reagents [6]. In the first step, $Al_2Be_3[Si_6O_{18}]$ is heated above its melting point (1923 K) followed by quenching with water. The quenched melt is treated with hot sulfuric acid, resulting in the formation of $BeSO_4$ (eq. 2.7.1). This solution is extracted with di-2-ethylhexylphosphoric acid (DEHPA) in kerosene. This organic solution is re-extracted with aqueous ammonium carbonate solution. Heating to 343 K leads to precipitation of co-extracted aluminum and magnesium, which are filtered off. Further heating to 368 K precipitates basic beryllium carbonate (2 $BeCO_3 \cdot$ $Be(OH)_2$). The basic carbonate is heated in an autoclave to 438 K to eliminate CO_2 and produce beryllium hydroxide ($Be(OH)_2$). Alternatively, beryl can be mixed with sodium hexafluoridosilicate (Na_2SiF_6) and sodium carbonate (Na_2CO_3) and heated to 1043 K. During this reaction, sodium tetrafluoridoberylate (Na_2BeF_4) is formed (eq. 2.7.2). Na_2BeF_4 can be separated from the other reaction products by dissolving it in water followed by precipitation via NaOH, again forming $Be(OH)_2$.

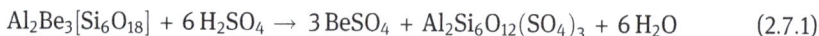

$$Al_2Be_3[Si_6O_{18}] + 6\,H_2SO_4 \rightarrow 3\,BeSO_4 + Al_2Si_6O_{12}(SO_4)_3 + 6\,H_2O \qquad (2.7.1)$$

$$Al_2Be_3[Si_6O_{18}] + 2\,Na_2SiF_6 + Na_2CO_3 \rightarrow 3\,Na_2BeF_4 + Al_2O_3 + 8\,SiO_2 + CO_2 \quad (2.7.2)$$

$Be(OH)_2$ is a suitable starting material for the production of metallic Be. This can be achieved e.g. by a carbothermic reduction of BeO (eq. 2.7.3) formed by a heat treatment of $Be(OH)_2$ in the presence of elemental chlorine, similar to the synthesis of elemental titanium (Chapter 2.3.2).

$$BeO + C + Cl_2 \rightarrow BeCl_2 + CO \qquad (2.7.3)$$

$BeCl_2$, when mixed with the same amount of NaCl, can finally be electrochemically reduced on a Ni electrode. Small Be pieces are obtained by this process, which can be melted to larger compact pieces above 1423 K.

However, the dominant industrial process for the production of beryllium metal is the Schwenzfeier process. Here, $Be(OH)_2$ is added to ammonium hydrogen difluoride to produce ammonium tetrafluoridoberyllate according to eq. (2.7.4). Subsequently $[NH_4]_2[BeF_4]$ is thermally decomposed into beryllium fluoride between 1173 and 1273 K (eq. 2.7.5). Beryllium fluoride is then chemically reduced with magnesium in graphite crucibles at 1173 K to metallic beryllium (eq. 2.7.6) [7].

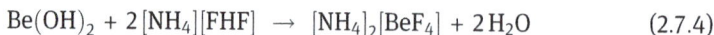

$$Be(OH)_2 + 2\,[NH_4][FHF] \rightarrow [NH_4]_2[BeF_4] + 2\,H_2O \qquad (2.7.4)$$

$$[NH_4]_2[BeF_4] \rightarrow BeF_2 + 2\,NH_3 + 2\,HF \tag{2.7.5}$$

$$BeF_2 + Mg \rightarrow Be + MgF_2 \tag{2.7.6}$$

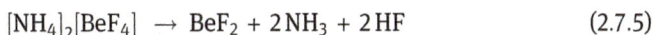

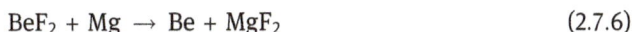

Beryllium and its compounds are regarded as toxic and carcinogenic. Therefore, thorough occupational safety measures need to be in place. To minimize exposure and to control exposure levels, administrative, engineering and medical measures are employed. Of these the most effective ones are engineering measures, like enclosure of the beryllium work, utilizing wet processes wherever possible, ensuring good ventilation and effective removal of airborne particles with high-efficiency particulate air (HEPA) filters. Additionally, personal protective equipment (PPE) is mandatory. Since the most significant exposure route is inhalation, respirators are the most effective form of PPE. On the administrative level, it is necessary to assign specific areas for beryllium work, which must be clearly labeled. In addition, the employer needs to make sure that all workers are regularly trained and that good work practice is in place. With these measures, beryllium exposure can be reduced to a safe level. To monitor this, regular medical checks of the workforce should be undertaken if tasks with high exposure risk are carried out. Generally, this is done via the beryllium lymphocyte proliferation test (BeLPT). If the BeLPT shows beryllium exposure, further and more extensive monitoring of the affected worker is carried out to assure early detection of the development of chronic beryllium disease [8].

2.7.1.3 Applications
Due to the low number of electrons in the metal (4 e⁻) and the good mechanical properties, metallic beryllium is used for several technical applications. These can be roughly grouped into four main applicational areas: 1) mechanical, 2) thermal and physical, 3) special and 4) nuclear properties [4]. The most prominent application is probably the use of metallic Be as X-ray tube window (Figure 2.7.2), since it has a very low interaction with X-ray radiation and a good stability with respect to the evacuated tube. For the same reason, beryllium is also widely used for the casings of γ-ray detectors and as material for refractory lenses for high energy X-ray beams, e.g. at the European Synchrotron Radiation Facility (ESRF) [9]. The vacuum detector tubes in particle accelerators like the large hadron collider (LHC) at CERN are built of pure beryllium due to its high permittivity to elementary particles [10]. Due to its low molar mass, beryllium efficiently scatters neutrons elastically and reduces their speed. It is therefore applied as a neutron reflector and moderator in nuclear reactors and weapons [8]. ^{9}Be, the only naturally occurring beryllium isotope, reacts in an α,n reaction according to eq. (2.7.7) and can therefore be used as a laboratory neutron source (Figure 2.7.3).

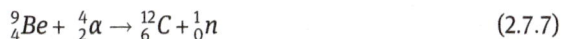

$$^9_4Be + \,^4_2\alpha \rightarrow \,^{12}_6C + \,^1_0n \tag{2.7.7}$$

Figure 2.7.2: Fine structure X-ray tube. The X-ray tube windows made of elemental Be can be seen on the right side. They are held in place by the brass-colored metal joints.

Figure 2.7.3: Solid beryllium casing for α-emitters and corresponding beryllium lid, used as a laboratory neutron source.

When irradiated with protons or X-rays beryllium emits neutrons, according to eqs. (2.7.8) and (2.7.9), the resulting neutron beam can be finely focused and therefore, this method for neutron generation is applied in cancer therapy [9].

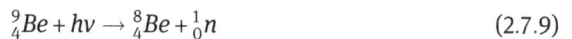

$$^{9}_{4}Be + ^{1}_{1}H \rightarrow ^{9}_{5}B + ^{1}_{0}n \tag{2.7.8}$$

$$^{9}_{4}Be + h\nu \rightarrow ^{8}_{4}Be + ^{1}_{0}n \tag{2.7.9}$$

With fast neutrons, ^{9}Be undergoes a $n,2\,n$ reaction (eq. 2.7.10) [9]. This, in combination with the excellent physical properties (high melting point, mechanical strength and thermal conductivity), makes beryllium an ideal material for the neutron breeding blanket in nuclear fusion reactors like ITER and JET [11, 12].

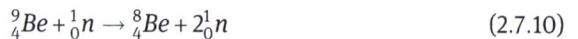

$$^{9}_{4}Be + ^{1}_{0}n \rightarrow ^{8}_{4}Be + 2^{1}_{0}n \tag{2.7.10}$$

The same properties enable its use in various parts of jet-propelled air- and space crafts or rockets and missiles such as brakes, nose shells or heat sink devices. A detailed usage on the example of NASAs Space Shuttle Orbiter is given in [13]. Another application is the use of beryllium for the fabrication of the primary, secondary and tertiary mirrors for deep space telescopes such as the James Webb Space Telescope [14]. Due to the high stability of beryllium metal at extremely high temperatures, it has also been tested as material for plasma limiters in nuclear fusion reactors (e.g. JET) [15]. Due to its low density paired with its high mechanical strength and stiffness, beryllium is also widely employed as a material for gyroscopes [16], high-end speakers [17] and for light-weight motor blocks in motor sports.

2.7.2 Be alloys

2.7.2.1 Applications
Due to the high cost of beryllium, the applications of beryllium-containing alloys are limited. However, the addition of beryllium to copper, aluminum or nickel results in a significant increase in the strength of the obtained alloy, especially when compared to the pure metals. This renders these alloys irreplaceable for many applications where the superior properties are crucial.

Beryllium copper alloys for example are widely applied in oil and gas exploration and extraction due to mechanical properties comparable to steel combined with the fact that they are non-sparking (Figure 2.7.4). The latter property is indispensable in areas with explosion hazards. Furthermore, these alloys are non-magnetic. Therefore, all tools for strong magnetic fields (e.g. magnetic resonance imaging) are made of beryllium copper alloys. Due to the high fatigue strength combined with high electric conductivity, these alloys are also widely applied in connectors and switches in consumer electronics, computers, automotive applications as well as in the medical and aerospace industries [8]. Nickel beryllium alloys show extremely high resistance against periodic temperature fluctuations even up to very high temperature and still retain their good spring characteristics. Therefore, these are widely applied as Belleville washers in fire protection sprinkler heads, thermostats, bellows and test and burn-in sockets [18]. Aluminum beryllium composites exhibit superior specific stiffness and more design flexibility compared to titanium, aluminum or other Al-composites. Their extrudability, formability and ease of machining are also good, which is in contrast to pure beryllium metal. Therefore, aluminum beryllium composites are widely used in aeronautics and space applications, where the relatively high costs are insignificant [19].

2.7.2.2 Alloys
Probably the most important beryllium-based alloys are beryllium copper (BeCu), also referred to as copper beryllium, beryllium bronze or spring copper. These

Figure 2.7.4: Tools made of non-sparking Be-Cu alloy. The tools were kindly made available by Feuerwehr Münster. Foto by Thomas Fickenscher.

alloys contain, in contrast to the e.g. aluminum-based alloys, only low amounts of the respective alloying element, in this case 0.5–4 wt% Be. They can be grouped into alloys with high strength (upper end of Be content) and high conductivity (lower end of Be content plus additional alloying elements). A single heat treatment (annealing or age hardening) leads to curing and yields the highest strength of all copper-based alloys and an electrical conductivity above the one of bronze. Due to the limited solubility of Be in Cu, in alloys with 1.6 to 2 wt% Be processed by age hardening between 500 and 700 K, the formation of the γ-phase, a Be-rich compound stable below 870 K, is observed. The γ-phase is the main contributor to the high strength due to precipitation hardening. Annealing between 980 and 1030 K leads to the dissolution of Be and the γ-phase. Quenching keeps the Be in the solid solution and therefore generates a soft and ductile material, which can be machined and subsequently age hardened. Alloy 25 for example is treated for 2 to 3 h at 590 K for maximum strength. For the highly conductive alloys, small amounts of Be (0.15–0.7 wt%) are added to the copper, along with e.g. Ni. In these alloys, the majority of the beryllium is found as intermetallic beryllium compounds. From a structural point of view, only a small amount of these intermetallics has been structurally characterized. Crystallographic data for the three cubic phases CuBe (CsCl type, $Pm\bar{3}m$) [20], $CuBe_2$ ($MgCu_2$ type, $Fm\bar{3}m$) [20] and NiBe (CsCl type, $Pm\bar{3}m$) [21] have been reported. Overall, the knowledge on Be intermetallic is rather scarce, the current state of knowledge has been summarized in two recent reviews [22, 23]. During solidification, coarse Be intermetallics are formed which hamper the grain growth during annealing. The age-hardening step, in contrast, leads to finer distributed precipitates with a significant impact on their strength. Parts made from beryllium copper alloys can be cast, forged, extruded, rolled and soldered. Besides their good machinability, these materials exhibit a good corrosion, wear and abrasion resistance, are non-magnetic and non-sparking. Usually, BeCu alloys are produced as wrought alloys. In contrast to the copper-based beryllium materials, where rather small amounts of Be are used, alloys based on the Al-Be system exhibit Be contents between 10% and 75%. It is interesting to note, that with increasing Be content, the melting point increases in an S-shape from Al (934 K) to Be (1560 K), eutectic at 3.1% Be and $T_{\mathrm{mp,eut.}} = 917$ K. Furthermore, no intermetallic compounds are found in the respective binary phase diagram [24]. Known by their trade name AlBeMet (Materion), they form metal matrix composites which combine the high strength and low density of beryllium with the fabrication and mechanical property behavior of aluminum [25]. Especially with respect to their significantly enhanced mechanical stability, their good vibrational dampening properties, their high thermal conductivity and stability as well as their reduced weight [19], these materials are used in air and space born vehicles. AlBeMet AM162 is one of the most commonly used and sold alloys. It contains 62% Be and 38% Al, leading to a density of only 2.1 g cm^{-3}.

References

[1]   Emsley J. The Elements, Clarendon Press, Oxford University Press, Oxford, New York, 1998.
[2]   Gordon P. J Appl Phys 1949, 20, 908.
[3]   Aldinger F, Jönsson S. Z Metallkd 1977, 68, 62.
[4]   Naik BG, Sivasubramanian N. Min Proc Extr Metall Rev 1994, 13, 243.
[5]   Foley NK, Jaskula BW, Piatak NM, Schulte RF. Beryllium. In: Schulz KJ, DeYoung JJH, Seal II RR, Bradley DC (Eds.) Professional Paper, Reston, VA, 2017. http://pubs.er.usgs.gov/publication/pp1802E
[6]   Hinz I, Koeber K, Kreuzbichler I, Kuhn P, Seidel A. The production of beryllium. In: Hinz I, Kugler HK, Wagner J. (Eds.) Be Beryllium: Supplement Volume A1 The element. Production, Atom, Molecules, Chemical Behavior, Toxicology, Springer Berlin Heidelberg, Berlin, Heidelberg, 1986.
[7]   Fröhlich P, Lorenz T, Martin G, Brett B, Bertau M. Angew Chem Int Ed 2017, 56, 2544.
[8]   Walsh KA. Beryllium Chemistry and Processing, ASM International, Materials Park, OH, 2009.
[9]   Everest DA. The Chemistry of Beryllium, Vol. 1, Elsevier, Ort, 1964.
[10] Veness R, Ramos D, Lepeule R, Rossi A, Schneider G, Blanchard S Installation and commissioning of vacuum systems for the LHC particle detectors, Proceedings of 23rd Particle Accelerator Conference, Vancouver, BC, Canada, 2009, https://accelconf.web.cern.ch/pac2009/index.htm.
[11] Raffray AR, Federici G, Barabash V, Pacher HD, Bartels HW, Cardella A, Jakeman R, Ioki K, Janeschitz G, Parker R, Tivey R, Wu CH. Fusion Eng Des 1997, 37, 261.
[12] International Atomic Energy Agency. Summary of the ITER Final Design Report, ITER EDA Documentation Series No. 22, IAEA, Vienna, 2001, http://www.jp-petit.org/NUCLEAIRE/ITER/ITER-EDA-DS-22.pdf.
[13] Norwood LB. J Spacecr Rockets 1985, 22, 560.
[14] Dunbar B. The "Not So Heavy Metal Video": James Webb Space Telescope's Beryllium Mirrors, 2011, https://www.nasa.gov/topics/technology/features/webb-beryllium.html, 15.06.2021.

[15] Stacey WM. Fusion: An Introduction to the Physics and Technology of Magnetic Confinement Fusion, 2nd edition, Wiley-VCH Verlag GmbH & Co. KGaA, Weinheim, 2010.

[16] Mackenzie DA. Inventing Accuracy – A Historical Sociology of Nuclear Missile Guidance, MIT Press, Cambridge, MA, 1990.

[17] Materion, Truextent Beryllium for Audio Domes, Cones and Assemblies, https://materion. com/products/beryllium-products/truextent-acoustic-beryllium, 18.06.2021.

[18] Materion, Materion Alloy 360, https://materion.com/products/high-performance-alloys /nickel-beryllium-alloy-360, 23.06.2021.

[19] Ropp RC. In: Ropp RC (Ed.) Chapter 6 – Group 13 (B, Al, Ga, in and Tl) Alkaline Earth Compounds in Encyclopedia of the Alkaline Earth Compounds, Elsevier, Amsterdam, 2013.

[20] Patskhverova LS. Sov Phys J 1969, 12, 646.

[21] Shen Y-S, Griffiths LB. Metall Trans 1970, 1, 2305.

[22] Amon A. Rare earth and actinide beryllides: Structural chemistry and physical properties. In: Pecharsky VK, Bünzli J-C (Eds.) Handbook on the Physics and Chemistry of Rare Earths, North-Holland/Elsevier, Amsterdam, Chapter 312, Vol. 59, 2021, 1–44. in https://doi.org/10.1016/ bs.hpcre.2021.04.001.

[23] Janka O, Pöttgen R. Z Naturforsch 2020, 75b, 421.

[24] Massalski TB, Okamoto H, Subramanian PR, Kacprzak L. Binary Alloy Phase Diagrams, 2nd edition, ASM International, Ohio, U.S.A., 1990.

[25] Materion, AlBeMet$^{(R)}$ Technical Data Sheet, https://materion.com/-/media/files/beryllium/al bemet-materials/maab-032albemettechnicaldatasheet.pdf, 07.07.2021.

2.8 Metals for implants and prosthesis

Detlef Behrend, Mareike Warkentin

Metals and metal alloys for medical applications have a very long history in mankind. First cases with good success with gold and ivory combinations for teeth replacements were reported by the old Egyptians [1, 2]. Some selected examples of current applications are listed in Table 2.8.1 and presented in Figures 2.8.1–2.8.12.

Table 2.8.1: Applications of metal and alloys in medicine.

Medical product/device	Metal/alloy(s)
Joint endoprosthesis	
Hip, knee, shoulder, foot	CrCoMo alloys, $TiAl_6V_4$ (316 L steel)
Heart valve cages	316 L steel, CrCoMo
Osteosynthetic systems	316 L, $TiAl_6V_4$, CrCoMo
Spinal cages	$TiAl_6V_4$, CrCoMo
Skull reconstruction plates	AgCu alloys
Pace maker and defibrillator housings	$TiAl_6V_4$
Electrodes for cochlea implants	PtIr
Ossicles	Au, Pt
Orofacial wires, springs and screws	316 L steel, CrCoMo, X5 CrNiMo18.8, $TiAl_6V_4$
Orthodontic systems	316 L steel, CrCoMo, $TiAl_6V_4$
Endodontic posts	316 L steel, CrCoMo, $TiAl_6V_4$, Au alloys
Dental filling materials	Au alloys, amalgam (AgCuSnCu)
Dental crowns and bridges	Au alloys, CrCoMo
Deep brain stimulating electrodes	PtIr, Au
Dental implants	$TiAl_6V_4$, CrCoMo, 316 L
Vascular stents	316 L steel, CrCoMo, Au, Mg and Fe alloys Shape memory alloys (NiTi)
Hernia networks	$TiAl_6V_4$

Acknowledgment: We thank Dipl.-Ing. Claudia Oehlschläger for the excellent photographic documentation of selected implant specimens.

https://doi.org/10.1515/9783110733143-015

Medical product/device	Metal/alloy(s)
Wound closing staples	316 L steel, CrCoMo, X5 CrNiMo18.8
Surgical instruments	316 L steel, CrCoMo, X5 CrNiMo18.8, $TiAl_6V_4$

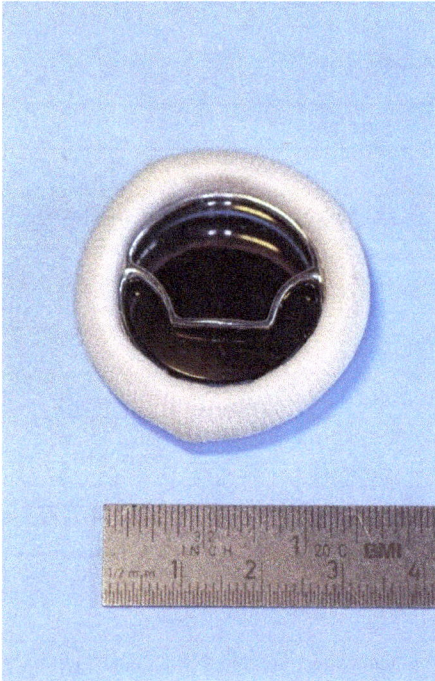

Figure 2.8.1: Cage of an artificial heart valve (Co-Ni-Cr-W alloy) of the Björk-Shiley type, disc is made from pyrolytic carbon, white material is a woven suture ring made of PET.

In comparison to other medical materials, metals own a high strength, but on the other hand tend to corrosion, especially tribocorrosion, in electrolytes with a very high corrosion potential like blood and other body fluids. Blood owns a corrosion capacity approximately three times higher than sea water.

In particular, heavy metal ions (Cr, Co, Nb, Ta) in higher blood plasma concentration may result at first to slight sensibility and later strong allergic reactions [6–8]. That, in combination with wear particles, initiates an osteolytic cascade and leads to aseptic relaxation or displacement of mechanical cyclic loaded prosthesis.

Figure 2.8.2: Pacemaker (VEB Transformatoren- und Röntgenwerk, Germany) – body (black) is made of TiAl$_6$V$_4$ and the header is made of epoxy resin.

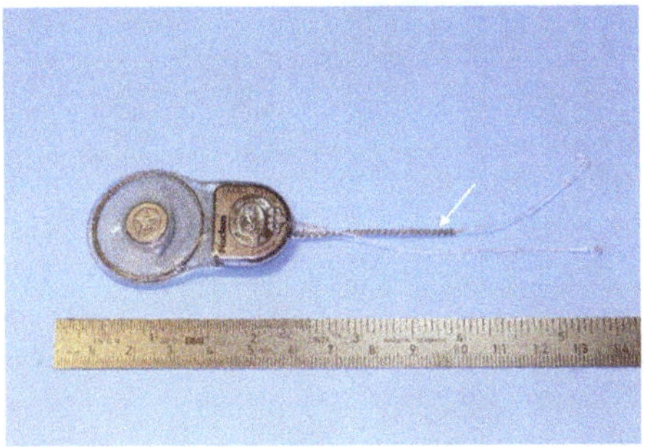

Figure 2.8.3: Cochlear^{TM} Nucleus® implant system CI00000 (Cochlear^{TM}, Australia) – cochlear electrode (Ø 200 µm) in a silicon housing (white arrow), rings made from PtIr.

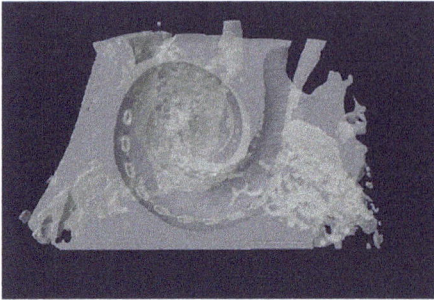

Figure 2.8.4: Micro CT of an implanted experimental endosteal cochlear electrode in a human temporal bone (semitransparent).

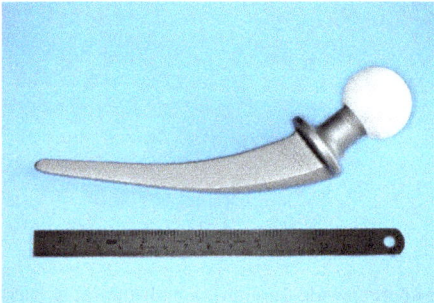

Figure 2.8.5: Artificial hip prosthesis (KERAMED 04 Standard, 135 mm, CS3711, Keramed Medizintechnik GmbH, Germany) made of $TiAl_6V_4$ with corresponding ball head, which is made from ZrO_2 (KERAMED R 5755 M).

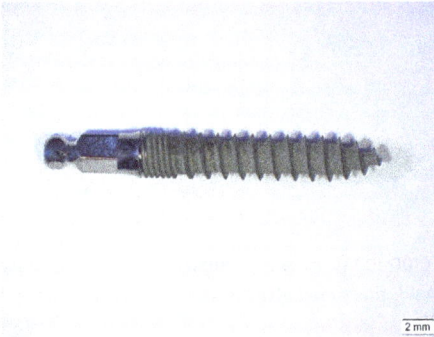

Figure 2.8.6: Ball head dental implant Champion® (Champions-Implants GmbH, Germany) made of Titanium Grade 4.

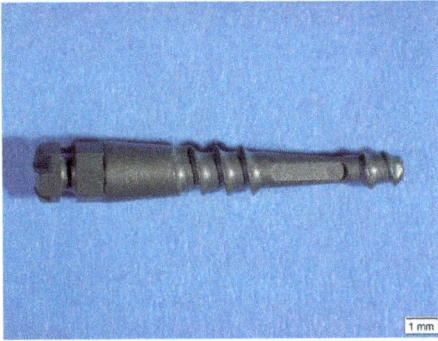

Figure 2.8.7: Experimental orthodontic mini screw (TiAl$_6$V$_4$) – a construction with optimized primary stability but less atraumatic risk for the jawbone.

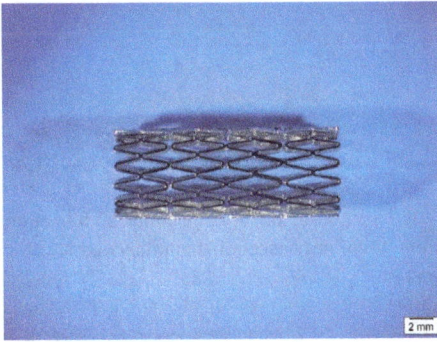

Figure 2.8.8: Balloon expanding cardiovascular stent – Tenax® (BIOTRONIK, Germany) made of 316 L with a a-SiC: H coating.

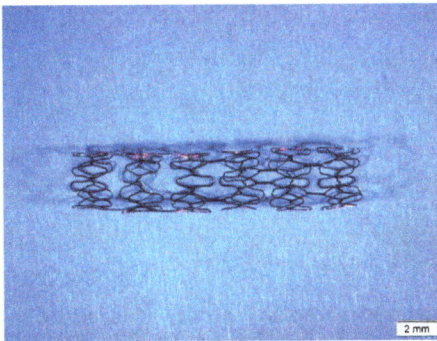

Figure 2.8.9: Self-expanding cardiovascular stent (NiTi Alloy, BIOTRONIK, Germany).

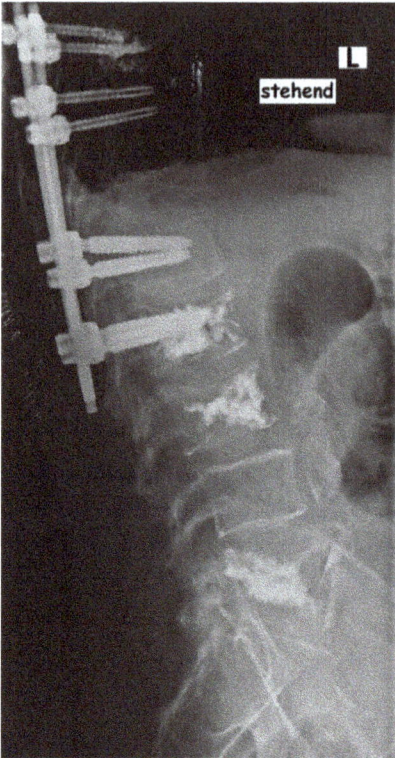

Figure 2.8.10: X-ray picture of a spinal stabilization system (titanium, Aesculap Implant Systems, Germany).

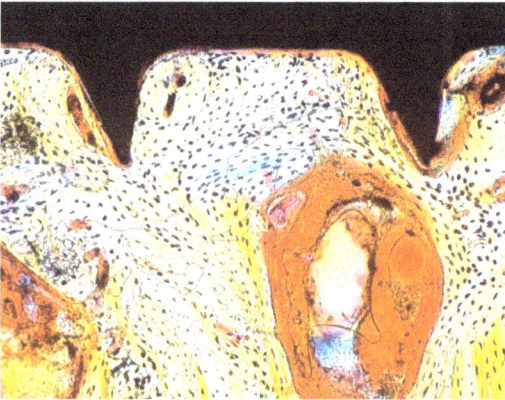

Figure 2.8.11: Thin polished microscopic slice of a canine jawbone with implanted orthodontic mini screw (LOMAS 1.4 × 7 mm, MONDEAL Medical Systems GmbH, Germany) and histologic differentiation (Giemsa and toluidine blue staining) of the hard tissue.

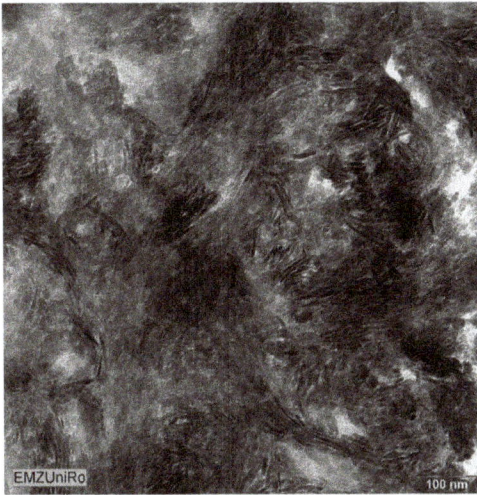

Figure 2.8.12: TEM Micrograph of a dentine section – hydroxyapatite nanocrystals (dark) in a collagen matrix.

Another problem is the long-term biocompatibility. This concerns surgical steel 316 L, e.g. X2 CrNiMo 18.10 with a content of 12.5 wt% of nickel, memory or superelastic alloys with even 50 wt%. This includes a high danger of systemic nickel allergic reactions.

One of the most interesting phenomena is the long-term biostability of titanium and titanium alloys in vivo. Titanium and titanium alloys exhibit low standard electrode potentials (−1.21/−1.77 V) and within microseconds (VROMAN effect [3]) they form TiO_2 on their surface.

Combined with perovskite ($CaTiO_3$) on the titanium surface and the interface with bone, these implants lead to an excellent osseointegration (Figure 2.8.10), because the osteoblastic cells recognize similar properties like the natural mineral hydroxyapatite ($Ca_5(PO_4)_3[OH]$) as nanocrystals, embedded in the collagen matrix of bones (Figure 2.8.12) [4, 5].

Bone remodeling is additionally promoted by the piezoelectric behavior of both perovskite and hydroxyapatite [5].

In future, further developments in the area of bioresorbable metal alloys based on magnesium and iron own a high potential for temporary implants.

References

[1]   Medicina antiqua: Codex Vindobonensis 93, Akad. Verlagsanstalt Graz, 1996. ISBN 3-201
      01659-4.
[2]   Rüster D. Alte Chirurgie, Verlag Gesundheit, Berlin, 1999. ISBN 3-333010291.
[3]   Vroman L. J Colloid Interface Sci, 1986, 111, 391–401.
[4]   Rott GA, Zhang F, Schlichting J, Haba Y, Kröger W, Burkel E. Dielectric properties of porous
      calcium titanate (CaTiO$_3$). Biomed Technol, 2010, Suppl 1, 55. DOI: 10.1515/BMT.2010.546
[5]   Wirthmann AJA, Paulmann C. Wechselwirkung von Knochen und Titan – Neue Einblicke in die
      ungewöhnliche „Hochzeit" von Knochen und Metall. Z Zahnaerztl Implantol, 2014, 30,
      288–300.
[6]   Cheng X, Dirmeier SC, Haßelt S, Baur-Melnyk A, Kretzer JP, Bader R, Utzschneider S, Paulus
      AC. Biological reactions to metal particles and ions in the synovial layer of mice. Materials,
      2020, 26(13), 1044. DOI: org/10.3390/ma13051044.
[7]   Markhoff J, Grabow N. Wear and corrosion in medical applications. Curr Dir Biomed Eng,
      2020, 6, 20203112.
[8]   Hembus J, Ambellan F, Zachow S, Bader R. Establishment of a rolling-sliding test bench to
      analyze abrasive wear propagation of different bearing materials for knee implants. Appl Sci,
      2021, 11, 1886. DOI: 10.3390/app11041886

2.9 Precious metals

Egbert Lox, Oliver Niehaus

Precious metals consist of the platinum group metals, gold and silver. Platinum group metals (PGMs) themselves consist of six silver-white metals: platinum, palladium, rhodium, ruthenium, iridium and osmium. They were discovered thousands of years after gold (4600 BC) and silver (3000 BC) and they are in the meantime indispensable in many industrial applications.

2.9.1 Primary production and recycling

Today, gold is mined in more than 100 countries and most of it is used for jewelry and investment. Only a minor fraction, about 10–15% is used in industrial and technological applications where its corrosion resistance, good electrical conductivity and good reflectivity for infrared light are an asset. Examples of such applications are mobile phones, computer motherboards, electric connectors and space-related applications such as coating of satellite components, mirrors of space telescopes and visors of astronaut helmets.

Silver is found all over the earth either in the elemental form, as an alloy or in some minerals. Nowadays silver is mainly produced as a byproduct of copper, gold, lead and zinc refining. Around half of the amount of silver mined is used in the form of investments, silverware and jewelry. The other half is used in applications in which its key properties of sensitivity to light, low contact resistance, the high electrical and thermal conductivity and anti-microbial effects are utilized. Examples of these applications are solar panels, rapid charging stations, circuit boards, some types of batteries, photography and water purification. The unique properties of silver and silver compounds make them essential in areas such as sustainable energy technologies and healthcare.

The six platinum group metals (PGMs) occur together in nature alongside nickel and copper. Mineable deposits of PGMs are very rare, with annual production amounting to around 450 tons, several orders of magnitude lower than many common metals (see Figure 2.9.1). Due to their economic value and higher quantities, platinum and palladium are the most important metals of the PGMs. The other four, rhodium, ruthenium, iridium and osmium, are mined as by-products of platinum and palladium.

Primary production of PGMs represents the transfer of metal from below ground resource to above ground material stock and happens mainly in South Africa, Russia and North America. Extraction, concentration and refining of PGMs require complex, costly and energy-intensive processes that may take several months to produce the final metal from the first time PGM-bearing ore is broken in a mine. For example, in South Africa, PGM-bearing ores generally have a low PGM content between 2 and 6

https://doi.org/10.1515/9783110733143-016

Weight of metals (in tonnes) mined world-wide in 2017

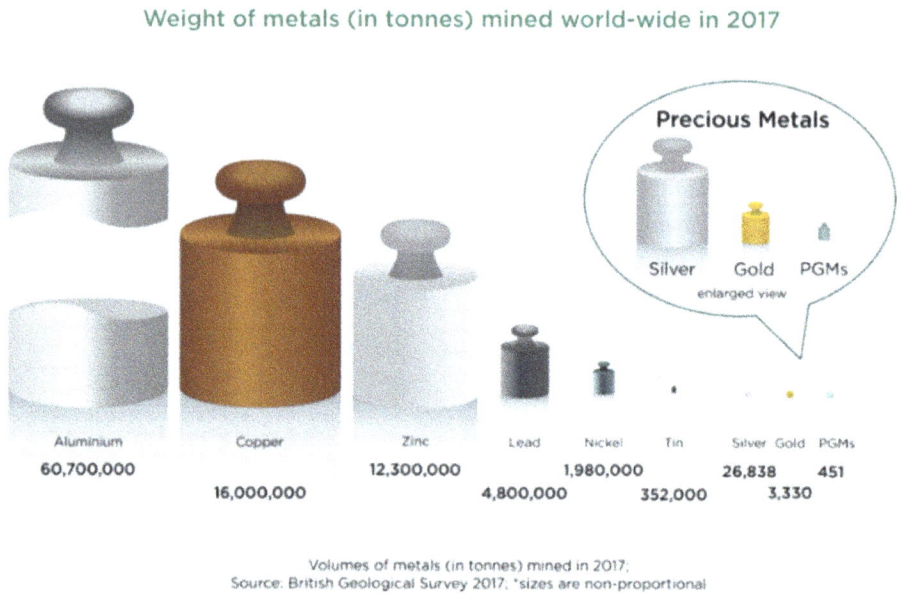

Precious Metals

Silver   Gold   PGMs
enlarged view

Aluminium	Copper	Zinc	Lead	Nickel	Tin	Silver	Gold	PGMs
60,700,000	16,000,000	12,300,000	4,800,000	1,980,000	352,000	26,838	3,330	451

Volumes of metals (in tonnes) mined in 2017;
Source: British Geological Survey 2017. *sizes are non-proportional

Figure 2.9.1: Annual primary production of precious metals compared to selected other technology metals (data for 2017, Source: IPA and British Geological Survey [1]).

grams per ton and it will usually take up to six months and between 10 and 40 tons of ore to produce one troy ounce of platinum (31.1035 g). PGM ore is mainly mined underground and to a minor extend in open-cast pits. It is blasted out of the ground; on surface, the ore is crushed and milled into fine particles. Wet chemical treatment known as froth flotation produces a concentrate which is dried and smelted in an electric furnace at temperatures above 1500 °C. A matte containing the valuable metals is transferred to converters to remove iron and sulfur. PGMs are then separated from the main base metals nickel and copper refined to a high level of purity using a combination of typically solvent extraction, electrochemical and ion-exchange techniques (see Figure 2.9.2).

The secondary production of PGMs includes the recycling of these metals from industrial applications and end-of-life products, as well as the recovery of metals from by-products and residues created in primary production (see Figure 2.9.3). Secondary production plays an important role in lowering the environmental footprint of global PGM production and contributes to a significant part, typically around 30% in the case of platinum, palladium and rhodium. The usage of PGMs is always related to high costs in comparison to alternative materials and is justified by their outstanding properties (see chapters below). Nonetheless, especially in industrial processes, recycling is economically crucial to justify the usage of PGMs. Secondary production is composed of two processes. First, PGM-containing materials are either smelted to form a molten metal matte or dissolved to bring the PGMs into a solution. Second, the

ENERGY | Ore Mining Operations

Mine Waste Rock

INPUTS
PROCESSES
WASTES

Ore

ENERGY

Ore Comminution

Ore Slurry

ENERGY
WATER
CHEMICALS | Flotation

Tailings

Concentrate

Emitted to air via stack at legislated levels

ENERGY
ADDITIVES | Smelting

Gas Cleaning — Sulphur Dioxide — Sulphuric Acid (used in the fertilizer industry)

Furnace Matte

ENERGY | Converting

CHLORINE ENERGY WATER CHEMICALS

ENERGY CHEMICALS

Refined PGMs

Converter Matte

ENERGY
CHEMICALS
SPENT
ELECTROLYTE | Leaching

Precious Metals Refining Leaching — Purification — Melting

Leach Solution Cu/Ni containing

Effluents Residues to recycling/tolling

ENERGY
CHEMICALS | Purification

Gas Cleaning

ENERGY

Ni/Cu/Co
Sodium Sulphate
Ammonium Sulphate
Effluents

Emissions to air

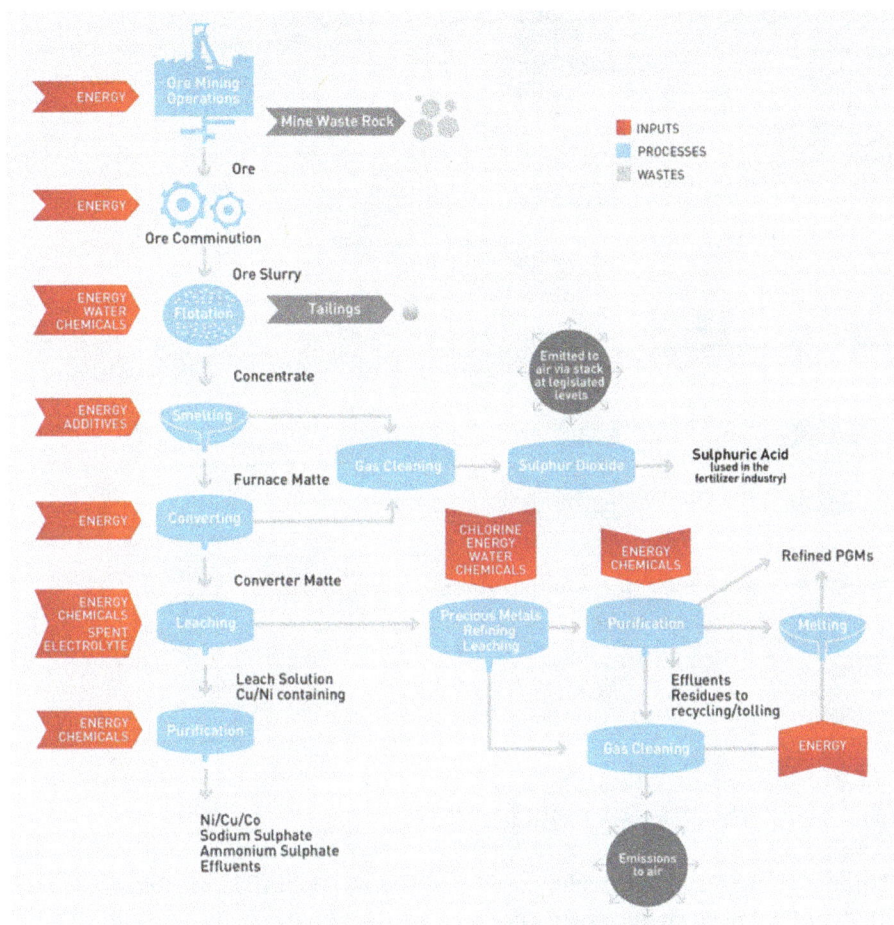

Figure 2.9.2: Generic flow chart for PGM primary production in a major metals refinery in South Africa (Source: IPA and Sibanye Stillwater Ltd [1, 2]).

PGM-enriched output from step one is then refined to recover the individual metals separately in a pure form identical to the output from primary production. PGM materials that are collected at a very high rate or enter secondary production directly from industrial processes, such as industrial catalysts, are recycled at a rate near the 95% maximum potential (close-loop recycling). For PGMs in consumer products, however, end-of-life recycling rates are lower, as high recovery requires an efficient recycling chain from collection to refining (open-loop recycling) [1–3, 16].

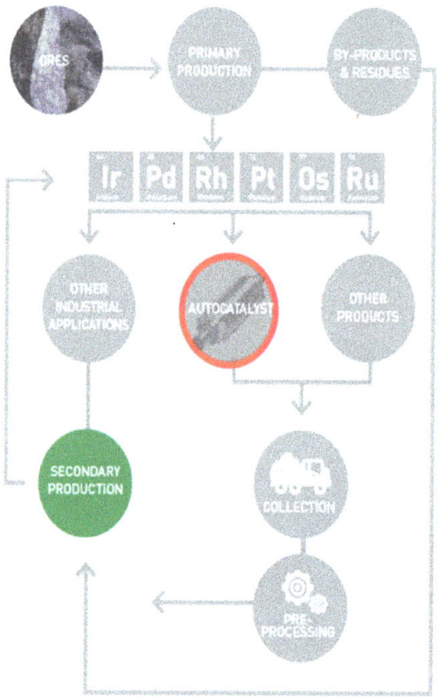

Figure 2.9.3: Secondary production in the PGM life cycle (Source: IPA [1]).

2.9.2 Main applications in industry

Platinum group metals are widely used in different application and segments (see Table 2.9.1). Typical characteristics of PGMs are their high density (especially Os, Ir, Pt), their high melting points and temperature stability, a great corrosion resistance, an excellent conductivity and oxidation resistance. All of them show excellent catalytical properties, especially as oxidation catalysts (see Table 2.9.2). In this chapter, a general overview about the industrial applications will be provided before deep diving into some major application areas.

The largest demand sector for PGMs is by far catalytic converters used in automotive applications (cars, trucks, motorcycles and off-road machinery). In 2020, those applications consumed almost 31% of the gross world demand for platinum, 85% for palladium and 93% for rhodium [13] (for further details, see Chapter 2.9.3).

The excellent catalytic properties of PGMs are additionally used in several industrial processes. Platinum gauzes often alloyed with Rh catalyze the production of nitric acid, and in the pharmaceutical industry it is used as a selective hydrogenation catalyst. Palladium acts as a catalyst in the petrochemical industry, in the production

Table 2.9.1: Summary of application fields of platinum group metals [4].

Fields of PGM applications	Pt	Pd	Rh	Ru	Ir	Os
Jewelry, decoration	▲	▲	■	■		
(Petro)chemistry	▲	▲	▲			
Pharmaceutical	■			■		
Medical engineering, dental	■	▲				■
Measurement engineering	■		■			
Glass industry, coating	▲	■	▲			
Electronic, electrical engineering	■	▲	■	▲		
Catalysis	▲	▲	▲	▲	■	■
Fuel cells, H$_2$ electrolysis	▲			■	▲	
Photovoltaics				■		
Superalloy	■			▲	■	

▲ Main applications or high significance in future
■ Additional applications

Table 2.9.2: Summary of physical properties of platinum group metals [5].

Properties	Pt	Pd	Rh	Ru	Ir	Os
Density (20 °C), g/cm^3	21.5	12.0	12.4	12.5	22.7	22.6
Mechanical hardness, HV, kg/mm^2	48	50	130	250–500	200	300–680
Tensile strength, MPa	125	190	420	500	500	
Melting point, °C	1772	1554	1966	2310	2410	3045
Thermal conductivity (20 °C), Wm^{-1}K^{-1}	73	75	150	117	147	87
Spec. electrical resistance (0 °C), µΩ cm	9.83	9.73	4.34	6.7	4.7	8.2

of terephthalic acid and in the purification of hydrogen peroxide. Many complex rhodium and ruthenium compounds have been developed as homogeneous catalysts, especially in the organic-chemical industry. Anodes for the electrolytic production of chlorine can be coated with ruthenium oxide, sometimes with a Pt-Ir alloy and more recently with an Ir-Ru alloy. Ruthenium catalysts can also be used to remove H$_2$S in oil refining, or ruthenium salts catalyze at low temperatures the production of ammonia (for further details see, Chapters 2.9.5 and 2.9.6).

Another high demand – even so not considered being industrially driven – is caused by jewelry, especially for gold and silver, but also roughly 30% of platinum is consumed in jewelry [1, 3]. Other PGMs are additionally used as alloying agent, e.g. palladium in white gold and ruthenium to impact hardness. In general, PGMs are often alloyed with each other or serve as an alloying agent to customize technical properties. A prominent example is the Pt/Ir alloy with a mass ratio of 9 to 1, which has an extremely low thermal expansion coefficient and is therefore be selected as standard meter.

The public interest in environmental protection as well as the economic importance (e.g. ESG criteria (environment, social and governance)) has increased significantly within the last years. PGMs are key contributors to this in multiple ways, beyond their application in exhaust depollution catalysts.

Driven by the change to become a CO_2-poorer society, the usage of hydrogen in multiple ways is expected to drastically increase. In fuel cells Pt/(Ir), electrodes will be widely used, while for the electrolysis of water, platinum is technically the best material, but for commercial reasons non-PGMs materials are also used (see Chapter 2.9.4). Potentially, the most remarkable property of palladium is its ability to absorb 900 times its own volume of hydrogen at room temperature, making palladium an efficient and safe hydrogen storage medium and purifier. As a thin membrane, palladium will allow hydrogen to permeate through, but block all other gases. Due to a better manufacturing ability, alloys of palladium with silver are used for those membranes, which are relevant in different chemical processes that require hydrogen purification. The use of ruthenium in alloys for aircraft turbine blades can reduce the CO_2 impact of air travel as their high melting point will allow for higher operational temperatures and, therefore, a more efficient burning of aircraft fuel.

PGMs make additionally a major contribution to waste reduction due to their high recyclability. Furthermore, their high melting point, high corrosion resistance as well as high density make goods containing PGMs more reliable and longer lasting. Ruthenium is a useful addition to reduce abrasion in electrical contact surfaces. In combination with its good conductivity, osmium (comparable to Pt and Pd) is a good alternative to gold as plating in electronic products in case a more durable product is required.

The abovementioned properties qualify PGMs also for products that must resist special conditions like required for laboratory ware or in the glass production. Platinum and rhodium, often alloyed, are used as a drawing die material in the fabrication of high-grade glass (e.g. for computer screens). Rhodium is also key to other industrial processes like the manufacture of thermocouples and laboratory crucibles. Electrodeposition of rhodium gives extremely hard coatings used in the manufacture of mirrors for optical instruments. Iridium is the most corrosion-resistant element and has the second highest melting point [7]. Consequently, it is used in high-temperature applications like iridium linings and fabrications in the glass industry because the metal stays inert in contact with molten glass. Another example is crucibles that are used to grow synthetic single crystals for LED and data-storage technologies. Osmium is roughly ten times harder than platinum and therefore it is predestined for applications in which frictional wear is a critical parameter. Most prominent examples are fountain pen nibs, styluses, needles and instrument pivots.

An excellent electric conductivity is another prominent property of PGMs lead-ing to important applications in the fields of electricity and electronics. Platinum alloy coatings are used in computer hard discs. Palladium can be a more effective and durable plating than gold in electronic components and is processed in multi-layer ceramic capacitors (MLCC), used in computers and mobile telephones. In-creasingly, ruthenium is being used in computer hard discs to increase the density at which data is stored. Ru-Pt and Ru-Pd alloys are used in electrical contacts for thermostats, relays and some hard disk drives.

Due to high corrosion resistance and low reactivity as elements, PGMs are in general considered as biologically compatible and used in multiple medical appli-cations. Platinum, often hardened by other PGMs like iridium, is used in pace-makers, medical probes and for other similar cases. Most amazingly, platinum compounds can inhibit the growth of cancerous cells. Since biological compatibility of PGMs has been mentioned, it should be stated that certain platinum salts are harmful. Iridium is as well widely used in health technologies combating cancer (light-activated cancer therapy), Parkinson's disease, heart conditions and even deafness and blindness [1, 4, 6, 16].

2.9.3 Catalytic converters

Burning gasoline or diesel in internal combustion engines results in exhaust emis-sions of mainly nitrogen (N_2), carbon dioxide (CO_2) and water vapor (H_2O), along with small quantities of the so-called criterion pollutants hydrocarbons (HC) origi-nating from unburnt and partially burnt fuel, carbon monoxide (CO) and particu-lates together with oxides of nitrogen (NO_x) from the (undesired) oxidation of the nitrogen in the air. The tailpipe emission of these criterion pollutants is subject of stringent emission legislations worldwide, which can be achieved by the incorpo-ration of catalysts, adsorbers and filters in the powertrain system, along with vari-ous exhaust gas composition sensors and electronic engine management devices.

The catalytic converter consists in most cases of a substrate (flow through or fil-ter) with a catalytic coating called washcoat on top or inside of the substrate walls. The nanotechnology used in catalytic coatings involves stabilized crystallites, wash-coat materials that maintain high surface area up to temperatures around 1000 °C (in gasoline applications), specifically designed PGM particles, improved oxygen storage components and novel coating processes to optimize the distribution of the coatings. A more fundamental chemical analysis of materials in automotive catalysts can be found in [8]. Catalysts generally need the exhaust line to heat up to a suitable operat-ing temperature, but with modern systems, appropriate system layout, thermal man-agement and optimized washcoat formulations this is reached within seconds after starting the engine. Various catalyst technologies exist for different engines and ap-plication types as shown in Table 2.9.3.

Oxidation catalysts are the original type of catalysts and were used from the mid-1970s for gasoline cars until ultimately replaced by three-way catalysts. They convert carbon monoxide (CO) and hydrocarbons (HC) to carbon dioxide (CO_2) and water but have little effect on nitrogen oxides (NO_x). Typically, platinum has been used for those catalysts because the high presence of catalysts poisons, such as sulfur and residual lead, would have caused a fast deactivation of palladium. Diesel oxidation catalysts (DOC) are nowadays a key technology for diesel engines where the high oxygen content of the exhaust precludes the use of three-way catalysts. These diesel oxidation catalysts (DOC) convert CO and HC but also decrease the mass of diesel particulate emissions by oxidizing some of the hydrocarbons that are adsorbed onto the carbon particulates. Platinum, palladium or a mixture of both are used depending on the maximum exhaust gas temperature and the required poisoning resistance. DOCs containing more Pt have a lower light-off temperature, however, a higher Pd content improves the temperature resistance.

Three-way catalysts (TWC) are the main technology used to control emissions from positive ignition engines, for example: gasoline, natural gas and liquified petroleum gas (LPG) engines. Their washcoats incorporate alumina, ceria and other inorganic oxides supporting combinations of the precious metals platinum, palladium and rhodium. Platinum and palladium mainly support the oxidation of carbon monoxide and hydrocarbons (Pd being the best catalytic converter for unburnt hydrocarbons under operating conditions of TWCs), whereas rhodium mainly catalyzes the reactions allowing the reduction of nitrogen oxides (see Figure 2.9.4). Three-way catalysts operate in a closed-loop system including an oxygen sensor to regulate the air-to-fuel ratio on these engines. The catalyst can then simultaneously oxidize carbon monoxide and hydrocarbons to carbon dioxide and water, while reducing NO_x to nitrogen and water or CO_2 [6].

Thermally durable catalysts with stability up to around 1000 °C exhaust gas temperature allow the catalytic converter to be mounted close to the engine. Advanced Ceria-based oxygen storage components stabilize the surface area of the washcoat, maximize the air-to-fuel "window" for three-way operation and help the oxygen sensors to indicate the "health" of the catalyst for on-board diagnostic (OBD) systems.

Diesel engines and lean-burn direct injection gasoline engines cannot apply conventional three-way catalyst technology to reduce NO_x. One performant option for this function is the selective catalytic reduction (SCR) technology, which was originally introduced on stationary power plants and stationary engines. Nowadays SCR is a key technology for all diesel applications. In a diesel after-treatment system, as shown in Figure 2.9.5, ammonia is used as a selective reductant, in the presence of excess oxygen, to convert up to 99% of NO and NO_2 to nitrogen. Different precursors of ammonia can be used; the most common option is a solution of 32.5% urea in water (e.g. AdBlue®) carefully metered from a separate tank and sprayed into the exhaust system where it hydrolyses into ammonia ahead of the SCR catalyst. Important for the efficiency of the SCR reaction is the NO/NO_2 ratio, a sufficient amount of NO_2 is provided by the oxidation catalysts upstream of the SCR catalysts.

Table 2.9.3: Summary of application fields of exhaust emission control catalysts [4, 6, 7].

Technology	Light duty		Heavy duty		NRMM	Motorcycle
	Diesel	Petrol	Diesel	Gas	Mostly diesel	Mostly petrol
Oxidation catalyst (OC)						▲
Three-way catalysts (TWC)		▲		▲		▲
Gasoline particulate filter (GPF)		▲		▲		
NO_x storage catalyst, lean NO_x trap (NSC, LNT)	▲	▲				
Diesel oxidation catalyst (DOC)	▲		▲		▲	
Diesel particulate filter (DPF)	▲		▲		▲	
Selective catalytic reduction (SCR)	▲		▲		▲	
Ammonia slip catalyst (ASC)	▲		▲		▲	

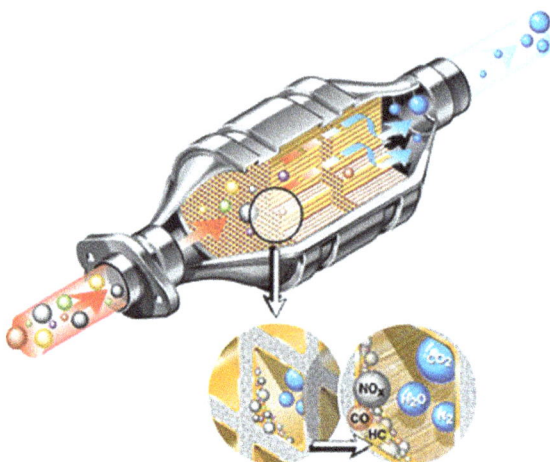

Figure 2.9.4: Illustration of a three-way catalyst in a canning and washcoat distribution on the surface of a flow-through substrate. The TWC removes NO_x, CO and HC, which may be accompanied by a gasoline particulate filter (GPF) (Source: Umicore) [8].

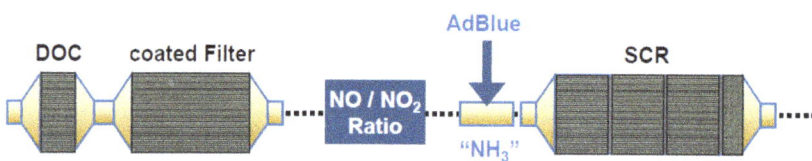

Figure 2.9.5: Basic configuration of a diesel after-treatment system. Function principle of the selective catalytic reduction technology to eliminate NO_x in lean burn applications (Source: Umicore).

The active components in SCR catalysts are mainly copper and vanadium, as well as to a lower extent Fe. In order to achieve highest NO_x conversion rates without exposing the environment with excess NH_3, SCRs are mostly followed by ammonia slip catalysts (ASCs), which are from a simplified perspective a combination of a DOC and a SCR catalyst.

A lesser-performant option to reduce NO_x is the NO_x adsorber, also known as lean NO_x trap (LNT) or NO_x storage catalyst (NSC), which operates in two modes. In the first one, during several minutes of lean exhaust gas compositions, it catalyzes the conversion of nitric oxide (NO) to nitrogen dioxide (NO_2) over the platinum part of the catalyst so that the NO_2 can be rapidly stored as nitrate on alkaline earth oxides. In the second mode, a brief return to rich operation for one or two seconds is enough to desorb the stored NO_x and convert them to N_2 over the incorporated conventional three-way catalyst using platinum and rhodium.

The catalyst functions that remove the gaseous criterion pollutants can be completed and combined with particulate filters to remove the fine particulate matter

(PM). Both the diesel particulate filters (DPF) – generally mounted between the DOC and the SCR/ASC catalysts – as well as the gasoline particulate filters (GPF) are based upon the wall-flow filtration mechanism, in which particulate matter is removed from the exhaust by physical filtration using a honeycomb structure similar to a flow-through substrate but with the channels blocked at alternate ends. The exhaust gas is thus forced to flow through porous walls between the channels and the particulate matter is deposited as a soot cake on the walls. Such filters are made of ceramic (cordierite, silicon carbide or aluminum titanate) materials. Ceramic wall flow filters almost completely remove the carbon particulates, including fine particulates of less than 100 nanometers (nm) diameter with an efficiency >95% in mass and >99% in number over a wide range of engine operating conditions. Dedicated washcoats improve additionally the filtration of smaller particles, which are essential for modern filter technologies (mainly GPF) to comply with the most stringent emission legislations [6, 9, 10].

2.9.4 Fuel cells and electrolyzers

Fuel cells are devices that convert hydrogen and oxygen into water to generate electricity. There exist several types of fuel cells but only one of them, called polymer electrolyte membrane or Proton Exchange Membrane (PEM) fuel cells, uses some platinum group metals. This type of fuel cell has the advantage to work at a relatively low temperature; it also quickly starts working and is robust with limited corrosion thanks to the solid electrolyte used. The applications range from portable power generation to transportation and personal mobility.

As shown in Figure 2.9.6, the PEM fuel cell consists of a membrane electrode assembly (MEA), which includes the polymer electrolyte membrane, the cathode and the anode at each side of the membrane respectively and gas diffusion layers (GDLs) on top of them.

The catalyst layers include nanometer-sized particles of mostly platinum dispersed on a high-surface-area carbon support, as shown in Figure 2.9.7. Sometimes, cobalt and nickel are also included, as well as supported iridium oxide. For those PEM fuel cells that would be used in vehicles for personal mobility, the total platinum amount used ranges from 60 to 25 g/vehicle, with a long-term target to further reduce this to about 10 g/vehicle. These catalyst particles are mixed with an ion-conducting polymer and coated on the membrane in a thin layer, as also shown in Figure 2.9.7. On top of these catalyst coatings, a microporous carbon and polymer containing gas diffusion layer (GDL) is applied to manage the transport of the reactants and products as well as the retention of some water, which is needed to maintain the conductivity of the membrane.

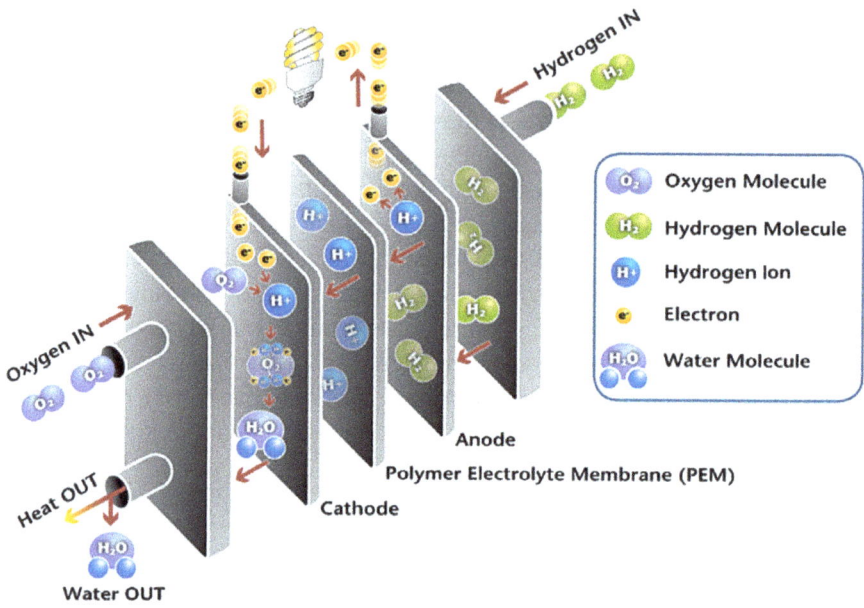

Figure 2.9.6: Schematic buildup of a PEM fuel cell (Source: Umicore).

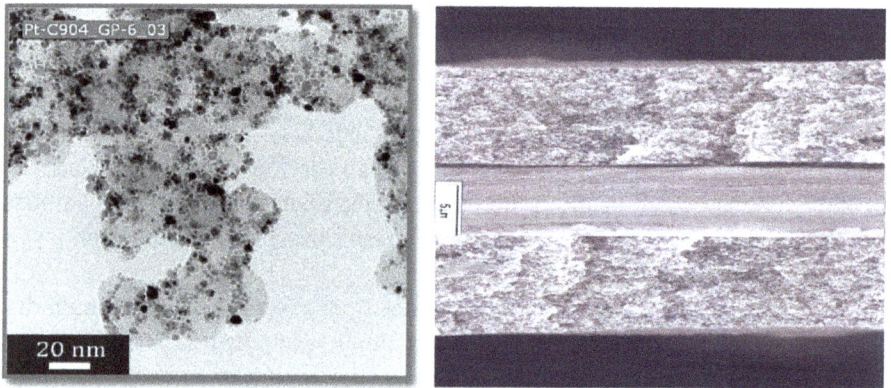

Figure 2.9.7: (left) Catalyst particle with platinum (black) on carbon support (grey). (right) Anode and cathode coatings on the proton exchange membrane (Source: Umicore).

Electrolyzers are devices that convert water into hydrogen and oxygen by means of electricity. Similar to fuel cells, different types of electrolyzers exist, amongst which the polymer electrolyte membrane electrolyzer is the one based upon PGMs. The subcomponents are similar to the ones of fuel cells, an anode and a cathode separated by a polymer electrolyte membrane. Water reacts at the anode to form oxygen and positively charged hydrogen ions, which move selectively through the membrane while the electrons flow into the electrical circuit. Hydrogen ions then combine at the cathode with electrons from the electrical circuit to form hydrogen (see Figure 2.9.8) [11].

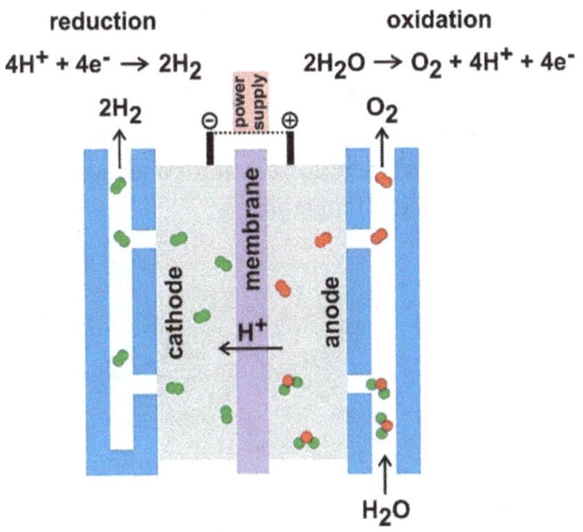

reduction   oxidation

$4H^+ + 4e^- \rightarrow 2H_2$ $2H_2O \rightarrow O_2 + 4H^+ + 4e^-$

Figure 2.9.8: Schematic functional principle of a PEM-based electrolyzer.

2.9.5 Heterogeneous catalysis in chemistry and petrochemistry

Various precious metals are used in a broad range of heterogeneous catalysts, other than automotive emission control catalysts, fuel cells and electrolyzers. One major application field is petrochemical operations, in which typically supported (doped) platinum catalysts are used for fossil oil reforming and isomerization reactions. For the reforming reactions, monometallic platinum or bimetallic platinum-rhenium, platinum-iridium and platinum-tin supported on alumina are widely used. For the isomerization reactions, mostly platinum supported on zeolites catalysts are used. For the synthesis of methanol from CO or CO_2 and hydrogen, supported palladium, rhodium, gold and silver catalysts are mentioned in literature.

For dehydrogenation reactions, large-scale industrial operations are based upon doped bimetallic platinum-tin supported on inorganic oxides such as alumina, zinc-aluminate and magnesium-aluminate. Similarly, specific hydrogenation reactions are catalyzed by a wide variety of supported PGM catalysts, especially palladium and platinum, sometimes combined with rhodium or ruthenium, are used [12].

2.9.6 Homogeneous catalysis in chemistry and pharmacy

Homogeneous catalysis (see also Chapter 7.1) is a young technology which has blossomed with the exploration of coordination chemistry, and already resulted in three Nobel Prizes in Chemistry [14–16]. The understanding of the mechanisms of catalytic cycles on a molecular basis has enabled the steady improvement of the catalysts used. Homogeneous catalysts can be designed very specifically for the target reaction by purposeful ligand choice corresponding to the active species promoting the targeted reaction in the catalytic cycle (see Figure 2.9.9). This leads to the well-known advantages when using homogeneous catalysts: high turn-over-numbers (TON), high chemo- and stereoselectivity, mild reaction conditions and good functional group tolerance – often combined all together.

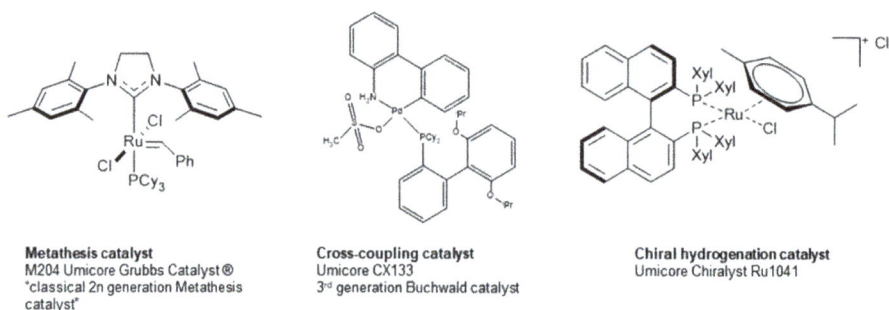

Metathesis catalyst
M204 Umicore Grubbs Catalyst®
"classical 2n generation Metathesis catalyst"

Cross-coupling catalyst
Umicore CX133
3ʳᵈ generation Buchwald catalyst

Chiral hydrogenation catalyst
Umicore Chiralyst Ru1041

Figure 2.9.9: Examples of modern tailor-made homogeneous PGM catalysts.

Those benefits go hand-in-hand with economic disadvantages of metal catalysis: high catalyst cost per kg and the need for recycling of the metal at the end of the catalyst functional life. Especially for precious metals-based catalysts, the specific catalyst cost is often outbalanced by the aforementioned advantages. The possibility to work on molecular basis (resulting often in high TON) enables a very low catalyst hold-up in the reactors making very efficient use of each "atom" of the precious metal.

Two different modes of homogeneous catalyst use exist. *In-situ processes* make use of simple metal-based precursors which are combined with the ligands before or within the process. This contrasts with the more and more familiar use of *tailor-made pre-formed catalysts,* which allows the stabilization of (often sensitive) ligands and thus the easier handling of catalyst dosing and precise ligand-to-metal stoichiometry.

The use of precious metals-based homogeneous catalysis can roughly (and arbitrarily, as all intermediate steps exist) be separated into two application areas – large-scale bulk chemical processes with strong focus on efficiency (TON, yield, chemoselectivity, etc.) and fine chemical approaches with a closer eye towards functional group tolerance, chemo- and stereoselectivity.

One of the earliest developed uses of precious metals catalysts is their application in carbonylation processes. Today, these account for some of the largest-scale implementations of homogeneous catalysis in the world.

Hydroformylation is a metal-catalyzed reaction originally developed by O. Roelen converting olefins into aldehydes by incorporation of carbon monoxide (CO). The products are mainly used as intermediates for other chemicals with a major downstream use as plasticizers for PVC or as detergents. While the original cobalt-based process required comparatively harsh conditions, a new rhodium-based technology was introduced by Union Carbide in the 1970s, enabling the conversion under much milder pressure and temperature regimes with high selectivity. This Rh-based process has undergone many improvements. The selectivity of the reactions (*n*- vs. *iso*-carbonylation) and the range of participating olefins was addressed by continuous progress in the design of phosphorus-based ligands (from PPh_3 originally towards e.g. chelating phosphite ligands in recent developments). Also one of the first biphasic industrial processes using homogeneous catalysts (Ruhrchemie/Rhône-Poulenc process) was a Rh-based hydroformylation process, where the catalyst in aqueous phase, through a water-soluble ligand, promotes the reaction at the liquid-liquid interface with the reaction partners in an organic phase. This concept enables a good separation of the used catalysts in liquid residues from products by distillation and can be used by many other precious metal-containing side streams (addressing one of the above mentioned challenges of homogeneous catalysis).

Palladium-based carbonylation processes of olefins have also been industrialized. The synthesis of methyl propionate, an intermediate for the manufacturing of methyl methacrylate (MMA) by carbonylation of ethylene in presence of water has been industrialized by Lucite using a Pd-phosphine-system as homogeneous catalyst. MMA is used, for example, in synthesis of acrylic glass. A second very important carbonylation process using precious metals for the preparation of bulk chemicals is the synthesis of acetic acid, where methanol is converted into acetic acid through an iodide-mediated reaction cycle. This process was originally introduced by Monsanto using RhI_3 or Rh(III) acetate as the source of active catalytic species $[Rh(CO)_2I_2]_2$. This process was then further developed towards an Ir/Ru-promoted process by BP (Cativa process) using $[Ir(CO)_2I_2]_2$ as the propagating species by which a significant increase in throughput can be achieved. Both processes are still in large-scale use today.

Besides the above mentioned carbonylation processes, olefins are converted on significant scale by addition of reactive H-X bonds. Hydrolyzation (addition of $H\text{-}SiR_3$) is the most prominent example for this type of reaction. This reaction can be used for synthesis of organosilanes, which finds multiple uses as chemical intermediates in downstream applications. A very large application of hydrosilylation is the cross-linking of polysiloxanes bearing vinyl-substituents with short-chain siloxanes as cross-linkers. This reaction is used for example in the synthesis of silicone or in large-scale paper-coating processes. Originally, H_2PtCl_6 was used as catalyst in those applications. With improved understanding of the catalytic pathways in the meantime, a number of different Pt-based catalysts were developed of which the "Karstedt catalyst", a Pt(0) species, has found a prominent role in many of those processes.

Besides the manufacturing processes for large-scale chemicals described above, a large and still growing number of fine chemical processes leverage the high flexibility and tunability of precious metals-based homogeneous catalysts. These advantages are indispensable to the synthesis of highly functionalized molecules applied in pharmaceuticals, agrochemicals or the electronic industry. The most prominent examples of these reactions are built upon the prize-winning technologies mentioned in the beginning of this chapter.

Pd-based cross-coupling reactions exemplify the connection of an aryl electrophile (typically an aryl halide) with a nucleophile. The differently named reactions vary by the nucleophile used. One of the most prominent reactions, the so-called Suzuki coupling, engages boronic acids or esters as nucleophiles. Other prominent examples use e.g. amines in presence of a base (Buchwald-Hartwig amination), Grignard reagents (Kumada coupling), zinc reagents (Negishi coupling) and so on. Originally, the reactions were based on simple *in situ* processes using Pd(II)-precursors and phosphine ligands. With better mechanistic understanding of turnover-limiting steps, the selection of the Pd(0)-propagating species now encompasses a multitude of phosphine and N-heterocyclic carbene ligands. Furthermore, the toolbox now includes stable Pd (II) precatalysts that can be readily activated under mild conditions, enabling fast low-temperature reactions, minimal catalyst loadings and, in many cases, conversion of very challenging substrates prone to unproductive side reactions (see Figure 2.9.10).

Another very prominent reaction in the fine chemical context is the enantioselective hydrogenation of olefins or ketones using catalysts with bulky chiral ligands coordinating to the metal center. With this comparatively "simple" reaction it is possible to introduce a new stereocenter into a molecule avoiding in many cases the need of chiral resolution or chromatography for purification. This allows expedient access to many molecules relevant in life science applications. As in the base of cross-coupling above, the development of a multitude of different ligands and preformed catalyst systems allows a larger scope of substrates to be addressed. In the meantime, a number of other enantioselective conversions have been achieved, even though hydrogenation remains by far the most prominent application.

Finally, a third reaction type promoted by homogeneous catalysts is the olefin metathesis reaction (see Figure 2.9.11). This so-called – "double-bond dance" – enables two olefin molecules to exchange their bonding partners. This reaction is promoted by specific Ru-carbene complexes which are highly tolerant to functional groups. Over time a range of catalysts was developed that exemplify different reactivities and

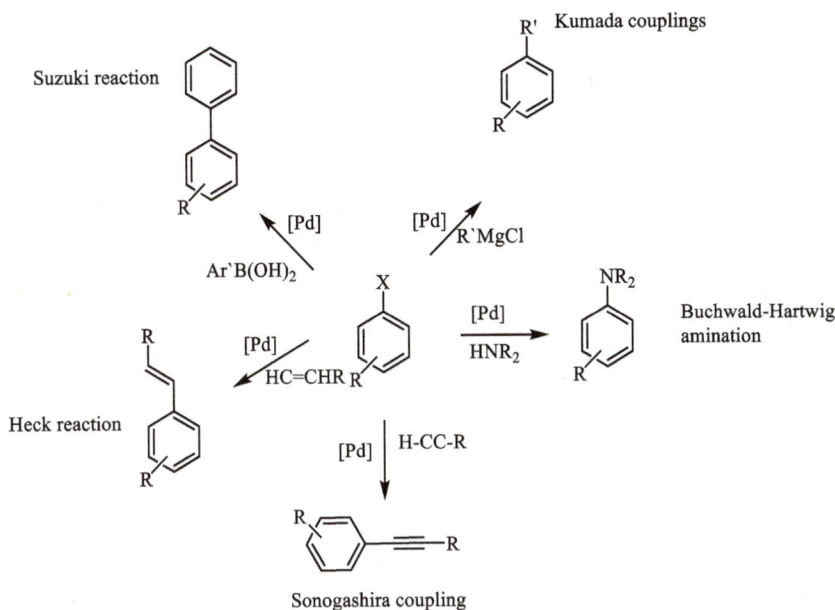

Figure 2.9.10: Well-known Pd-based CX-reactions used in many types of chemical and pharmaceutical industry.

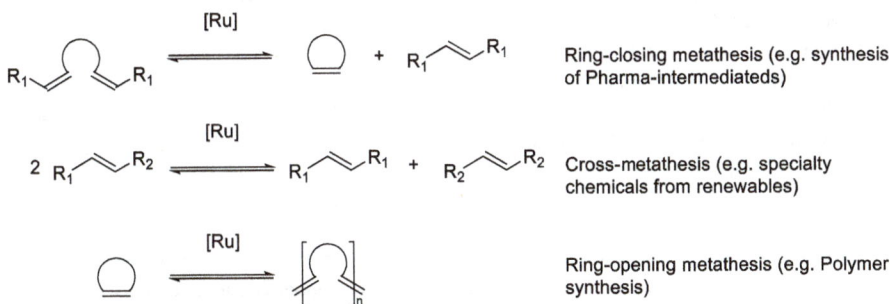

Figure 2.9.11: Main classes of metathesis reactions based on ruthenium catalysts.

selectivities to approach a number of synthetic challenges. The use of olefin metathesis is multifold. It can be used in pharmaceutical applications, e.g. through the generation of macrocycles via ring-closing metathesis as exemplified by the first clinical development campaign executed by Boehringer Ingelheim towards a potential Hepatitis C treatment. The reaction can also be used to generate highly resistive polymers by ring-opening metathesis of strained unsaturated cyclic monomers, as implemented e.g. by Materia Inc. with their poly-DCPD-based Proxima resin. Finally, it is used by cross-(or self-)metathesis of two olefins as in the conversion of seed oils or fatty acid esters in the synthesis of olefinic intermediates [17].

📖 References

[1]   Homepage of the International Platinum Group Metals Association, Schiess-Staett-Strasse 30, 80339 Munich, Germany, https://www.ipa-news.de/, accessed January 05, 2022.

[2]   Homepage of Sibanye Stillwater Limited, Cnr 14th Avenue & Hendrik Potgieter Road, Bridgeview House, Ground Floor (Lakeview Avenue) Weltevreden Park, 1709, South Africa, https://www.sibanyestillwater.com/, accessed January 05, 2022.

[3]   Homepage of the European Precious Metals Federation, Avenue de Tervueren 168, Box 6, 1150 Brussels, Belgium, https://www.epmf.be/, accessed January 05, 2022.

[4]   Hagelüken C. Metall, 2019, 10, 396–403.

[5]   Beck G, Beyer H-H, Gerhartz W, Haußelt J, Zimmer U. (Eds.) Edelmetall-Taschenbuch, Giesel Verlag GmbH, Isernhagen, 2001.

[6]   Homepage of Umicore NV/SA, Broekstraat 31, 1000 Brussels, Belgium, https://www.umicore.com, accessed January 05, 2022.

[7]   Homepage of Materials Performance, 15835 Park Ten Pl., Houston, TX 77084, USA, https://www.materialsperformance.com, accessed January 05, 2022.

[8]   Datye AK, Votsmeier M. Nat Mater, 2021, 20, 1049–59.

[9]   Homepage of the European Association for Automobile Emission Control, Boulevard Auguste Reyers 80, 1030 Brussels, Belgium, https://www.aecc.eu, accessed January 05, 2022.

[10] Deutschmann O, Grunwaldt J-D. Chem Ing Techn, 2013, 85, 595–617.

[11] Homepage of U. S. Department of Energy, 1000 Independence Ave. SW, Washington DC 20585, USA, https://www.energy.gov/eere/fuelcells/parts-fuel-cell, accessed January 05, 2022.

[12] Ertl G, Knoetzinger H, Weitkamp J. (Eds.) Handbook of Heterogeneous Catalysis, Wiley-VCH, Weinheim, 1997.

[13] Homepage of Johnson Matthey PLC, www.matthey.com, PGM market report May 2021, accessed January 05, 2022.

[14] Homepage of The Nobel Foundation, Sturegatan 14, Stockholm, Sweden, https://www.nobelprize.org/prizes/chemistry/2010/press-release/, Heck RF, Negichi E, Suzuki A. Pd-catalyzed Cross-coupling in organic synthesis, accessed January 05, 2022.

[15] Homepage of The Nobel Foundation, Sturegatan 14, Stockholm, Sweden, https://www.nobelprize.org/prizes/chemistry/2005/press-release/, Chauvin Y, Grubbs R, Schrock RR. Development of the metathesis method in organic synthesis, accessed January 05, 2022.

[16] Homepage of The Nobel Foundation, Sturegatan 14, Stockholm, Sweden, https://www.nobelprize.org/prizes/chemistry/2001/press-release/, Knowles WS, Nojori R, Sharpless B. Catalytic asymmetric synthesis, accessed January 05, 2022.

[17] Umicore. Precious, Materials Handbook, Vogel Communications Group GmbH & Co. KG, Würzburg, 2011.

2.10 Shape memory alloys

Stefan Engel, Oliver Janka

The class of shape memory materials (SMMs) contains compounds that all exhibit the so-called shape memory effect (SME). These materials can be deformed by an external force; however, they maintain the deformed shape only until the process is reversed by a specific stimulation. The SME describes the process in which such a deformed material returns to its original shape when subjected to heat or a magnetic field. During this process, significant forces are generated that can be utilized. This leads to a key criterion for the search for SMMs: the material must exhibit two phases with different crystal structures, connected by a diffusion-less phase transition. A second important factor is the temperature range in which these phase transitions occur since they determine where these materials can be used. Several classes of materials are known, e.g. metallic materials such as shape memory alloys (SMAs), high-temperature shape memory alloys (HTSMAs) and magnetic SMAs (MSMAs) but also thin film materials, shape memory polymers (SMPs) and some ceramic examples. This chapter will focus on shape memory alloys with nitinol as the most important technical representative [1].

2.10.1 Brief history

The shape memory effect was determined first in the binary Au/Cd system by Ölander in 1932 [2]. He investigated different compositions in this system and realized, that wires made with a Au content between 50 and 52.5% were elastic and "almost reminded of rubber". A similar effect was observed for Cu/Zn and Cu/Sn alloys in 1938 [3]. In 1949, Kurdyumov and Khandros described the thermoelastic equilibrium of martensitic transformations based on light microscopic investigations of the microstructure of cooled alloys, leading to an enhanced understanding of this process [4]. Only slightly later, the same observations were made in the Au/Cd system [5]. These observations sparked the interest of many scientists; however, manufacturing complexity and costs prohibited the practical application. The discovery of the ductile intermetallic compound NiTi and its alloys, named as "Nitinols" [6, 7] (NiTi alloys discovered at the Naval Ordnance Laboratory, White Oak, Maryland, USA) fueled the interest in these materials, especially, since some alloys showed "unusual mechanical vibration damping properties". In a subsequent publication [8], the authors stated that an annealed 55.4 wt.-% Ni wire with about 0.020 inch in diameter coiled to a tight helix retains its shape at room temperature indefinitely. However, submerging the wire in 65 °C hot water reverts it back to its original straight shape. When heated to 100 °C and coiled, the material can be straightened at room temperature and recurs to its coiled shape when heated to > 65 °C. Especially surprising is the fact, that

https://doi.org/10.1515/9783110733143-017

these processes happen in less than one second [7]. These discoveries, in turn, finally lead to practical applications and the commercialization of these materials. The first commercial SMA application was the CryoFit™ "shrink-to-fit" pipe produced by the Raychem Corporation for the F-14 fighter jet followed by the use in orthodontic bridge wires [7]. While in the first example, the memory abilities of the material are used, in the latter case, the generated forces of SMAs are utilized. A brief overview of todays applications of shape memory materials (SMM) will be given in Chapter 2.10.5.

2.10.2 Phase transitions

The phase transitions that occur in SMAs revolve around the respective austenitic and the martensitic phases. These phases are well known in the Fe/C system and are the main reason for the physical and mechanical properties of steel. While in the Fe/C system the austenitic phase crystallizes with a face centered cubic structure ($Fm\bar{3}m$) with C occupying the octahedral voids (defect NaCl type structure), HT-NiTi crystallizes with the cubic primitive space group $Pm\bar{3}m$ (CsCl type structure). However, the structures not being related, the high-temperature phase is still called austenitic. In the Fe/C system, the low-temperature phase also known as the martensitic phase crystallizes in the tetragonal crystal system with space group $I4/mmm$. Due to the group-subgroup relationship (*translationengleiche* transition of index 3) of the two space groups and structures, during the formation of martensite three different potential domains can be formed, leading to the formation of trillings. In contrast, in the NiTi system, from a group-subgroup point of view, several phase transitions must occur until the room temperature phase of NiTi (martensitic phase) in the monoclinic crystal system with space group $P2_1/m$ is formed (Figure 2.10.1). It has been shown that the transition from the CsCl type (B2) takes places via a trigonal / rhombohedral phase (R) till finally the monoclinic RT-NiTi phase (B19′) is formed. While the structures of the HT and the RT are well established, for the R-phase several different structures have been reported with space groups $P\bar{3}$ [9–11], $P3$ [11–13], $R3$ [14] and $P31m$ [15]. Alongside these, also phases like rhombohedral Ni_4Ti_3 ($R\bar{3}m$, Pu_3Pd_4 type) have been reported to be formed in Ni-rich samples [14, 16, 17]. However, since the transition is diffusion-less it must obey a group-subgroup relationship. Therefore, all structures that are reported with space groups not being direct subgroups of $Pm\bar{3}m$ are potentially incorrect.

2.10.3 Different shape-memory effects

As shown before (Figure 2.10.1), SMAs exhibit an austenitic and a (twinned / detwinned) martensitic phase with the latter one being the at room temperature stable one. Upon heating, the phase transformation from the martensitic to the austenitic

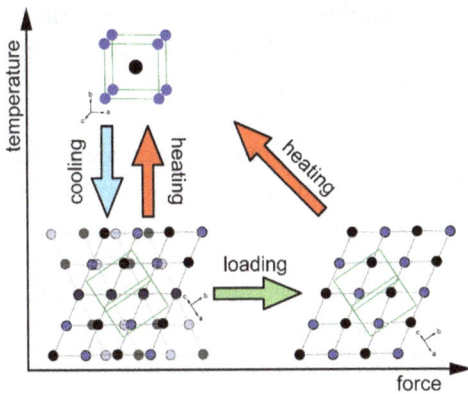

Figure 2.10.1: Fundamental processes in SMAs. Top left: austenitic NiTi phase (CsCl type); bottom left: twinned martensitic NiTi; bottom right: detwinned martensitic NiTi (monoclinic), formed upon the application of a load.

phase takes place. For a potential application, the austenite start (A_s) and end (austenite finish = A_f) temperatures of this transformation are crucial. As soon as A_s is reached, the material recovers its original shape, even under applied loads. When cooled, the transformation to the martensite structure starts at M_s and ends when the martensite-finish temperature M_f is reached. This process is called a one-way shape memory effect (OWSME). After application of an external force, the material stays in the deformed state und only recovers after heat is applied and therefore the transformation to the austenitic phase is triggered. Finally, the temperature above which a permanent deformation of the martensitic phase is introduced is called M_d. In contrast, materials with a two-way shape memory effect (TWSME) exhibit a reversible SME, meaning that they can remember their shapes both at high and low temperatures. However, several drawbacks (lower recovery strain, deterioration of strain at high temperatures, required 'training' of the material) hamper their technical use [18, 19]. Finally, pseudo- (PE) or superelasticity (SE) can be observed. Here, the SMA returns to its original shape when a load is applied while the material is at a temperature between A_f and M_d.

2.10.4 Classes of materials

The class of shape memory materials (SMMs) contains not only metallic shape memory alloys (SMAs) but also metallic high-temperature shape memory alloys (HTSMAs), metallic magnetic SMAs (MSMAs) or non-metallic shape memory polymers (SMPs). High temperature shape memory alloys (HTSMAs) [20] usually have transition temperatures above 100 °C, which can be achieved through addition of a third component, e.g. Au, Pd, Pt or less noble metals such as Zr or Hf. Besides increased costs for these materials, also a limited ductility and processing difficulties hamper their use. Magnetic shape memory alloys (MSMAs) are usually ferromagnetic materials (therefore also known as FSMAs), in which the energy is transferred through magnetic fields rather than heat. Due to the differences in the energy transfer mechanism, magnetic materials can operate at much higher frequencies. However, FSMAs are often brittle and can only operate at low temperatures, alongside the problems regarding their shaping. Finally, shape memory polymers (SMPs) have to be mentioned. Their SME originates from the glass transition of the polymer, which can be tailored for the specific application. Due to their lower costs, higher efficiency and sometimes even biodegradability, these materials have gained a lot of attention [21, 22].

2.10.5 Application fields

All the materials mentioned above can be divided into four categories based on their primary function. Materials that exhibit the shape memory effect (SME) can be used to generate motion and / or force, while materials with superelasticity (SE) can store the applied deformation energy. In the following sub chapters, examples for the different types of applications will be given.

2.10.5.1 Transportation & construction

The number of components in modern transportation vehicles have increased tremendously with the increasing demand for more comfort and safety. In quite a number of cases, SMA based actuators can be utilized for linear movements, e.g. in climate control flaps, fuel door actuators, windshield wipers or for the fixation of bearing on shafts. However, also the superelasticity and deformation storage ability can be used, e.g. in washers or bumpers. Also in the field of aviation, smart wings for planes or vibration-damped rotor blades in helicopters have been reported (Table 2.10.1).

Thermal actuators have also been employed in several constructive applications. Cu-Zn-Al based springs for example have been used as a bifunctional devices for temperature sensing and automatically operating greenhouse windows. Steam traps or air conditioning units operate similarly, while SMAs can also be used for example for static rock breakers (Table 2.10.2).

Table 2.10.1: Applications of SMAs in the automotive and aerospace industry.

Task	Description and use	
generate force	springs in thermal valves	[1, 23–26]
	fixation of bearings on shafts	[1, 24, 26]
	actuator for shifting in automatic automobile transmissions	[1, 25, 26]
	spring for pressuring the windshield wiper or door lock	[1, 25–27]
	wire to move a glass plate to measure dust built-up on mars rover	[28, 29]
	couplings for joining aircraft hydraulic tubing	[1, 24, 30]
store deformation	shape memory washer to reduce rattling noise	[1, 26]
energy	beams for energy absorption in automobiles	[26, 31, 32]
	smart wings (form of aircraft wing dependent on speed)	[1, 33–35]
	rotor blades of helicopters to reduce vibration	[1, 34, 36]

Table 2.10.2: Constructive applications of SMAs.

Task	Description and use	
generate force	wire for temperature control in greenhouses	[24, 37]
	joints for air conditioning tubing	[24]
	air conditioning louvres	[24, 34, 38]
	steam traps	[34]
	air flow controller, air damper in multifunction electric oven	[34]
	composites as static rock breaker (expansion force breaks rocks)	[34, 39]
	self-healing loose bolted joints in civil structures and components	[40]

2.10.5.2 MEMS and robotics

Actuators based on SMAs exhibit several advantages when it comes to miniaturization: the simplicity of the mechanism, their silent motion, their rather easy fabrication process or their low driving voltage [41]. Especially their excellent power to weight ratio (up to 100 W kg^{-1}) and the possibility to form thin films makes them excellent components for e.g. robotic devices [42], MEMS (micro-electromechanical systems) [43] or when designing artificial muscles [44, 45]. For MEMS, SMAs play an important role, as they can contribute with the mechanical part to this device (Table 2.10.3). Besides thermally activated valves, also micro pumps or micro-grippers have been realized. The range of robotic devices spans from jumper and crawler robots all the way to walker or fish [42, 46–49] and even flower robots [42, 50] (Table 2.10.4).

Table 2.10.3: Mini actuators and micro-electromechanical systems (MEMS) based on NiTi SMAs.

Task	Description and use	
generate force	springs in thermal valve	[44]
	plates in microvalve to lift the plunge	[44, 51–55]
	thin films inside micropump as diaphragm	[44, 52, 54–57]
	thin films for micro-clipper	[44, 58]
	in micro-grippers to open the jaws	[44, 55, 59–61]
	springs as linear actuator in optical microswitch	[44, 45, 54, 55, 62, 63]

Table 2.10.4: NiTi SMAs used in robotic devices.

Task	Description and use	
generate motion	springs/wires in crawler robots for caterpillar-like movement	[42, 64, 65]
generate force	coils for deforming in jumper robot movement	[42, 66]
generate force	coils to move flower stem and pedals in flower robot	[42, 50]
and movement	springs/bands for propulsion for fish robot	[42, 46, 47]
	wires for mimicking muscles in walker robots	[42, 48, 49]
	wires for mimicking agonist-antagonist muscles in robotic hands	[42, 67, 68]

2.10.5.3 Biomedical

Like already sketched before, depending on the potential application, the different properties of SMAs can be used. Besides generating force and motion, also the storage of energy can be usefully employed in biomedical applications [69]. Table 2.10.5 summarizes some examples for the different tasks. Endovascular stents made from nitinol for example can be inserted into the blood vessels in its collapsed form and extended when in place to prevent the failure of the blood vessel. Wires made from

Table 2.10.5: Biomedical applications of SMAs.

Task	Description and use	
generate force	endovascular stents to prevent collapse of blood vessels	[69–74]
	Ni-Ti wire as flow control for intravenous tubes	[75]
	SMA nails, staples, and plates for causing uniform constraint	[76–78]
	endosseous dental implants to provide stability	[79]
generate force &	archwire for braces to generate force and motion in the teeth	[80]
motion	suture anchor for reattachment ligaments to the bone	[81, 82]
store deformation	NiTi spacing disk to prevent movement of vertebrae	[83]
energy	cava filters to collect blood clots in cava vein	[84]
	hook shaped needle for breast cancer location	[75]

nitinol can be used for dental applications, e.g. for braces, while disks made of nitinol can be used to stabilize the spinal column and prevent, due to the high dampening properties, traumatic movement of the vertebrae during the healing process.

2.10.5.4 Miscellaneous

Finally, SMAs are also employed in consumer products (Table 2.10.6), different industrial applications and in fashion (Table 2.10.7). Besides industrially used couplings and fasteners for tubes and the as springs in fire check valves [24, 34, 85, 86], SMAs were also implemented in consumer products and fashion items. Their main goal is to increase the comfort of the user usually due to their pseudo-elastic behavior. The important combination of stiffness and super-elasticity was the key to success [34]. Typical applications are eye glass frames, frames for brassieres or antennas for mobile phones. The dampening effects of SMAs are also enhancing the properties of skis, tennis rackets or fishing lines [34].

Table 2.10.6: Consumer products and industrial applications utilizing SMAs.

Task	Description and use	
generate force	fishing lines	[34, 87]
	springs in coffee makers to control brewing temperature, anti-scald valve in sink, tub and shower	[24, 38, 86, 88]
	wires in skis to reduce vibration	[34, 89]
	fibers in tennis rackets to enhance structural damping	[34, 40]
	antenna for cellphones	[34]
	disc actuator to prevent the development of uncontrollable temperatures due to overcharge or short circuit in lithium ion cells	[86]
	couplings for pipes	[24, 34, 85]
	spring as fire check valve, closes flow of toxic and high flammable gases and vents air, when predetermined temperature is exceeded	[86]

Table 2.10.7: SMAs used in fashion.

Task	Description and use	
generate force	inserts for shoes	[34]
	eye glass frames	[24, 34]
	headband for headphones (for the MiniDisk Walkman by Sony®)	[34]
store deformation energy	underwire for brassieres	[24, 34, 90]
	core wire of wedding dress petticoat	[34]

References

[1]   Mohd Jani J, Leary M, Subic A, Gibson MA. Mater Des, 2014, 56, 1078.
[2]   Ölander A. J Am Chem Soc, 1932, 54, 3819.
[3]   Greninger AB, Mooradian VG. AIME Trans, 1938, 128, 337.
[4]   Kurdyumov GV, Khandros LG. Dokl Akad Nauk SSSR, 1949, 66, 211.
[5]   Chang LC, Read TA. AIME Trans, 1951, 189, 47.
[6]   Buehler WJ, Wiley RC. Noltr, 1962, 65.
[7]   Kauffman GB, Mayo I. Chem Educ, 1997, 2, 1.
[8]   Buehler WJ, Gilfrich JV, Wiley RC. J Appl Phys, 1963, 34, 1475.
[9]   Goryczka T, Morawiec H. J Alloys Compd, 2004, 367, 137.
[10] Khalilallafi J, Schmahl W, Toebbens D. Acta Mater, 2006, 54, 3171.
[11] Sitepu H. Textures Microstruct, 2003, 35, 185.
[12] Hara T, Ohba T, Okunishi E, Otsuka K. Mater Trans, 1997, 38, 11.
[13] Hara T, Ohba T, Otsuka K. J Phys IV, 1995, 05, C8.
[14] Saburi T, Nenno S, Fukuda T. J Less-Common Met, 1986, 125, 157.
[15] Fruchart D, Soubeyroux JL, Miraglia S, Obbade S, Lorthioir G, Basile F, Colin C, Faudot F,
      Ochin P, Dezellus A. Z Phys Chem, 1993, 179, 225.
[16] Dlouhy A, Khalil-Allafi J, Eggeler G. Philos Mag Lett, 2010, 83, 339.
[17] Tirry W, Schryvers D, Jorissen K, Lamoen D. Acta Crystallogr B, 2006, 62, 966.
[18] Schroeder TA, Wayman CM. Scripta Metall, 1977, 11, 225.
[19] Huang W, Toh W. J Mater Sci Lett, 2000, 19, 1549.
[20] Ma J, Karaman I, Noebe RD. Int Mater Rev, 2010, 55, 257.
[21] Xia Y, He Y, Zhang F, Liu Y, Leng J. Adv Mater, 2021, 33, 2000713.
[22] Liu C, Qin H, Mather PT. J Mater Chem, 2007, 17, 1543.
[23] Shaw G, Prince T, Snyder J, Willett M, Lisy F. US 7,587,944 B1, 2009.
[24] Duerig TW. Mater Sci Forum, 1991, 56-58, 679.
[25] Jani JM, Leary M, Subic A. Appl Mech Mater, 2014, 663, 248.
[26] Stoeckel D. Mater Des, 1990, 11, 302.
[27] Buchanan HC, Victor KR. US 5,062,175, 1991.
[28] Landis GA, Jenkins PP. J Geophys Res Planets, 2000, 105, 1855.
[29] Landis GA, Jenkins PP. Dust on mars: Materials adherence experiment results from mars
      pathfinder. Conf Record Twenty Sixth IEEE Photovoltaic Specialists Conf, 1997, 865.
[30] Hartl DJ, Lagoudas DC. Proc Inst Mech Eng G, 2007, 221, 535.
[31] Browne AL, Johnson NL. US 6,910,714 B2, 2005.
[32] Melz T, Seipel B, Sielhorst B, Zimmerman E. US 7,905,517 B2, 2011.
[33] Gonzalez L. Morphing Wing Using Shape Memory Alloy: A Concept Proposal, 2005, 1.
      https://citeseerx.ist.psu.edu/viewdoc/download?doi=10.1.1.513.8597&rep=rep1&type=pdf.
      Accessed January 3rd, 2022.
[34] Van Humbeeck J. Mater Sci Eng A, 1999, 273-275, 134.
[35] Zhu J-H, Zhang W-H, Xia L. Arch Comput Methods Eng, 2015, 23, 595.
[36] Epps JJ, Chopra I. Smart Mater Struct, 2001, 10, 104.
[37] Alazzawi S, Filio P. Design exploration study of a smart passive window based on NiTi bent
      actuator. IOP Conf Ser: Mater Sci Eng, 2021, 1076, 012085.
[38] Takaoka S, Horikawa H, Kobayashi J, Shimizu K. Mater Sci Forum, 2002, 394-395, 61.
[39] Benafan O, Noebe RD, Halsmer TJ. Shape memory alloy rock splitters (SMARS) – a non-
      explosive method for fracturing planetary rocklike materials and minerals. Nasa/tm, 2015-
      218832, 2015.
[40] Hurlebaus S, Gaul L. Mech Syst Signal Process, 2006, 20, 255.

[41] Ikuta K. Micro/miniature shape memory alloy actuator. Proc IEEE Int Conf Robotics Automation, 1990, 2156.

[42] Kheirikhah MM, Rabiee S, Edalat ME. A review of shape memory alloy actuators in robotics. In: Ruiz-del-solar J, Chown E, Plöger PG. (Eds.) RoboCup 2010: Robot Soccer World Cup XIV, Springer, Berlin, Heidelberg, 2011, 206.

[43] Choudhary N, Kaur D. Sens Actuators A: Phys, 2016, 242, 162.

[44] Kohl M. Shape Memory Microactuators, Springer-Verlag, Berlin, Heidelberg, New York, 2004.

[45] Fujita H, Toshiyoshi H. Microelectron J, 1998, 29, 637.

[46] Wang Z, Hang G, Wang Y, Li J, Du W. Smart Mater Struct, 2008, 17, 025039.

[47] Kyu-Jin C, Hawkes E, Quinn C, Wood RJ. Design, fabrication and analysis of a body-caudal fin propulsion system for a microrobotic fish. In: 2008 IEEE Int Conf Robotics Automation 2008, 706.

[48] Furuya Y, Shimada H. Mater Des, 1991, 12, 21.

[49] Menon C, Sitti M. Biologically Inspired Adhesion based Surface Climbing Robots. In: Proc 2005 IEEE Int Conf Robotics Automation, 2005, 2715.

[50] Hao L, Huang S-H, Park Y-S. Shape Memory Alloy based Flower Robot. In: 39th Int Symp Robotics, 2008, 888.

[51] Nath T, Raut G, Kumar A, Khatri R, Palani IA. Investigation on laser assisted actuation of shape memory alloy based micro-valve. In: 2015 Int Conf Robotics, Automation, Control Embedded Systems (RACE) 2015, 1.

[52] Kahn H, Huff MA, Heuer AH. J Micromech Microeng, 1998, 8, 213.

[53] Barth J, Megnin C, Kohl M. A bistable shape memory microvalve. In: 2011 IEEE 24th Int Conf Micro Electro Mechan Systems 2011, 1067.

[54] Fu Y, Du H, Huang W, Zhang S, Hu M. Sens Actuators A: Phys, 2004, 112, 395.

[55] Sun L, Huang WM, Ding Z, Zhao Y, Wang CC, Purnawali H, Tang C. Mater Des, 2012, 33, 577.

[56] Liu DH, Liao GH. Appl Mech Mater, 2011, 63-64, 800.

[57] Makino E, Mitsuya T, Shibata T. Sens Actuators A: Phys, 2001, 88, 256.

[58] Takeuchi S, Shimoyama I. Three dimensional SMA microelectrodes with clipping structure for insect neural recording, Technical Digest. In: IEEE Int MEMS 99 Conf. Twelfth IEEE Int Conf Micro Electro Mechanical Systems (Cat. No.99CH36291) 1999, 464.

[59] Lan -C-C, Lin C-M, Fan C-H. IEEE ASME Trans Mech, 2011, 16, 141.

[60] Just E, Kohl M, Miyazaki S. J Phys IV, 2001, 11, Pr8-559-64.

[61] Lee AP, Ciarlo DR, Krulevitch PA, Lehew S, Trevino J, Northrup MA. Sens Actuators A Phys, 1996, 54, 755.

[62] Barth J, Krevet B, Kohl M. Smart Mater Struct, 2010, 19, 094004.

[63] Bhattacharya B. Sens Transd, 2011, 4, 103.

[64] Young Pyo L, Byungkyu K, Moon Gu L, Jong-Oh P. Locomotive mechanism design and fabrication of biomimetic micro robot using shape memory alloy. In: IEEE Int Conf Robotics Automation Proc ICRA '04, 2004, 5007.

[65] Liu CY, Liao WH. A snake robot using shape memory alloys. IEEE Int Conf Robotics Biomimetics, 2004, 601.

[66] Sugiyama Y, Hirai S. Int J Rob Res, 2006, 25, 603.

[67] Cho KJ, Rosmarin J, Asada H. SBC hand: A lightweight robotic hand with an SMA actuator array implementing C-segmentation. Proc IEEE Int Conf Robotics Automation, 2007, 921.

[68] Yang H, Xu M, Li W, Zhang S. IEEE Trans Ind Electron, 2019, 66, 6108.

[69] Mantovani D. Jom, 2000, 52, 36.

[70] Machado LG, Savi MA. Braz J Med Biol Res, 2003, 36, 683.

[71] Pelton AR, Stöckel D, Duerig TW. Mater Sci Forum, 2000, 327-328, 63.

[72] Petrini L, Migliavacca F. J Metall, 2011, 1. DOI: doi.org/10.1155/2011/501483.

[73] Gil FJ, Planell JA. Proc Inst Mech Eng H, 1998, 212, 473.

[74] Tominaga R, Kambic HE, Emoto H, Harasaki H, Sutton C, Hollman J. Am Heart J, 1992, 123, 21.

[75] Fischer H, Vogel B, Welle A. Minim Invasive Ther Allied Technol, 2004, 13, 248.

[76] Kramer M, Muller CW, Hermann M, Decker S, Springer A, Overmeyer L, Hurschler C, Pfeifer R. J Mech Behav Biomed Mater, 2017, 75, 558.

[77] Li Q, Zeng Y, Tang X. Australas Phys Eng Sci Med, 2010, 33, 129.

[78] Braun JT, Ogilvie JW, Akyuz E, Brodke DS, Bachus KN. Spine, 2004, 29, 1980.

[79] El H, Kan H, Adilp lu I, Aktas G, Kocadereli I. Cumhur Dent J, 2011, 14, 119.

[80] Miura F, Mogi M, Ohura Y, Hamanaka H. Am J Orthod Dentofac Orthop, 1986, 90, 1.

[81] Moneim MS, Firoozbakhsh K, Mustapha AA, Larsen K, Shahinpoor M. Clin Orthop Relat Res, 2002, 402, 251–59.

[82] Krumme J, McDevitt D. WO2009029468A2, 2011.

[83] Petrini L, Migliavacca F, Massarotti P, Schievano S, Dubini G, Auricchio F. J Biomech Eng, 2005, 127, 716.

[84] Wolf F, Thurnher S, Lammer J. Rofo, 2001, 173, 924.

[85] Tabesh M, Boyd J, Atli KC, Karaman I, Lagoudas D. J Intell Mater Syst Struct, 2017, 29, 1165.

[86] Wu MH, Schetky LM. Industrial applications for shape memory alloys. Proc Int Conf Shape Memory Superelastic Techn, 2000, 171.

[87] Van Humbeeck J. Adv Eng Mater, 2001, 3, 837–50.

[88] Asai M, Suzuki Y. Mater Sci Forum, 2000, 327–28, 17.

[89] Scherrer P, Bidaux JE, Kim A, Månson JAE, Gotthardt R. J Phys IV, 1999, 09, Pr9-393-400.

[90] Gök MO, Bilir MZ, Gürcüm BH. Procedia Soc Behav Sci, 2015, 195, 2160.

2.11 Bulk metallic glasses

Stefan Engel, Oliver Janka

Metallic glasses (MG) are materials consisting of amorphous metals or alloys. All these materials have metallic character, meaning fair electrical and thermal conductivity and the typical silvery metallic luster. The production of metallic glasses involves the prevention of the formation of a crystalline structure. Without an ordered crystal structure, metallic glasses become very hard, even harder than steel, smooth and possess a very low Young's modulus. Bulk metallic glasses (BMG) can be used as material in consumer products and fashion items, in industrial applications, in biomedical, MEMS and technology, as well as in transportation, aerospace applications and defense. The properties of the glassy metals can be influenced, depending on the selection of the elements. BMGs containing Fe, Co and Ni are often soft-magnetic and used primarily as magnetic cores and sensors. Ti- and Zr-based BMGs possess very high hardness and low Young's modulus and are typically used as springs, diaphragms and coatings.

2.11.1 Brief history

$Au_{75}Si_{25}$ was the first composition for that was reported by Klement, Willens and Duwez in 1960 to give a diffraction pattern of an amorphous metallic glass [1]. It was obtained via rapid quenching from ~1300 °C to room temperature. The initial requirement to rapidly cool the specimen led to a limitation regarding their shape as typically only wires, ribbons or foils could be produced. Therefore, the thickness of the samples was limited to a few hundred micrometers in at least one direction. The required cooling rates drastically decreased from initially 10^6 K s^{-1} to only 100–1000 K s^{-1} with the finding of the Pd-T-Si glasses (T = Cu, Ag, Au), leading to the possibility to produce amorphous spheres [2]. Using a melt-spinning device, the first iron-based glass $Fe_{40}Ni_{40}B_{20}$ was developed in 1976 [3]. In the 1980s, it was commercialized as *Metglas*. The group of Turnbull was finally able to produce the first bulk metallic glasses. They could obtain ingots with 5 mm diameter of $Pd_{40}Ni_{40}P_{20}$, which increased their critical casting thickness to 1 cm during processing in a boron oxide flux [4]. The discoveries of Inoue and Johnson finally enabled critical casting thicknesses well above 1 cm by conventional molding due to their findings of glassy systems that needed only cooling rates of 1 to 100 K s^{-1} [5]. $Zr_{65}Al_{7.5}Ni_{10}Cu_{17.5}$ for example could be cast up to 16 mm thickness [6], while $Zr_{41.2}Ti_{13.8}Cu_{12.5}Ni_{10}Be_{22.5}$ could be cast up to 10 cm [7]. The alloy was commercialized as Vitreloy 1 (Vit1). More details regarding the history of bulk metallic glasses can be found in these review articles [5, 8, 9].

https://doi.org/10.1515/9783110733143-018

2.11.2 General understanding, synthesis and structure

To synthesize metallic glasses, one has to understand the required conditions. Initially, a rapid quenching (10^5 to 10^6 K s^{-1}) of the respective melt and clean conditions to avoid crystallization due to contact with nucleates were thought to be necessary. These high cooling rates can be achieved by various ways, e.g. by melt spinning, in which the molten alloy is quenched by ejecting it onto the flat rim of a rotating copper wheel [10, 11]. By this procedure, bands and ribbons can be produced. For the production of wires, the melt can be ejected into water [12]. Alternatively, an amorphization of the material can be achieved by other techniques such as deposition from the liquid (electro- [13] or other deposition methods [14, 15]) or gas phase (evaporation [16] or sputtering [17]) or by grinding and mechanical alloying [18, 19].

Achieving the glassy/amorphous state should therefore be the easiest, when the temperature interval between the melting point (T_m) and the glass transition temperature (T_g) is small. Since T_m can strongly be influenced by the chemical composition, a formation of glasses is rather likely near minima in the respective phase diagrams, which often are eutectic points. Solutes furthermore increase the chance of forming a glassy state due to the stabilization of the liquid. However, incorporation can only be achieved when the differences of the atomic diameters of the constituent elements is <10%. While nearly all metals can be utilized for the formation of metallic glasses, palladium and iron seem to be elements that enable an easier access to the amorphous state. Aluminum, in contrast, tends to be rather difficult as the main component for the formation of glasses. Due to its important physical properties (e.g. low density, see Chapter 2.3), significant research has been conducted. For >80 at% Al, finally, the system Al-T-Ln, with T being a transition metal and Ln being a lanthanide, arose. Here, compositions such as $Al_{90}Fe_5Ce_5$ or $Al_{85}Ni_{10}Y_5$ should be noted. As a palladium-based glass ($Pd_{40}Ni_{40}P_{20}$) could be produced with low cooling rates (< 1 K s^{-1}), new research interest sparked. This led to the discovery of Ln-Al-T and Mg-T-Ln glasses, which sometimes do not even need special clean conditions. From this point, a rapid development set in, leading to observation of the "confusion principle" [20] by adding more and more elements to the system. These different components hamper, especially due to their different sizes, the crystallization of the system and destabilize competing crystalline phases.

The absence of crystallization in turn leads to the formation of an amorphous material. For "classical" oxidic silicate-based glasses, the structural understanding is that there is a short-range ordering (SiO_4 tetrahedra); however no long-range ordering as found e.g. in silicate minerals. For metallic glasses, also no long-range ordering can be observed, due to the absence of Bragg reflections in diffraction experiments. However, there had been a vivid discussion about whether these materials are indeed amorphous or microcrystalline and if they can really be regarded as glasses that continuously form from the melt. The latter question could be answered by annealing experiments that showed that the properties of metallic glasses change in the same way as the ones of conventional glasses [21]. Furthermore, the direct formation from the melt could be proven [22]. However, in comparison with oxidic glasses, metallic glasses are less stable and tend to recrystallize at lower temperatures [8].

Although no long-range ordering exists, short- and medium-range ordering based on size effects and chemical affinity is expected to be pronounced in these materials. Their high atomic packing density alongside the different interactions between the constituent elements greatly influences the respective structure and properties. Since the structure of the glass exhibits no periodicity, every sample, even with the same chemical composition, exhibits its very own atomic arrangement. Due to this fact, it is neither nearly impossible nor necessary to determine the structure of BMGs. However, getting an idea about the three-dimensional arrangement of the atoms via experiments in combination with computational methods leads to a significantly better understanding of the formation and properties of the respective material. Since no structure determination is possible, other methods have to be employed to describe the glassy solid. This can be done via pair distribution function (PDF), coordination number (CN), the chemical short-range order, common neighbor analysis (CAN) and others. The data for these methods can be obtained via X-ray/neutron diffraction, extended X-ray absorption fine structure (EXAFS), electron microscopy (EM), nuclear magnetic resonance (NMR) spectroscopy and of course computational methods. This leads to a variety of different structural models that can be used for the description of metallic glasses. One of the pioneering models is the one of dense randomly packed hard spheres. This model discusses the random packing of identical hard spheres based on possible holes in the structure and the constraint not to produce crystallinity/periodicity. A very detailed review published in 2011 summarizes all these methods as well as the respective structural approaches [23].

2.11.3 Properties and applications

Like normal, crystalline metals, intermetallic compounds and alloys (bulk), metallic glasses are of silvery appearance and exhibit the typical metallic reflectivity. In contrast to normal metals, however, they usually exhibit lower ductility and fatigue strength but a significantly enhanced hardness, higher tensile and elastic strains, high elastic deformation as well as improved wear and corrosion resistance and scratch resistant surfaces with a persistent metallic shine [24, 25]. These properties originate from their amorphous solid state with an absence of grain boundaries. Therefore, both electrical and thermal conductivity behave differently compared to classical metals. Because of the non-ordered structure, the thermal conductivity is lower due to increased phonon scattering [9]. Also, the electrical conductivity behaves differently. While in metals, the conductivity increases with decreasing temperature, in metallic glasses a rather low conductivity can be observed already at room temperature, rendering them "bad metals" [26]. Fe-, Co- and Ni-containing glasses usually retain their ferromagnetic properties, while others are non-ferromagnetic.

The good mechanical properties led to a commercialization in different areas, e.g. sporting goods (golf clubs, baseball bats, ski edges) or fashion products (Table 2.11.1). The enhanced hardness is also employed in a large diversity of medical (razors, microsurgery scissors, scalpels, blades; Table 2.11.2), industrial (wear resistant coatings, optical mirrors, high-performance springs; Table 2.11.3) and transportation applications (fibers in tires or valve springs; Table 2.11.3). But also the ferromagnetic properties are of great importance for commercial applications. Magnetic ribbons can be used for various sensors, transformers (Table 2.11.4) or information storage and microgeared motor parts (Table 2.11.5).

Table 2.11.1: Applications of BMG in consumer and fashion products.

Product and task	Reference
Golf club heads	[9, 27, 28]
Baseball bats, bicycle spokes, edges in skis or skates	[9, 28, 29]
Casing for electronics, cell phones, watches, pens	[26, 27, 29]
Tooling, e.g. knife edges, self-healing anti-scratch coating	[9, 26, 30]
Rings, spectacle frames, jewelry	[9, 26]
Brazing foil	[8]
Wires for musical instruments	[9]

Table 2.11.2: Medical applications of BMG.

Product and task	Reference
Surgical instruments: tweezers, razors, microsurgery scissors, scalpels, blades, thin-film coatings with antimicrobial effects	[9, 27, 31]
Implants for bone/teeth replacements	[9, 27, 29]
Bone substituent (CaMgZn)	[32]
Medical probes, skin-based drug delivery patches	[30]
Self-expandable stent	[33]
Thin-film coating on medical needles and guide wires to reduce slide resistance	[30]

Table 2.11.3: Defense, automotive and aerospace applications of BMG.

Product and task	Reference
Reinforced fibers in automobile tires	[8]
Alloy diaphragms (Zr-Ni-Al-Cu) as pressure control	[29, 34]
Automobile valve spring	[29]
Kinetic energy penetrator	[9, 29, 35]
Solar wind collector in NASA's Genesis project	[9]
Aircraft fasteners, missile components in fins, bodies, gimbals and nosecones	[29]

Table 2.11.4: Industrial applications of BMG.

Product and task	Reference
Thin ribbons of magnetic nanocomposites and purely glassy materials in transformer cores and shieling	[27]
Thin-film coating for diamond abrasive blades or as diffusion barrier	[30]
Coating material (FeCrMoCB) for cast iron and steel	[34]
Optical mirror device (ZrCuAlNi)	[29]
High performance springs	[9, 26, 27]
Thin ribbons as magnetic sensors	[27]
Metallic glass foams	[26]
Catalysts, e.g. in fuel cells, PtCuNiP nanowire for oxidation of methanol and ethanol	[8]
Transducers and sensors, e.g. force sensor, shock sensor, extensometer, accelerometer, displacement transducers, torque transducer, field sensor of flux-gate type, rotation sensors, position inducers	[36]
Glassy tubes as Coriolis mass flowmeter	[25, 29, 34]
Glassy diaphragms in pressure sensor	[25]
Transformers and load bearing applications	[37]

Table 2.11.5: MEMS and technological applications of BMG.

Product and task	Reference
High-precision parts, e.g. Al_3Ti hinges, cogs, shafts, spring actuators, linear comb-drive actuators	[9, 26, 27]
Information storage and reproduction	[26]
Templates for nanoimprinting	[38]
Electrodes in batteries and supercapacitors	[38]
MEMS relays, RF dielectric switches, springs	[9]
Thin-film coating for diaphragms in acoustic devices	[30]
Thin-film metal-insulator-metal diodes, detectors and sensors	[30]
Microgeared motor parts, precision optics, micromachines	[9, 29, 34]
liqualloy sheet as electromagnetic noise suppression sheet in electromagnetic instruments as digital still cameras, in radio-frequency identification system	[34]
FeCrPBSi as powder core for power inductors in laptops	[34]
Material for digital master disks	[9]

References

[1] Klement W, Willens RH, Duwez POL. Nature, 1960, 187, 869.
[2] Chen HS, Turnbull D. Acta Metall, 1969, 17, 1021.
[3] Liebermann HH, Graham C. IEEE Trans Magn, 1976, 12, 921.
[4] Kui HW, Greer AL, Turnbull D. Appl Phys Lett, 1984, 45, 615.
[5] Telford M. Mater Today, 2004, 7, 36.
[6] Inoue A, Zhang T, Nishiyama N, Ohba K, Masumoto T. Mater Trans JIM, 1993, 34, 1234.

[7]   Peker A, Johnson WL. Appl Phys Lett, 1993, 63, 2342.
[8]   Greer AL. Science, 1995, 267, 1947.
[9]   Axinte E. Mater Des, 2012, 35, 518.
[10]   Liebermann HH. Mater Sci Eng, 1980, 43, 203.
[11]   Narasimhan MC. US 4,142,571, 1979.
[12]   Hagiwara M, Inoue A. In: Liebermann HH. (Ed.) Rapidly Solidified Alloys: Processes-Structures -Properties-Applications, CRC Press, Dekker, New York, 1993.
[13]   Brenner A, Couch DE, Williams EK. J Res Natl Bur Stand, 1950, 44, 109.
[14]   Watanabe T, Tanabe Y. Mater Sci Eng, 1976, 23, 97.
[15]   Yeh XL, Samwer K, Johnson WL. Appl Phys Lett, 1983, 42, 242.
[16]   Buckel W. Z Phys, 1954, 138, 136.
[17]   Leamy HJ, Dirks AG. J Phys D: Appl Phys, 1977, 10, L95.
[18]   Koch CC, Cavin OB, McKamey CG, Scarbrough JO. Appl Phys Lett, 1983, 43, 1017.
[19]   Schwarz RB, Petrich RR, Saw CK. J Non-Cryst Solids, 1985, 76, 281.
[20]   Greer AL. Nature, 1993, 366, 303.
[21]   Greer AL. In: Liebermann HH. (Ed.) Rapidly Solidified Alloys: Processes-Structures-Properties-Applications, CRC Press, Dekker, New York, 1993.
[22]   Kui HW, Turnbull D. Appl Phys Lett, 1985, 47, 796.
[23]   Cheng YQ, Ma E. Progr Mater Sci, 2011, 56, 379.
[24]   Russell AM, Lee KL. Structure-Property Relations in Nonferrous Metals, 2005.
[25]   Nishiyama N, Amiya K, Inoue A. J Non-Cryst Solids, 2007, 353, 3615.
[26]   Ashby M, Greer A. Scr Mater, 2006, 54, 321.
[27]   Löffler JF. Intermetallics, 2003, 11, 529.
[28]   Johnson WL. MRS Bull, 1999, 24, 42.
[29]   Suryanarayana C, Inoue A. Bulk Metallic Glasses, CRC Press, Boca Raton, London, New York, 2011.
[30]   Yiu P, Diyatmika W, Bönninghoff N, Lu Y-C, Lai B-Z, Chu JP. J Appl Phys, 2020, 127, 030901.
[31]   Chu JP, Huang JC, Jang JSC, Wang YC, Liaw PK. Jom, 2010, 62, 19.
[32]   Wang YB, Xie XH, Li HF, Wang XL, Zhao MZ, Zhang EW, Bai YJ, Zheng YF, Qin L. Acta Biomater, 2011, 7, 3196.
[33]   Praveen Kumar G, Jafary-Zadeh M, Tavakoli R, Cui F. J Biomed Mater Res B Appl Biomater, 2017, 105, 1874.
[34]   Inoue A, Takeuchi A. Acta Mater, 2011, 59, 2243.
[35]   Nagireddi S, Majumdar B, Bonta S, Diraviam AB. Trans Indian Inst Met, 2021, 74, 2117.
[36]   Hernando A, Vazquez M, Barandiaran JM. J Phys E: Sci Instr, 1988, 21, 1129.
[37]   Li J, Doubek G, McMillon-Brown L, Taylor AD. Adv Mater, 2019, 31, e1802120.
[38]   Liu L, Hasan M, Kumar G. Nanoscale, 2014, 6, 2027.

3 Technical glasses

3.1 Ultra-strong glasses and glass-ceramics and bioactive materials

Hellmut Eckert, Martin Letz

3.1.1 Introduction

For several centuries, glasses have been the eyes of science empowering us and guiding our insight into both the microscopic world and the macroscopic universe. Today, with a plethora of uses ranging from the domestic realm to high-end technologies they are not only pervading all aspects of our everyday life but are at the forefront of fundamental and applied research and development. Optical fibers have become the highways of our modern information age. Fast ion-conducting glasses and glass-ceramics ensure the high power densities required for batteries used in electromotion. Ultra-stable and impact-resistant glasses and glass-ceramics assure personal safety. Bioactive glasses and scaffolds activating osteoproduction genes promote bone and tissue healing. Catalytically active porous glasses functionalized at their surfaces help to protect the environment. It is widely held by scientists and engineers that our technological society is currently entering a new era, appropriately termed the "glass age". This development has been recently recognized by the United Nations who have declared the year 2022 officially the Year of Glass.

For transforming a promising solid material into an application-relevant functional device, many of its physical properties need to be optimized simultaneously. For this objective, glasses are the perfect materials base: we can modify the physical properties of a glass over wide regions by adjusting its chemical composition, by adding new constituents and even by changing the way they are made, processed and thermally treated after their synthesis. Thus glasses offer a vast parameter space for fine-tuning a material for its desired application. An excellent summary of the current state of the art has been recently published in the Springer Handbook of Glass, summarizing in 52 Chapters the important areas of research and development [1].

Because of the complexity involved with technological multi-parameter optimization, even today much of this work is still being done empirically, by trial and error (melting and testing) which is costly, time-consuming and obviously extremely inefficient. At the current stage of the art, this situation is about to change: the ultimate goal of predicting an adequate glass composition for a given application is coming within reach. Three key developments are responsible for this paradigm shift: (a) results from

Acknowledgment: This work was supported by FAPESP through the CEPID program, process number 2013/07793-6 and CNPq and the Deutsche Forschungsgemeinschaft.

https://doi.org/10.1515/9783110733143-019

60 years of modern fundamental research on glasses have been organized in a comprehensive database (SciGlass) [2], (b) highly effective artificial intelligence methods (e.g., neural networks) have been developed for mining this database [3] and (c) powerful theory-based and computational techniques are becoming increasingly reliable for the prediction of structures and atomic transport processes from molecular dynamics simulations and the calculation of physical properties from density functional theory [4]. All of these three research directions continue to be under active development, requiring further extensive experimental testing by fundamental research directed at composition-structure-function relationships. However, already at the present stage, they comprise an enormously influential arsenal of techniques awaiting use in the functional design of glasses for desired applications. In short, we are currently on the threshold of an exciting and challenging decade of research and innovation in glass science.

3.1.2 The glassy state: definition and characteristics

3.1.2.1 Thermodynamic aspects

Numerous definitions of "glass" can be found in the literature, and efforts of improving the definition are continuing [5]. Glasses are non-crystalline solids, originating from the freezing of supercooled liquids. This classifies them as non-equilibrium, non-stationary states [6–8]. The situation is best illustrated on the basis of a state diagram, in which a thermodynamic state function X (X = molar volume, enthalpy or entropy) is plotted as a function of temperature (Figure 3.1.1). If crystallization is kinetically inhibited, the supercooled liquid state persists below the freezing point T_m, and state functions continue to decrease monotonically with decreasing temperature as long as the system can respond sufficiently rapidly to the change in temperature by adjusting its nanoscopic molecular/atomic structure. This process is called relaxation and is driven by molecular dynamics. However, as the temperature keeps decreasing, molecular motion keeps slowing down, until the system ultimately reaches the characteristic *glass transition region* where the cooling rate exceeds the rate of molecular relaxation. In this regime, molecular/structural adjustment towards the stationary supercooled liquid or – ultimately – to the thermodynamically favored crystalline state can no longer occur. The temperature where this is observed, known as the glass transition temperature, T_g, manifests itself in a change of slope, dX/dT. The actual value of T_g detected in this way depends on the observable and the rate with which it is being monitored, as well as the thermal history of the glass under investigation, and the concept of fictive temperature, T_f has been introduced (see Figure 3.1.1). Obviously, T_f depends on the cooling rate and can be changed by annealing at temperatures below T_g. The most common observables are the heat capacity (measured via differential scanning calorimetry), the viscosity or the thermal expansion coefficient. Most importantly, however, T_g depends on the structural organization of the glass, which is determined by its chemical

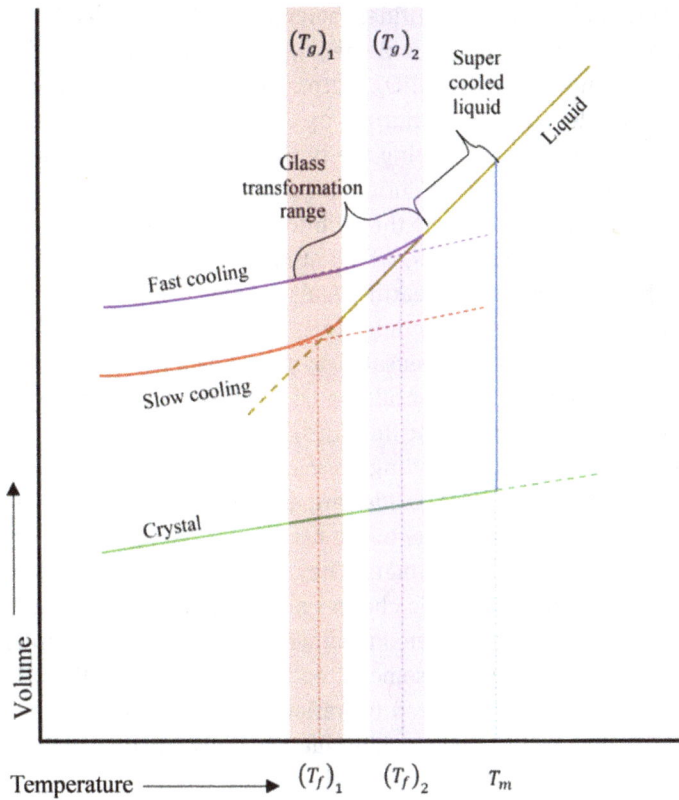

Figure 3.1.1: Molar volume versus temperature diagram for a glass-forming liquid including definition of the fictive temperature, T_f. Reproduced with permission in adapted form from [8].

composition. Broadly speaking, in glasses held together by covalent forces, T_g increases with increasing strength of the chemical bonds and their overall density.

3.1.2.2 Chemical aspects

As implied by the foregoing description, the glass-forming ability of a given material is just a matter of cooling rate. Basically, the temperature interval between T_m and T_g must be crossed fast enough to suppress crystallization. For each material specific *time-temperature-transformation (TTT) diagrams* (see below) have been established, from which the necessary cooling rates can be read off [5–7]. Under standard laboratory conditions (cooling rates < 100 K s^{-1}) a great variety of materials can be prepared in the glassy state, including metals, organic and inorganic molecular solids, salts and compounds with extended structures predominantly characterized by covalent bonding. Glasses prepared from the latter group of materials will be at the focus of this book contribution. They are based on the single- and

multicomponent oxides, chalcogenides (sulfides, selenides, tellurides) and halides (fluorides, chlorides) of the main group elements, the so-called *network formers*. The most common ones are silica, SiO_2, boron oxide, B_2O_3, phosphorus
oxide, P_2O_5, germanium oxide, GeO_2 and tellurium oxide, TeO_2. Frequently, more
than one network former is present, modifying the physical properties of glasses
due to special *network former mixing effects*, originating from specific interactions
between the different components. Most of the technologically relevant glasses
also contain the so-called *network modifiers*, which are generally oxides, chalcogenides or halides of the alkali or alkaline-earth metals as well as of the trivalent
rare earths. Incorporation of such compounds transforms the network into a
macro-polyanion, whose charges are compensated by the network modifier cations, which interact with bridging and non-bridging O, S, Se, Te, F or Cl species.
Some covalent oxides (as well as sulfides or fluorides) can act as either network
former or network modifier species, depending on glass composition. The most
prominent examples for such *intermediate oxides* are aluminum and gallium oxides which in many glass systems participate in the network, forming anionic tetrahedral units bridging to other network formers. They can also occur in the five-
and six-coordinated states, acting as cationic charge compensators for anionically
functionalized networks and then are considered modifiers. Likewise, the rare earths
are usually considered modifiers, although some exceptions may occur in fluoride
glasses. A similarly dual role may be adopted by various transition metal oxides
(TiO_2, V_2O_5, Nb_2O_5, MoO_3, WO_3) and post-transition metal oxides (ZnO, CdO, SnO,
SnO_2, PbO, Sb_2O_3, Bi_2O_3) which by themselves cannot be produced in the glassy state
under usual melt-cooling conditions, but are frequently encountered in multicomponent oxide glasses.

3.1.2.3 Structural aspects

Owing to the lack of translational symmetry, structural models developed for glasses
are usually cast in terms of interatomic distance statistics, the so-called *pair correlation
functions*, which are specified for each possible pair of atoms occurring in the material.
The most radical reference state is that of a random packing of spheres, which may be
appropriate for the structural description of some metallic glasses. Randomly packed
spheres have no type of ordering at all, not even short range ordering. Abundant experimental evidence shows, however, that this extreme model is not applicable for the
more common covalent network glasses, which can be considered random assemblies
of well-defined coordination polyhedra. This is the famous *continuous random network
model* of Zachariasen [9], which despite numerous challenges during the past decades,
has withstood the test of time. Experimental validation has been recently obtained by
atomic resolution microscopy, see Figure 3.1.2. The excellent agreement between the
recent experimental result and Zachariasen's anticipation 80 years earlier is truly impressive [10].

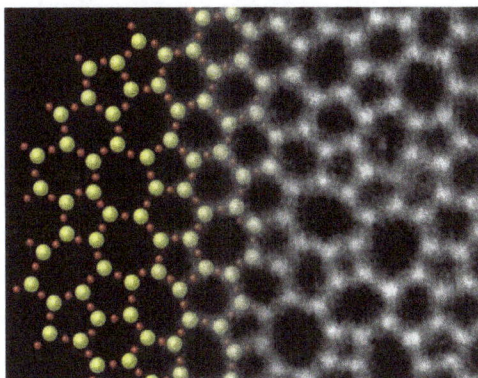

Figure 3.1.2: Two-dimensional representation of the continuous network model of Zachariasen and experimental data from atomic resolution imaging via scanning tunneling microscopy of silica thin films on a Ru(0001) surface. Reproduced with permission in adapted form from reference [10].

Obtaining a comprehensive picture of glass structure then involves a detailed description of how the assembly of coordination polyhedra leads to loss of correlation as a function of distance. It is therefore convenient to discuss this subject on different length-scale domains: (a) *short-range order* involving only the first atomic coordination spheres (distance region 0.15–0.3 nm) describing the bonding of the central atom X and the bridging or non-bridging oxygen, chalcogen or halogen atoms bonded to them, (b) *intermediate-range order* comprising distance correlation in the second coordination sphere (0.3–0.5 nm), (c) *medium-range order* (0.5–1 nm) involving the formation of larger structural units (including clusters, chains and rings), (d) *mesostructure* (1–200 nm) and, finally, (f) *microstructure* (>200 nm). Although possibly appearing somewhat arbitrary these separate domains are closely related to the informational content of the experimental techniques available to probe glass structure, including X-ray and neutron diffraction, vibrational spectroscopy, nuclear magnetic resonance, extended X-ray absorption fine structure (EXAFS) and molecular dynamics simulations. While all of these methods respond most sensitively to changes and details concerning the first coordination spheres, they also contain information on the second and higher coordination spheres. On the other hand, the usual transmission and scanning electron microscopies cover the regimes of meso- and microstructure. The region within 0.5 to a few nanometers distance may be considered "least-known territory" as there is no ideally suited physical examination technique for deriving specific structural information in this length scale domain. Therefore, structural concepts describing medium-range order in glasses usually emerge from extrapolation, via computational modeling and/or energy minimization, respecting constraints posed by the information on first and second coordination spheres, meso- and microstructure as well as knowledge deduced from macroscopic physicochemical observables.

Zachariasen's continuous random network model is based on covalently bonded, well-defined structural units. Figure 3.1.3 summarizes such network former units (NFU-s) as encountered in the most common glass systems based on SiO_2, B_2O_3, P_2O_5 and GeO_2. Analogous structural units can be drawn for sulfide, oxysulfide, selenide, nitride and halide glasses. From X-ray and neutron diffraction experiments, it is now well-known that the typical bond lengths of these units are very similar to those encountered in the structures of corresponding crystalline model compounds (Figure 3.1.4). The second coordination sphere comprises information such as (a) distance correlations between network formers, (b) distance correlations between network formers and network modifiers and (c) distance correlations between the network modifiers. In covalent network glasses, the existence of intermediate-range order follows from the distance and angle-dependent interaction potentials defining chemical bonds. The corresponding three-body interaction potential $V_{abc}(\theta)$ clearly depends on the O–X–O bond angle θ. If $V(\theta)$ has a sharp minimum, such as is the case for the tetrahedral O–Si–O bond angles in silica glasses, then there is a rather narrow distribution of second-nearest neighbor distances. This is indeed the case for the O–O next nearest neighbor correlations in most oxide glasses. On the other hand, $V(\theta)$ functions with broad and shallow energy minima result in a wider distribution of θ, and hence a wider distribution of next-nearest neighbor distances. This is the case for the Si–Si pair correlation function in silicate glasses. Therefore, characterizing such bond angle distributions is an important part of describing intermediate range order in glasses. Thus, the atomic arrangement in the second coordination spheres is neither expected nor found to be random [11]. A related aspect of network-former connectivity concerns the formation of preferred linkages in mixed network-former glasses, owing to chemical preferences. The connectivity of individual NFU-s also involves two torsional angles, defining the spatial relations between atoms that are within the third and fourth coordination spheres of each other [12]. Information about their distribution can be extracted from molecular dynamics simulations. Ordering phenomena beyond the second coordination sphere involve the formation of larger structural motifs such as clusters, chains or rings, which again are a consequence of the directionality of chemical bonding. This distance regime is called "medium-range order", which can be identified in chemically very diverse glass systems [13]. In highly polymerized glasses, a common way of characterizing such effects is specifying ring size distributions [14].

Further, in glasses containing network modifiers, the distribution of these modifiers is not necessarily random, and clustered arrangements have been postulated in some systems on the basis of such simulations. This is the modified random network model proposed by Greaves for alkali silicate glasses, leading to the postulate of ion-conduction channels [15]. It is important to realize that such larger-scale segregation effects will generally also manifest themselves at the level of the second coordination sphere, resulting in non-random network former – network modifier correlations as well as network modifier – network modifier correlations [13].

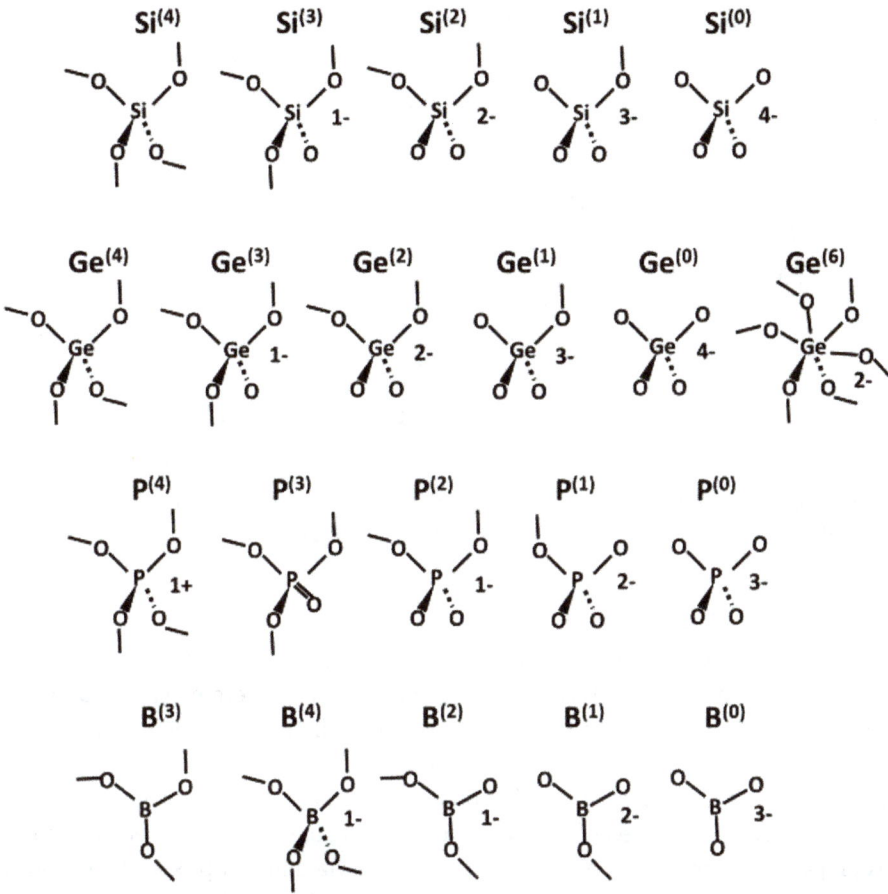

Figure 3.1.3: Common network-former units (NFU-s) occurring in oxide glasses.

Nanoscale segregation effects of the abovementioned kind are the consequence of phase separation tendencies, which are usually kinetically suppressed at and below the liquidus temperature, resulting in glasses that appear more homogeneous than expected on the basis of equilibrium thermodynamics. For such systems, the thermal history of the glass is an important control parameter, by means of which the size distribution of nanosegregated domains, and hence the macroscopic physical properties of the glass can be manipulated. As phase separation phenomena frequently lead to subsequent crystallization, an understanding of medium-range order also provides an important bridge toward the elucidation and elaboration of functional glass ceramics [16–18].

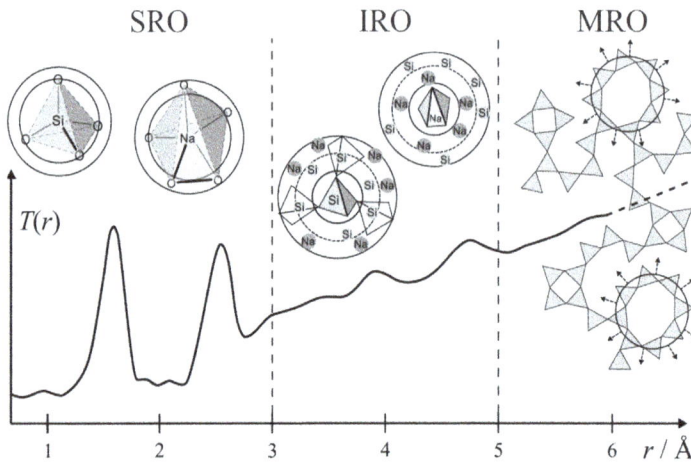

Figure 3.1.4: Spatial correlations in glasses characterizing interatomic distance distribution typical for short-range (SRO), intermediate-range (IRO) and medium-range order (MRO).

3.1.2.4 Glass ceramics

The controlled crystallization (devitrification) of glasses and the use of nucleation agents is known since the 1920s [19], and metal colloids were crystallized out of glasses in the 1940s to obtain photosensitive glasses [20]. In the 1950s, scientists at Corning Glassworks discovered glass ceramics with near zero thermal expansion [21], which became a new scientific sub-discipline in their own right, called glass ceramics (GC) [22]. According to a recently updated definition all inorganic, non-metallic materials prepared by controlled crystallization of glasses via different processing methods are called GC. They contain at least one type of functional crystalline phase and a residual glass. The volume fraction crystallized may vary from ppm to almost 100% [23].

The conversion of a glass into a glass ceramic is shown in a time-temperature transformation (TTT) diagram as illustrated in Figure 3.1.5. Glasses can be prepared if the cooling rate q is larger than the critical cooling rate $q_c = (T_L - T_N)/\delta t$, where T_L is the liquidus temperature and T_N, the *nose temperature*, is the temperature at which the time δt to achieve a crystalline fraction of 10^{-6} is shortest. Cooling rates $q < q_c$ result in uncontrolled crystallization. The most important part of the definition of a glass-ceramic is the time/temperature control exercised during the annealing process as the microstructure of the ceramic can be influenced in this fashion. Most commonly single- and double-stage annealing protocols are being used. Classical theory differentiates between two distinct processes involved in the formation of glass-ceramics: (1) the spontaneous formation of crystal nuclei (nucleation), which can only occur if aggregates of a certain critical size are formed, and (2) the diffusion-controlled addition of more material from the melt to these nuclei (growth). The rates of nucleation and growth depend differently on temperature: typically, the nucleation rate shows a maximum at a lower temperature than the growth rate. Figure 3.1.5 shows typical one- and two-stage annealing protocols leading to ceramics B and C with different microstructures. For example, if a material with small crystallites is desired, the sample is first heated at an initial (lower) temperature where the nucleation rate is high but the growth rate is still low. In this way, many crystal nuclei are produced, whose subsequent growth (induced by heating at the higher temperature) is then limited by the amount of material available in melt surrounding these nuclei. The characterization of a glass-ceramic involves (1) the fraction of crystalline material, (2) the number and identity of the crystalline phases present, (3) their morphology and size distributions and (4) the degree of porosity [23]. As mentioned above, these characteristics are determined by the chemical composition of the base glass and the specific annealing protocol chosen.

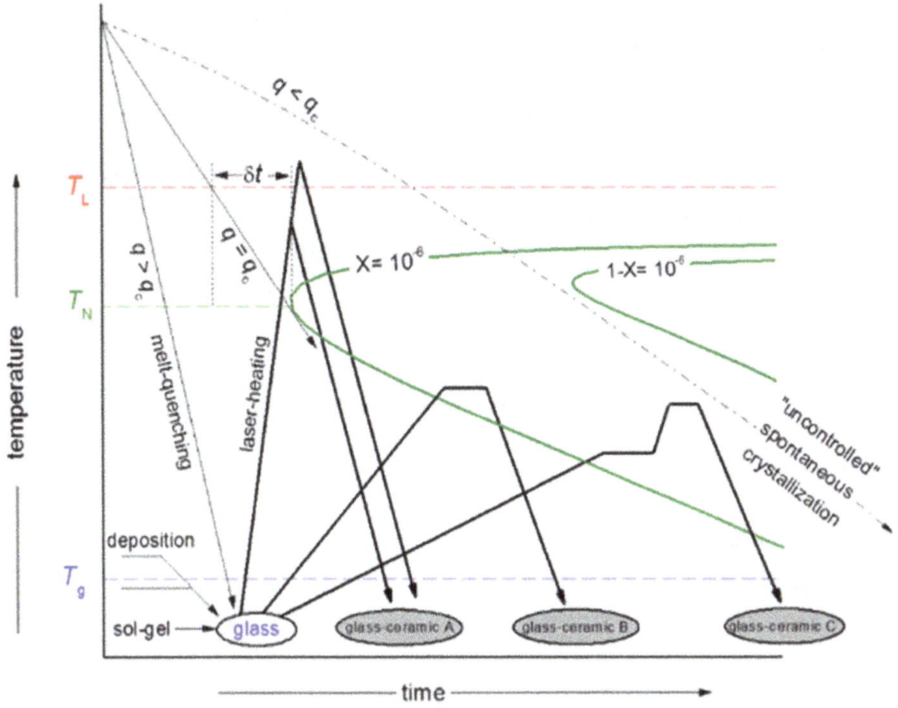

Figure 3.1.5: TTT diagram illustrating the conversion of a glass into a glass-ceramic. See text for further explanation. The glass-ceramic A is obtained during melt cooling, whereas glass-ceramics B and C are converted by single- and double-stage heat treatments, respectively. Reproduced with permission from [23].

3.1.3 Economic significance of glasses

3.1.3.1 Historical trajectory and current high-market-share uses

While natural glasses (obsidian) have been in use as tools and weapons since the stone ages, the first man-made glass dates back to ~3500 BC (Egypt). Glass panels for windows appeared in ancient Rome during the first century AD, but their first uses in Northern European castles can be traced back to the end of the first millennium. Up to the seventeenth century, glass remained mostly an object of admiration and symbol of power (Hall of Mirrors in Versailles). The order for the glass mirrors in Versailles, which was the founding history of today's company Saint Gobain, paved the transition from a royal manufacture towards industrial production and broke the Venice monopoly on mirrors in Europe. The early invention of eyeglasses (thirteenth century) was followed by the use of glasses as essential components of scientific instrumentation (mirrors, lenses in microscopes and telescopes) only several centuries later, leading to today's innumerous technological commodities enhancing daily life. Table 3.1.1 summarizes the currently economically most important application areas. There are large, high-volume markets for glasses such as window glasses and container glasses for e.g. jars and bottles, with production volumes of several thousand tons per year in a single manufacturing site. Also glass wool (see also Chapter 1.2) products as insulation material for buildings have a large market. Most of these high-volume markets are dominated by soda-lime glass (composition approximately 74 wt % of SiO_2, 13 wt% of Na_2O and 10 wt% of CaO [24], which makes up around 90% of

Table 3.1.1: Economically significant glass applications.

Volume markets:
Flat glass (soda-lime)
Container glass (soda-lime)

Special glass markets:
Optical glasses
Flat glass (non-soda-lime, e.g. fire-resistant glazing, cover glasses)
Container glass (non-soda-lime, e.g. pharmaceutical packages)
Glass fibers for telecommunication (fused silica)
Glass fibers for illumination
Laser and filter glasses
Glass powders in electronic packaging
Dental glass powders
Glass-ceramics (e.g. cooking stoves)
Glasses for solid-state memory

Emerging markets:
Ion conducting glasses and glass-ceramics (e.g. for batteries)
Glasses for bone and tissue healing

the total volume of produced glasses. Until the 1950s, the main method to produce flat glasses (for windows, etc.) utilized a glass blowing technique. With this method a long, hollow cylinder was produced, which was cut and flattened at temperatures at which the glass was still flexible. This method limited the available sizes of windows. Also, the flattening process left a characteristic blurring of the glasses, which can be sometimes found in historical buildings. In 1954, Pilkinton [25] revolutionized the production method for flat glasses with the invention of the float process (Figure 3.1.6). In this process, the molten glass is poured on a bath of liquid tin and gradually stretched and slowly cooled down until the glass solidifies. In this way, large formats with very good homogeneity can be produced.

The availability of large glass panels produced in this manner revolutionized modern architecture, which is not thinkable without large window fronts and large-area glass structures. Today, the world market for floated flat glass is around 270 billion USD/year [26] (status 2020). A further important innovation in glass manufacturing was the invention of the ribbon process by Corning in the 1920s [27]. In the ribbon machine, a ribbon of molten glass is moved across a sequence of openings in an underlying plate. Gravity sags the glass down into a mold in which a hollow glass form is produced by blowing air in from above. The ribbon machine revolutionized the manufacturing of light bulbs and illuminated the world. In 1926, a ribbon machine was already able to produce 300 light bulbs per minute [28]. Although light bulbs are meanwhile nearly completely replaced by LED illumination, similar processes are still in use for container glasses. The total world market for container glasses

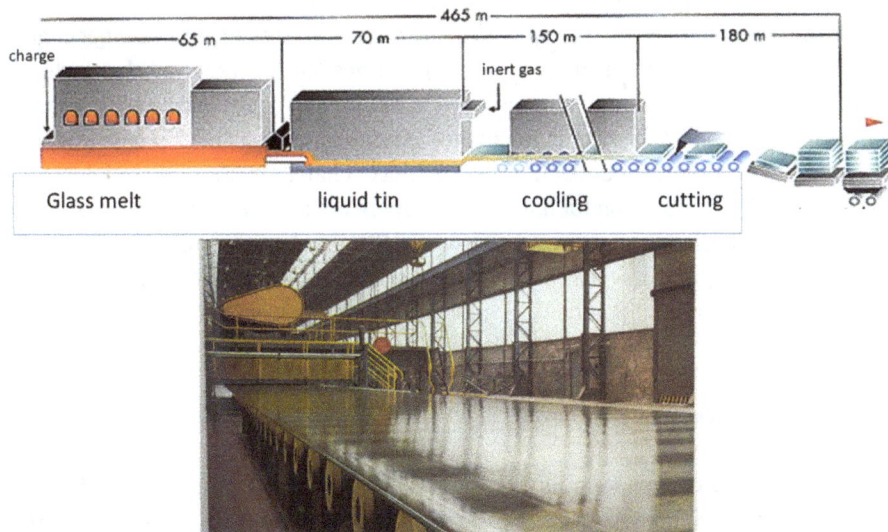

Figure 3.1.6: Schematic of the float glass production process and photograph of a glass layer transported on the cooling tape.

is of the order of 59 billion USD/year (status 2020) [29]. Within the container glass market volume, there is a market included for pharma packaging which has a volume of 4–7 billion USD/year. A second glass type besides soda-lime glass, which makes up a certain part of the glass volume markets are borosilicate glasses. They have an approximate glass composition of 80 wt% SiO_2, 13 wt% B_2O_3, 4 wt% Na_2O and 2–3 wt% Al_2O_3 [30]. The glass was invented by Otto Schott in the years 1887–1893. The introduction of boron oxide into the glass composition leads to glasses with increased chemical durability, lower thermal expansion coefficients and reduced alkali ion content. In history, one of the first applications of borosilicate glasses was for the upcoming gas lanterns, which needed highly temperature stable glasses. Also an, at that time, important problem in the accurate measurements of temperatures was solved using the new boron-containing glass types [31]. But also for cooking ware (brand names like Pyrex [32] or Jenaer Glas [33]), chemical laboratory equipment (brand name Duran) [34] and pharmaceutical packages (brand names like Fiolax) [35] borosilicate glasses find their applications today and will have a certain market share in future.

3.1.3.2 Markets for specialty glasses

There are a large number of glass compositions produced in lower volumes for numerous specialized applications. Clearly, the largest range of compositions is needed for the optical glass markets (mirrors, lenses, optical fibers). The worldwide market volume is around 4 billion USD/year (status 2020). In the area of optical glasses, an important application concerns color correction. Glass applications in electronic devices have significantly decreased due to the replacement of radio tubes by transistors and of cathode ray tubes by flat panel displays, meanwhile even with LED backlighting. The strong reduction in the volume of glass needed has required the development of new manufacturing formats on the millimeter scale. Another important market has opened for chemically toughened glasses for robust smart phone and tablet covers. Chalcogenide glasses find use as phase change memory materials using the glass-to-crystal transition for data storage. There are also fast growing markets for glasses in electronic packaging and sensor or MEMS (micro-electromechanical system) applications, and low-loss substrate materials with very smooth surfaces and good metal adhesion for 5G and 6G technology with frequency bands up to 170 GHz.

3.1.4 Glass science and technology

Glass science as a discipline in its own right could arguably be dated to the famous publication of the volume *Ars vitrea experimentalis* by **Johannes Kunckel** (1630–1703) [36], which summarized the results from century-old experimentations by Venetian glass makers. Understanding glass and improving its performance requires a multidisciplinary approach, as, for instance, documented by the successful collaboration of the chemist **Otto Schott** (1851–1935), the physicist **Ernst Abbé** (1840–1905) and the engineer **Carl Zeiss** (1816–1888) during the nineteenth century. Understanding the glass transition itself remains one of the deepest unsolved problems in theoretical solid state physics. The famous citation of P. W. Anderson (Nobel Prize 1977) remains true today:

> The deepest and most interesting unsolved problem in solid state theory is probably the nature of the glass transition. This could be the next breakthrough in the coming decade. The solution of the problem of spin glass in the late 1970s has broad implications in unexpected fields like neural networks, computer algorithms, evolution and computational complexity. The solution of the more important and puzzling glass problem may also have a substantial intellectual spin-off. Whether it will help to make better glass is questionable. [37]

On the side of knowledge-driven fundamental research, glass scientists attack fundamental research questions dealing with the thermodynamic and kinetic foundations of glass formation, relaxation and crystallization towards glass-ceramics. They want to understand and predict glass-forming ability, and how and why crystallization is suppressed upon fast cooling and how it can be introduced in a controlled way for enhancing physical and chemical properties of glass-ceramics. On the application-oriented side, glasses present unique combinations of mechanical, optical, thermal and electrical properties that can be varied over an expansive range, by altering chemical compositions and processing parameters. The quest for controlling physicochemical properties of glasses and tailoring them to their specific applications motivates detailed investigations of the various structural aspects describing the glassy state of matter. Armed with such knowledge new glasses and glass-ceramics with enhanced physical properties can be designed. Examples covered in the present review and the following contribution to this volume include high compressive strength (for impact-resistant protection devices), tailorable hydrolytic kinetics (for bioactive glass-ceramics optimally adapted in their physiological environment), high ionic conductivity (for high energy density batteries) and efficient photonic response for more powerful lasers and sensing devices.

3.1.4.1 Glass property predictions

Glass-forming ability (GFA) is a property of utmost importance in glass science and technology. In this respect, glass scientists look for suitable predictors, based on the various characteristic temperatures published for a wide range of glass compositions: glass transition temperature (T_g), the onset of crystallization temperature (T_x), the temperature of the crystallization peak (T_p) and the liquidus temperature, (T_L). A recent systematic investigation has shown that the Weinberg parameter $K_w = (T_x - T_g)/T_L$ holds the strongest predictive power of GFA also called glass stability (GS) [38]. Recently, a new GFA parameter has been proposed, based on the quotient $\eta(T_L)/T_L^2$ where $\eta(T_L)$ is the viscosity at the liquidus temperature. Vitrification is favored by a high viscosity at the liquidus temperature, $\eta(T_L)$, and a low value of the liquidus temperature. This parameter bears a great advantage over all other indicators in that there is no need to form a glass to evaluate the GFA.

Against this theoretical background, the crystallization resistance was assessed experimentally in various binary and ternary oxide glasses [39]. Pure oxides rank in the order $B_2O_3 > SiO_2 > GeO_2 > TeO_2 > Al_2O_3$. For the best glass formers B_2O_3, SiO_2 and GeO_2, the addition of small amounts of modifier oxides (Na_2O, CaO, etc.) drastically reduces the GS, whereas for the conditional glass formers TeO_2 and Al_2O_3 the GFA is increased by it. Above 50 mol% modifier oxide, the liquidus temperature is the main factor controlling GFA. Glass-forming compositions are usually found around eutectica in the phase diagrams.

The intrinsic limitation of purely mining experimental data sets lies with (a) the availability of large data sets, (b) the quality/reliability of the data contained therein and (c) the consistency (or not) of the trends uncovered with the laws of physics. These restrictions can be overcome by not simply relying on published data bases of experimental studies but by including results from molecular dynamics (MD) simulations. An illustrative recent example is given in Figure 3.1.7 [40]. In this case the "experimental dataset" originated from high-throughput MD, shown in part (a), while part (b) shows the prediction of an artificial neural network model trained based on the data present in Figure 3.1.7a. In this case, the artificial neural network is able to successfully capture the complex, non-linear evolution of the Young`s modulus as a function of composition while filtering out the intrinsic noise of the simulation data. Figure 3.1.7 shows that the predicted data agree very well with the experimental data for a series of compositions $(CaO)_x(Al_2O_3)_{40-x}(SiO_2)_{60}$. Note that even though no experimental data is available for glasses wherein [CaO] < [Al_2O_3] due to the poor glass-forming ability of such compositions, the combination of MD simulations with machine learning allows for an extrapolation of glass property values.

To move forward from systematic trial-and-error to data- and simulation-driven strategies, data-driven predictive models (neural networks, NN) are being combined with genetic algorithms to design glass compositions with desired non-trivial combinations of properties. In a recent study aimed at developing a novel glass having a combination of a low glass transition temperature ($T_g < 500$ °C) and high refractive index ($n_d > 1.7$) the NN algorithm was trained with data sets of more than 40,000 compositions with 38 different elements. Two potential candidate compositions were suggested by the combined algorithm and produced in the laboratory, closely matching the results the software had calculated [41]. Also, the algorithm yielded fair predictions for extremely low (≤ 450 K) and extremely high (≥ 1150 K) T_g values [42]. Recently, the machine-learning approach has been applied for a broader range of physical properties, such as T_L, elastic modulus, thermal expansion coefficient and Abbé number [43]. These developments constitute important milestones on the road of machine learning-guided design of novel glasses, ready to be used.

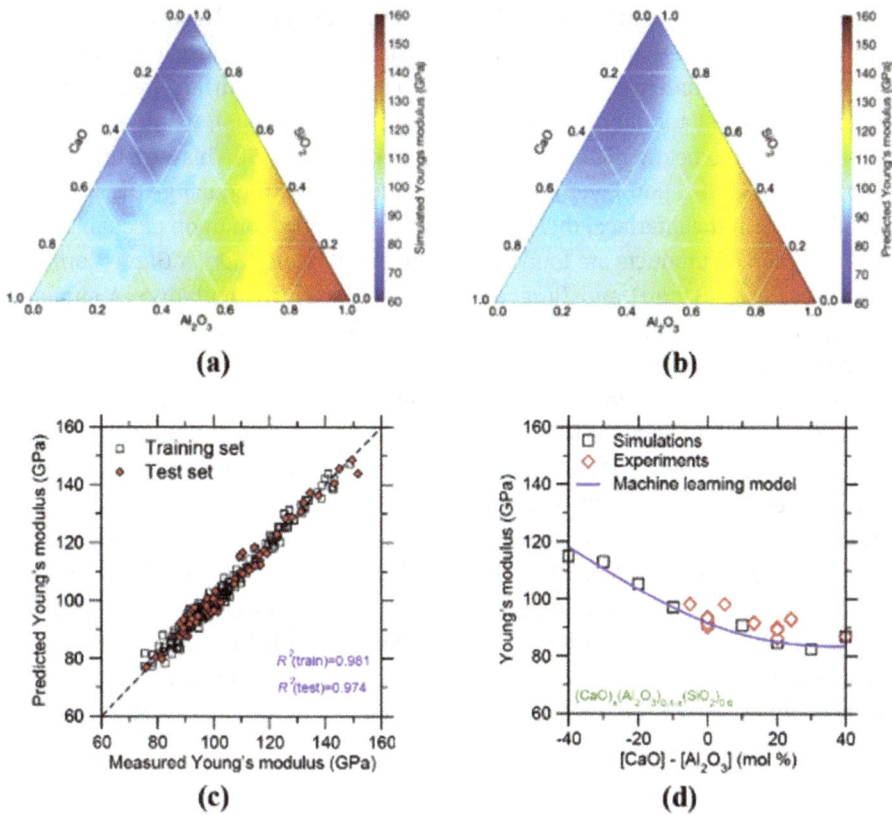

Figure 3.1.7: Ternary diagram showing the Young modulus values as a function of composition in the CaO-Al$_2$O$_3$-SiO$_2$ glass system (a) computed by high-throughput molecular dynamics (MD) simulations and (b) predicted by artificial neural network (ANN). (c) Comparison between the Young modulus values predicted by the ANN model and computed by MD simulations. (d) Comparison between the Young modulus values computed by MD simulations and predicted by ANN with selected available experimental data for the series of compositions (CaO)$_x$(Al$_2$O$_3$)$_{40-x}$(SiO$_2$)$_{60}$. Reproduced with permission from [40].

3.1.4.2 Mechanically ultra-strong glasses and glass-ceramics

Recent years have seen increasing demands for stronger and more damage-resistant glasses. Strength, toughness and elastic properties of glass are currently the main materials limitations impeding the further development of high-capacity telecommunication and fiber technologies, crack-resistant displays, solar modules and planar lighting devices and numerous other applications. Physical quantities to be optimized are strength, fracture toughness, elastic modulus and crack initiation resistance. Increasing the mechanical strength of glasses comes down to reducing the concentrations of defects occurring on different length scales [44]. Thermal annealing at temperatures below, but close to T_g, serves to heal many defects and results in significantly increased crack and scratch resistance. Another common toughening strategy is the exchange of network modifier cations at the surface by their larger homologues (most commonly sodium ions by potassium ions), using exposure to molten salt baths below the glass transition temperature [45]. This leads to the formation of a mixed-cation layer typically tens of μm thick, producing stress buildup and plasticity at the interface, thereby allowing crack deflection upon mechanical impact. Commercial products are found under the brand names *Gorilla Glass* (Corning), *Xensation Cover* (Schott) and *Dragontrail* (Asahi Glass) in circulation. A similarly processed glass was produced until 1990 in the GDR and marketed under the trade name *Superfest*.

Indentation deformation is a new experimental approach for studying the mechanical properties of glasses [46, 47]. Owing to some intrinsic degree of plasticity of the glass, indentation leads to densification (reduction in total volume), which can be quantified by measuring the volume recovery after annealing near T_g. Figure 3.1.8 shows a good correlation of crack initiation resistance with the extent of densification achieved in this manner [47]. The structural changes effected by densification have been successfully monitored using Raman spectroscopy of glassy silica [48]. Figure 3.1.9 shows that densification produces a significant intensity loss of the main band near 430 cm^{-1} relative to the D1 and D2 bands at 470 and 600 cm^{-1}. This effect has been interpreted to indicate a significant reduction of the average Si–O–Si angle upon indentation.

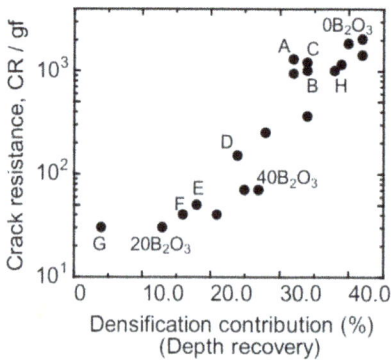

Figure 3.1.8: Relation between crack initiation resistance and glass densification upon nanoindentation for some borate and borosilicate glasses. Reproduced with permission from reference [47].

The most common approach towards enhancing the mechanical properties of glasses is their partial crystallization towards glass-ceramics. Transparency can be maintained if the dimensions of the crystalline phase are significantly smaller than the visible light wavelength range (400–800 nm). Because of their unique thermal stability, low density, chemical durability, high strength and shock resistance glass-ceramics are also ideally suited for a wide range of applications from ballistic protection [49] to restorative dentistry [50]. The intrinsic fracture toughness and flexural strengths of

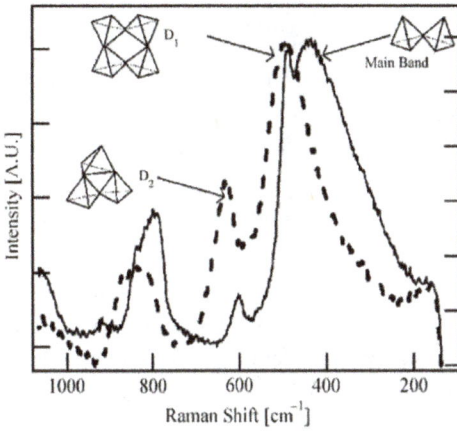

Figure 3.1.9: Raman spectra of glassy SiO_2 before (solid curve) and after indentation (dashed curve), including peak assignments to structural features. Reproduced with permission from reference [48].

glass-ceramics originate from their uniform microstructures, i.e. narrow grain size distribution, special crystal shapes and lack of voids. The common feature of these glass-ceramics is their fibrous microstructure formed by crystals with large aspect ratios, which promote crack deflection and toughening. An important glass-ceramic for this application is lithium disilicate ($Li_2Si_2O_5$), which has received particular interest as material for CAD/CAM manufacture technology in the area of dentistry [50]. Other ultrastrong important glass-ceramic compositions are based on magnesium aluminosilicates, in particular cordierite, $2MgO\text{-}2Al_2O_3\text{-}5SiO_2$ [51, 52]. A further promising new development in the area of glass-ceramics is partially selected patterning of crystals with a single crystal-like architecture by continuous wave (CW) and femtosecond (fs) pulsed laser irradiation [53]. When applying appropriate conditions highly oriented crystals with different morphologies such as straight/bending/spiral lines and two-dimensional planar shapes can be patterned using heat-driven-type CW laser irradiations with a scanning mode. Such ceramics are not only useful from the viewpoint of increased mechanical stability but also of great interest owing to their unique nonlinear optical properties. Recently femtosecond laser pulses have been used to inscribe optical waveguides inside a magnesium aluminum silicate (MAS) precursor glass and glass-ceramic [54], which has shown satisfactory mechanical properties to be applied as ballistic armor. Its good optical waveguiding and mechanical properties indicate that this new transparent glass-ceramic might be adequate for photonic devices that require high mechanical strength.

3.1.4.3 Bioactive glasses and glass-ceramics

Research and development of bio glass-ceramics over the last 40 years have greatly enhanced the quality of life. The replacement and restoration of hips, knees, eyes, ears, teeth, etc. ("human spare parts") has evolved into a multi-million-dollar business. Certain silicate-based glasses and glass-ceramics of the system SiO_2-CaO-Na_2O-P_2O_5 fulfill the necessary requirement of biocompatibility for such applications: the ability to bond to and integrate with living bone in the body without forming fibrous tissue around them or promoting inflation or toxicity [55]. Modern strategies utilize three-dimensional structures, termed "scaffolds", fabricated from a suitable bioglass with high porosity and pore interconnectivity. The scaffolds provide a passive structural support for bone cells and stimulate osteoblastic cell proliferation and differentiation. Various levels of bioactivity can be distinguished: *Osteoconduction* describes the ability of an inert scaffold to allow the formation of new bone, while *osteoproduction* denotes the ability of a bone graft to directly produce bone. It is attributed to the existence of bone-forming cells within the graft. Finally, the activation, proliferation and differentiation of host bone-producing cells is called *osteoinduction* [56]. In particular, the discovery of osteoinduction has resulted in a new therapeutic approach, employing bioactive ceramics as biomineralization scaffolds for inducing bone regeneration rather than as materials to substitute bone tissues [57].

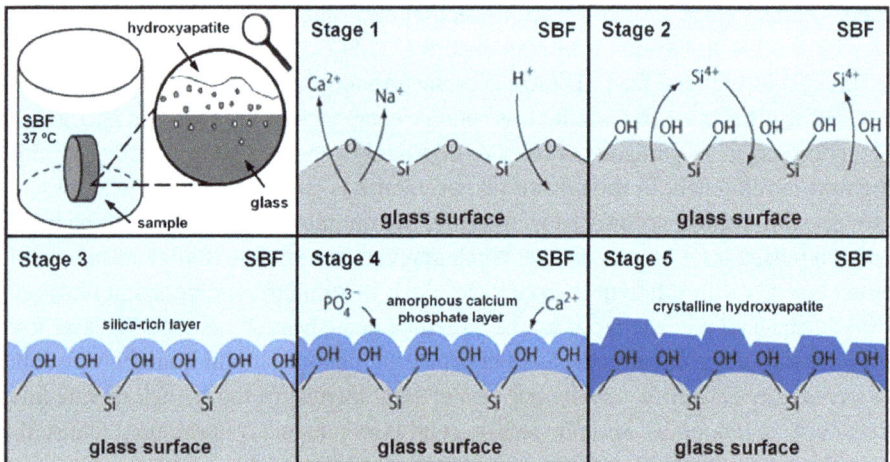

Figure 3.1.10: Mechanistic scheme for osteoconduction in bioactive ceramics.

Figure 3.1.10 summarizes the current view of the mode of action of bioactive glasses promoting the formation of hydroxyapatite. In contact with the body fluids, the glassy material undergoes a sequence of reactions:

In the first step, leaching of Na^+ and Ca^{2+} cations is followed by the formation of Si–OH units at the glass surface. The resulting silica gel surface layer incorporates calcium and phosphate ions from the solution, forming an amorphous calcium phosphate surface layer. Following the subsequent incorporation of carbonate into this layer, a nanocrystalline carbonate apatite layer (the mineral phase of bone and teeth) is formed in situ, which bonds to bone and teeth and for some compositions even to soft tissue.

Figure 3.1.11 gives an overview on the range of relevant bioactive compositions in the Na_2O-CaO-SiO_2 glass system containing a fixed amount of P_2O_5. Particularly strong bioactivity is found over a range of compositions near the calcium metasilicate–sodium metasilicate join. The reactivity of these bioglasses can be controlled by the $(Na_2O + CaO)/SiO_2$ ratio, the Na_2O/CaO ratio and the phosphorus content. Small amounts of P_2O_5 are considered to be critical to stimulate the mineralization of hydroxyapatite carbonate. The most prominent bioactive glass is the 45S5 glass, whose name signifies a silica content of 45 wt% and a Ca/P molar ratio of 5:1. Its exact molar composition is $46.1SiO_2$-$26.9CaO$-$24.4.Na_2O$-$2.6P_2O_5$. The mechanical stability of the scaffold can be increased by partial crystallization to form glass-ceramics. For example, Biosilicate® is the designation of the particular glass-ceramic composition ($23.75Na_2O$-$23.75CaO$-$48.5SiO_2$-$4P_2O_5$ (wt%)). Under controlled double-stage heat treatments, this material can be engineered to produce one (1P) or two crystalline phases (2P): a sodium–calcium silicate ($Na_2CaSi_2O_6$) or both $Na_2CaSi_2O_6$ and a sodium–calcium phosphate ($NaCaPO_4$) phase [58].

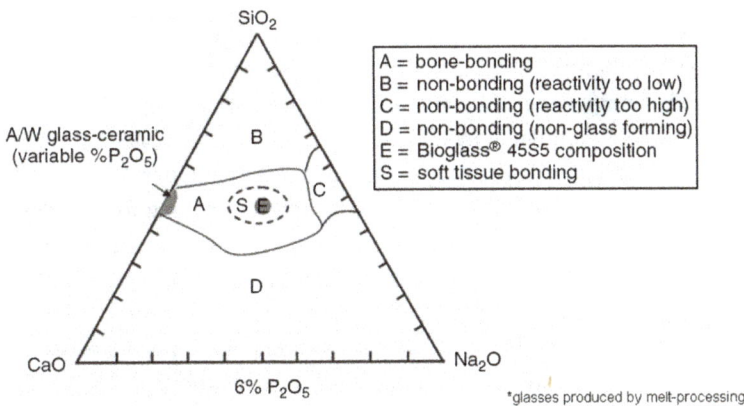

Figure 3.1.11: Composition diagram of the CaO-Na_2O-SiO_2 ternary system containing 6 wt% P_2O_5, indicating the various levels of bio-activity for bone and tissue healing.

Figure 3.1.12 summarizes the various biological responses triggered by the ions released from the bioglasses. In the areas of both hard and soft tissue engineering an important challenge is the development of efficiently vascularized networks. Indeed, bioactive glasses promote angiogenesis [59], opening up new horizons in

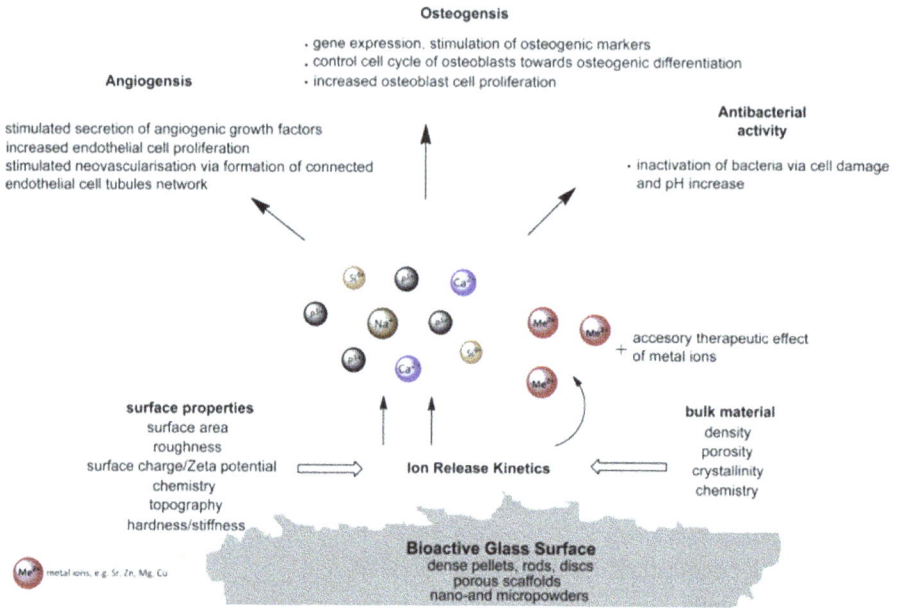

Figure 3.1.12: Summary of the biological effects trigged by ion release from bioactive glasses into body fluids. Reproduced with permission from reference [59].

biomedical tissue engineering. In view of these new applications, "bioactivity" is no longer defined as the ability of inducing carbonated hydroxyl apatite formation.

Dissolution kinetics, angiogenesis activity and other functional properties of bioactive glasses can be further modulated by the addition of other oxides [59–65]. Particularly important is the partial substitution of CaO by its homologues MgO or SrO [60–61]. Modified bioactivity is also encountered in fluoride-containing formulations [62], leading to the mineralization of apatite or mixed fluoro-hydroxyapatites. Boron may be included to impede crystallization, and formulations containing boron, gallium or silver ions are sometimes used to impart antimicrobial activity to the material [63–65]. The addition of alumina, while generally reducing the overall bioactivity, increases the mechanical stability of the material [66]. Following a similar motivation, TiO_2, ZrO_2, Nb_2O_5 and other transition metal oxides are frequently included in the formulation. In all cases the structural effects of such additives are of interest to rationalize modified functional properties, i.e. dissolution, release and mineralization rates, biocompatibility, porosity/pore structure, mechanical stability, and activity of angiogenesis.

Owing to their large surface areas and porosities sol-gel-derived ceramics generally have faster osteoproduction rates, faster bone bonding and excellent degradation and resorption rates compared to melt-quenched bioglasses [67, 68]. In addition, the incorporation of supramolecular chemical preparation strategies (using non-ionic and polymer templates) has resulted in mesoporous and hierarchically ordered bioglasses, which offer significantly increased bioactivity [69, 70], along with the opportunity of hosting active agents for drug delivery. Furthermore, owing to the large sizes of osteo-blastic cells, imparting macroporosity (pore sizes in the μm range) to the material are particularly important for promoting effective cell-bioceramic interactions leading to osteoblastic cell adhesion, proliferation, and differentiation. Hierarchically ordered sol-gel-derived bioceramics with both meso- and macroporosity are particularly attractive, because the mesopores of such sol-gel-derived materials can be loaded with anti-inflammatory agents, drugs or other biomolecules that contribute to the tissues' heal-ing processes. To be able to flexibly meet the diverse demands of specific applications, one aims at the production of samples for which dissolution kinetics and bactericidal activity can be actively tuned via composition and processing parameters. Tailorable ion dissolution rates are relevant, for example, in the controlled delivery of Sr to pro-mote bone regeneration, to prevent infection in bone and dental implants and en-hance contrast in X-ray imaging by replacing Ca [61]. Optimized sol-gel preparation procedures include details regarding the chemical nature and purity of the molecular precursors used, volumes and concentrations of the solutions, the use of co-solvents, complexation agents or additives/templates, the pH value, the temperature and dura-tion the of hydrolysis and polymerization reactions and annealing protocols, all of which control compositional homogeneity, surface area, and pore structure and vol-ume. Standard functional characterization involves (1) measuring the rates of ion-specific dissolution in simulated body fluid (SBF) [71], (2) monitoring amorphous cal-cium phosphate precipitation by IR-attenuated reflectance (ATR) spectroscopy and (3) tracking the conversion of the latter to crystalline hydroxyapatite carbonate by ATR and X-ray diffraction. In addition, the interaction of the scaffolds with live tissue can be characterized on explanted material from the body.

References

[1]   Springer Handbook of Glass, Musgraves JD, Hu J, Calvez L. (Eds.) 2019. ISBN 978-3-319-93726-7.
[2]   Priven AI, Mazurin OV, Liška M, Galusek D, Klement R, Petrušková V. Adv Mater Res 2008,39–40,147–52.
[3]   Liu H, Fu Z, Yang K, Xu X, Bauchy M. J Non Cryst Solids 2021,557,119419.
[4]   Mauro JC, Tandia A, Vargheese KD, Mauro JC, Smedskjaer MM. Chem Mater 2016,28,4267–77.
[5]   Zanotto ED, Mauro JC. J Non Cryst Solids 2017,471,490–95.
[6]   (a) Zallen R. The Physics of Amorphous Solids 1998, Wiley VCH, Weinheim, Germany. ISBN 9780471299417; (b) Feltz A. Amorphous Inorganic Materials and Glasses, Wiley VCH, Weinheim, Germany, 1993, ISBN 1560812125.
[7]   Doremus RH. Glass Science, Wiley VCH, Weinheim, Germany, 2nd edition, 1994, ISBN 0471891746.
[8]   Varshneya AK. Fundamentals of Inorganic Glasses, Gulf Professional Publishing, Elsevier, Amsterdam, 1994. ISBN 9780080571508.

[9] Zachariasen WH. J Am Chem Soc 1932,54,3841–51.
[10] Lichtenstein L, Büchner C, Yang B, Shaikhutdinov S, Heyde M, Sierka M, Włodarczyk R, Sauer J, Freund HJ. Angew Chem Int Ed 2012,51,404–07.
[11] Wright AC. J Non Cryst Solids 1990,123,129–48.
[12] Yuan X, Cormack AN. J Non Cryst Solids 2003,319,31–43.
[13] Eckert H. In: Dronskowski R, Kikkawa S, Stein A. (Eds.). Handbook of Solid State Chemistry, Wiley VCH, Weinheim, 2017, 93–137 and references therein.
[14] Yuan X, Cormack AN. Comp Mater Sci 2002,24,343–60.
[15] Gurman SJ. J Non Cryst Solids 1990,125,151–60.
[16] Greaves GN. J Non Cryst Solids 1985,71,203–17.
[17] Deubener J. J Non Cryst Solids 2005,351,1500–11.
[18] Zanotto ED, Tsuchida J, Schneider JF, Eckert H. Int Mater Rev 2015,60,380–94.
[19] Singer F. Patentschrift 570148, Reichspatentamt 1929.
[20] Stookey SD. J Am Ceram Soc 1949,32,246–49.
[21] Stookey SD. Ind Engin Chem 1959,51,805–08.
[22] Höland W, Beall GH. Glass Ceramic Technology, Wiley-VCH, Weinheim, Germany, 2019. ISBN 9781119423690.
[23] Deubener J, Allix M, Davis MJ, Duran A, Höche T, Honma T, Komatsu T, Krüger S, Mitra I, Müller R, Nakane S, Pascual MJ, Schmelzer JWP, Zanotto ED, Zhou S. J Non Cryst Solids 2018,501,3–10.
[24] Ashby MF. (Ed.) Materials and the Environment, 2nd edition, Elsevier, Amsterdam, 2013. ISBN 978-0-12-385971-6
[25] Pilkington LAB, Bickerstaff K. US2911759, 1959.
[26] Görz G, Schmid U, Schneeberger J, Braun T. (Eds.). Handbuch der Künstlichen Intelligenz, 6. Auflage, De Gruyter, Berlin, Germany, 2021. ISBN 978-3-11-021808-4
[27] Dyer D, Gross D. The Generations of Corning, Oxford, 2001, 119.
[28] Innovations in Glass, Corning, New York: The Corning Museum of Glass, 1999.
[29] https://www.mordorintelligence.com/industry-reports/global-glass-bottles-containers-market-industry. Accessed February 8, 2022.
[30] www.glassproperties.com. Accessed February 8, 2022.
[31] Kappler D, Steiner J. SCHOTT 1884-2009 vom Glaslabor zum Technologiekonzern, Schott, Mainz, 2009, 21.
[32] https://www.corning.com/worldwide/en/products/life-sciences/resources/brands/pyrex-brand-products.html. Accessed February 8, 2022.
[33] https://www.jenaerglas-shop.de/. Accessed February 8, 2022.
[34] https://www.schott.com/de-de/products/duran-p1000368. Accessed February 8, 2022.
[35] https://www.schott.com/de-de/products/pharmaceutical-tubing-p1000372/technical-details. Accessed February 8, 2022.
[36] Kunckel J. Ars Vitraria Experimentalis, Oder Vollkommene Glasmacher-Kunst. Frankfurt (Main), 1679. In: Deutsches Textarchiv. https://www.deutschestextarchiv.de/kunckel_glas macher_1679, Accessed February 8, 2022.
[37] Anderson PW. Science 1995,267,1615.
[38] Jiusti J, Cassar DR, Zanotto ED. Int J Appl Glass Sci 2020,11,612–21.
[39] Jiusti J, Zanotto ED, Feller SA, Austin HJ, Detar HM, Bishop I, Manzani D, Nakatsuka Y, Watanabe Y, Inoue H. J Non-Cryst Solids 2020,550,120359.
[40] Yang K, Xu X, Yang B, Cook B, Ramos H, Bauchy M. Condens Matter Phys 2019,1901,1–20.
[41] Cassar DR, Santos GG, Zanotto ED. Ceramics Int 2021,47,10555–64.

[42] Alcobaça E, Mastelini SM, Botari T, Pimentel BA, Cassar DR, de Carvalho ACPDLF, Zanotto ED.
      Acta Mater 2020,188,92–100.
[43] Cassar DR, Mastelini SM, Botari T, Alcobaça E, de Carvalho ACPDLF, Zanotto ED. Ceramics Int
      2021,47,23958–72.
[44] Wondracek L, Mauro JC, Eckert J, Kühn U, Horbach J, Deubener J, Rouxel T. Adv Mater
      2011,23,4578–86.
[45] Green J, Tandon R, Sglavo VM. Science 1999,283,1295–97.
[46] Januchta K, Smedskjier M. J Non Cryst Solids X 2019,1,100007.
[47] Yoshida S. J Non Cryst Solids X 2019,1,100009.
[48] Perriot A, Vandembroucq D, Barthel E, Martinez V, Grosvalet L, Martinet C, Champagnon B. J
      Am Ceram Soc 2006,89,596–601.
[49] Gallo LS, Villas Boas MOC, Rodrigues ACM, Melo FCL, Zanotto ED. J Mater Res Technol
      2019,8,3357–72.
[50] Ritzberger C, Apel E, Höland W, Peschke A, Rheinberger VM. Materials 2010,3,3700–13.
[51] Costa Oliveira FA, Franco JA, Cruz Fernandes J, Dias D. Adv Appl Ceram 2002,101,14–21.
[52] Boccaccini DN, Leonelli C, Romagnoli M, Pellacani GC, Veronesi P, Dlouhy I, Boccaccini AR.
      Adv Appl Ceram 2007,106,142–48.
[53] Komatsu T, Honma T. J Solid State Chem 2019,275,210–22.
[54] Ferreira PHD, Fabris DCN, Villas Boas MOC, Bezerra IG, Mendonça CR, Zanotto ED. Optics
      Laser Technol 2021,136,106742.
[55] Hench LL. J Am Ceram Soc 1991,74,1487–510.
[56] Montazerian M, Zanotto ED. RSC Smart Mater 2017,23,27–60.
[57] (a) Hench LL, Xynos ID, Polak JM. J Biomater Sci Polym Ed 2004,15,543–62; (b) Pneumaticos
      SG, Triantafyllopoulos, GK, Chatziioannou S, Basdra EK, Papavassiliou AG. Trends Molec Med
      2011, 17, 215–22.
[58] Crovace MC, Chinaglia CR, Peitl Filho O, Zanotto ED. J Non Cryst Solids 2016,432,90–110.
[59] Kargozar S, Baino F, Hamzehlou S, Hill RG, Mozafari M. Trends Biotechnol 2018,36,430–43.
[60] Watts SJ, Hill RG, O'Donnell MD, Law RV. J Non Cryst Solids 2010,356,517–24.
[61] O' Donnell MD, Hill RG. Acta Biomater 2010,6,2382–85.
[62] Brauer DS. Angew Chem 2015,54,4160–81.
[63] Rivadeneira J, Gorustovich A. J Appl Microbiol 2016,122,1424–37.
[64] Ylänen H. Bioactive Glasses: Materials, Properties and Applications, 2nd edition, Woodhead
      Publishing Series in Biomaterials, Elsevier, Amsterdam, 2017. ISBN 978-0-08-100936-9.
[65] Boccaccini AR, Brauer DS, Hupa L. (Eds.). Bioactive Glasses: Fundamentals, Technology and
      Applications, Royal Society of Chemistry, Cambridge, 2016. ISBN 1782622012.
[66] Hench LL. J Am Ceram Soc 1998,81,1705–28.
[67] Hoppe A, Güldal NS, Boccaccini AR. Biomater 2011,32,2757–74.
[68] Arcos D, Vallet-Regi M. Acta Biomater 2010,6,2874–88.
[69] Wei GF, Yan XX, Yi J, Zhao LZ, Zhou L, Wang YH C, Yu CZ. Micropor Mesopor Mater
      2011,143,157–65.
[70] Han X, Li X, Lin H, Ma J, Chen X, Bian C, Wu X, Qu F. J Sol Gel Sci Technol 2014,70,33–39.
[71] Kokubo T, Kushitani H, Sakka S, Kitsugi T, Yamamuro T. J Biomed Mater Res 1990,24,721–34.

3.2 Special glasses for optical and electrical device applications

Martin Letz, Hellmut Eckert

3.2.1 Introduction

Technologically relevant special glasses can cover a large number of material properties. The reason for that is that nearly all chemical elements from the periodic table can be incorporated in a glass composition. This gives a nearly infinite number of possibilities to tune and optimize particular material properties. The range of accessible material properties becomes even larger if glass-ceramics also are included. In glass-ceramic, a glass is molten and in a second and independent step, a crystal phase is grown using a well-defined time–temperature profile. In this way, ferroelectric phases [1], paraelectric phases [2], semiconducting phases [3] or even piezoelectric phases [4] can be obtained embedded in a residual glass phase.

3.2.2 Glasses and glass-ceramics for electrical device applications

There are numerous examples in electronics where glasses play a crucial role. Electronic packaging becomes increasingly important since multiple chiplets need to be combined into one single heterogeneous package. Here, glass substrates are one possible material solution. Glass sealings play a major role for the safety of high power electronics. For example, no airbag igniter would work reliably over decades even in different climate zones without a highly optimized glass-to-metal seal. Glass-to-metal sealings in high-frequency applications with defined electrical impedances are important for telecommunication applications like 5G/6G or radar-based sensors. Also MEMS (micro-electro mechanical systems) applications are realized with glass as substrates and based on glass wafer technology. In LTCC (low-temperature co-fired ceramics) applications, glasses with low softening temperature are essential to adjust the firing temperatures of different green tapes. Glass wafers as carriers or also as core substrates are already important for the semiconductor industry. In the following we discuss a few of such examples in more detail.

Acknowledgment: H.E. acknowledges support by FAPESP through the CEPID program, process number 2013/07793-6 and CNPq and the Deutsche Forschungsgemeinschaft.

https://doi.org/10.1515/9783110733143-020

3.2.2.1 Fast ion-conducting glasses and glass-ceramics

The development of efficient devices for energy storage, conversion and transmission is a key priority of societal efforts towards more sustainable energy economies. At their focus lies the development of high-energy and high-power batteries based on both lithium ion [5] and solid oxide fuel cell (SOFC) [6] technologies. High-performance batteries require solid electrolytes based on fast ion-conducting materials. Figure 3.2.1 summarizes the current state of the art in this field, in terms of the temperature dependence of ionic conductivity. Glassy materials feature an important part in this picture. For example, extremely high lithium ion conductivities up to 10^{-2} S cm^{-1} at room temperature are observed in various (oxy)sulfide and chalcohalide systems [7, 8]. Similar values are also found in some low-dimensional invert glasses combining lithium orthophosphate, molybdate, and iodide components [9]. The practical implementation of such glasses in high-energy batteries is complicated, however, by their low hydrolytic stability and their tendency to crystallize. Another issue limiting practical applications is their low oxidative and reductive stability at the interfaces with the cathode and anode compartments of electrochemical cells. High ionic conductivity is also featured by many oxidic crystalline superionic compounds with disordered cation substructures. For application of polycrystalline electrolytes in solid-state battery devices, fine powders of these crystals have to be well compacted to maximize interparticle contacts. Even so, ineffective ion transport across grain boundaries presents a problem and reduces device performance. Other serious transport limitations arise at the electrolyte–electrode interfaces. The advantageous features of both stable oxidic superionic crystals and favorable interfacial properties of glasses can be combined in an ideal way by using glass-ceramics processing, where the superionic crystalline phase is nucleated from a glassy precursor material. A case in point are the highly conductive ceramic electrolytes based on the Na-super ion conductor (NA-SICON) structure [9] presenting electrical conductivities in excess of 10^{-2} S cm^{-1} at room temperature [10]. Figure 3.2.2 sketches some structural details. The prototype material, $AT_2(PO_4)_3$ (A = Li, Na, T = Ge, Ti), possesses two sites for the mobile monovalent A cations, one of which (the $A2$ site) is unoccupied. This material can be rendered highly conductive by aliovalent substitution of tetravalent T atoms by a trivalent species T' (such as Al^{3+}, Ga^{3+}, Cr^{3+} or Sc^{3+}) and an additional A^+ cation for charge compensation leading to the composition scheme $Li_{1+x}T'_xT_{2-x}(PO_4)_3$. The extra monovalent cation now occupies the $A2$ sites, creating a three-dimensional conduction pathway imparting superionic conductivity to the material (see Figure 3.2.3). The analogous result can be obtained by substituting triply charged orthophosphate by quadruply charged orthosilicate anions. Lithium-containing NASICON glass-ceramics are already in commercial use in the form of membrane separators in lithium/air batteries offering potential energy densities of up to 11140 kWh kg^{-1} [11–15].

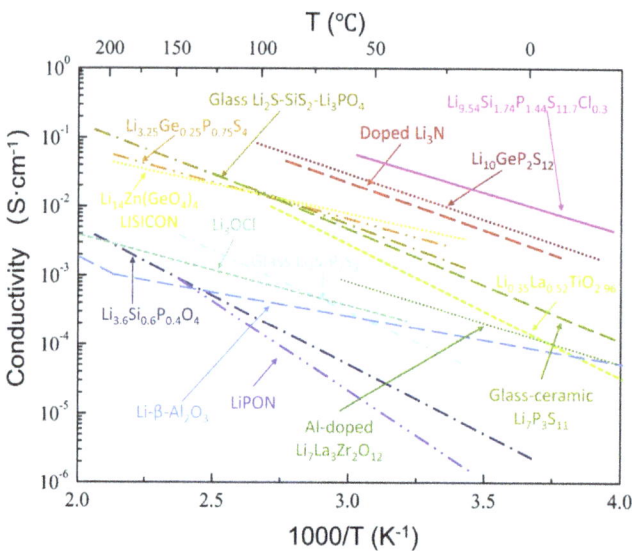

Figure 3.2.1: State of the art in fast ion conduction: DC ionic conductivity as a function of temperature. Reproduced with permission from reference [8].

In recent years, the development of analogous Na-batteries has been fueled by the much higher natural atomic abundance of this element (500 times that of lithium) and its much more widely spread geographical distribution. Intense research during the past few decades has shown that the best sodium ion conductors are usually not the isostructural analogues of the corresponding lithium compounds. Rather, a dedicated effort has to be made to develop the solid-state chemistry of new ion-conducting materials. Nevertheless, promising results have again been achieved with glass-ceramics based on the NASICON structure. Such systems present a much wider compositional variety than the corresponding Li-ion conductors. The literature reports more than 100 systems. The relevant crystal chemical parameters influencing ion-transport properties have been discussed in recent reviews [16, 17]. Na-conducting NASICONs that have attracted interest for solid-electrolyte applications in sodium batteries generally have higher alkali contents than the Li-NASICONs.

An important prototype material is $Na_3Zr_2(SiO_4)_2PO_4$, benefiting again from the advantages of glass-ceramic processing [18]. Alloying with sodium niobium phosphate glasses enhances their conductivities even further [18]. Finally, an extremely promising glass-ceramic system can be derived from the highly conductive $Na_5YSi_4O_{12}$ (N5) phase, which gives a σ value of 10^{-1} S cm^{-1} at 300 °C and an activation energy of 25 kJ mol^{-1} [19, 20]. The ionic conductivity is affected by the presence of two lower-conducting phases, however, of stoichiometries $Na_3YSi_3O_9$ (N3) and $Na_9YSi_6O_{18}$ (N9). Minimization of these competing phases is a severe challenge for defining the ceramization protocol.

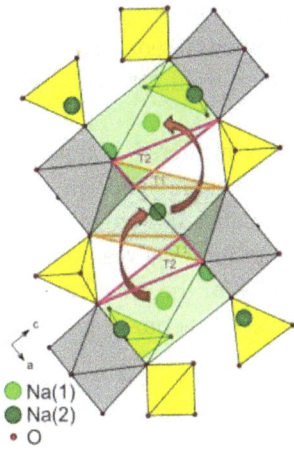

Figure 3.2.2: Alkaline ion environments (green colored) in NASICON solid electrolytes. In the unsubstituted material, the A2 sites are unoccupied. Aliovalent $T(IV) \leftrightarrow T'(III) + A(I)$ substitution leads to an occupation of the A2 sites. Tetrahedral phosphate ions are shown in yellow and GeO_6 octahedra are shown in gray color. Reproduced with permission from reference [15].

- Na(1)
- Na(2)
- O

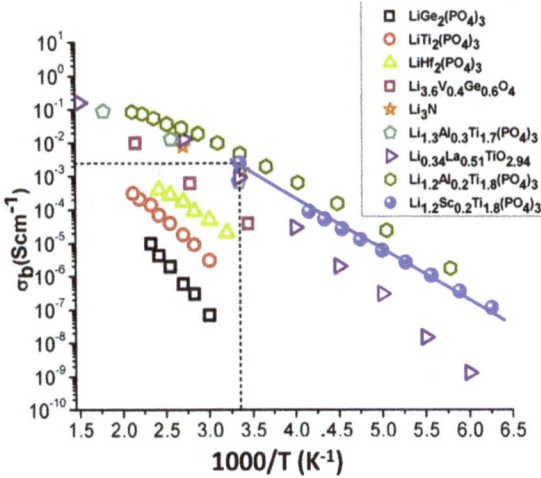

- $LiGe_2(PO_4)_3$
- $LiTi_2(PO_4)_3$
- $LiHf_2(PO_4)_3$
- $Li_{3.6}V_{0.4}Ge_{0.6}O_4$
- Li_3N
- $Li_{1.3}Al_{0.3}Ti_{1.7}(PO_4)_3$
- $Li_{0.34}La_{0.51}TiO_{2.94}$
- $Li_{1.2}Al_{0.2}Ti_{1.8}(PO_4)_3$
- $Li_{1.2}Sc_{0.2}Ti_{1.8}(PO_4)_3$

Figure 3.2.3: DC ionic conductivity as a function of temperature in Li-containing NASICON glass-ceramics and some reference materials. Reproduced with permission from reference [8].

3.2.2.2 Glasses for solid-state memory devices

The glass transition itself can be used for data storage. In optical memory materials as they are used in rewrite-able CD or DVD's the amorphous and crystalline states of the glass-forming chalcogenide system Ge-Te-Sb form the states "0" and "1" of a binary data storage. For a review on these materials, see [21]. In the chalcogenide system the amorphous state with low reflectivity and high electrical conductivity can be easily distinguished from the crystalline state with high reflectivity and low electrical conductivity. This strong contrast in material properties between amorphous and crystalline phase together with a fast crystallization kinetic make such systems applicable for rewriteable memory devices. In this case a fast crystallization kinetic means that a long laser pulse or electrical pulse, where long means an order of 10^{-8} s, with relatively low intensity should be sufficient to turn an amorphous region into a crystalline structure. On the other hand a short pulse, where short means $\ll 10^{-8}$ s, with high intensity will create an amorphous region.

The field of solid-state memory devices is usually not considered as a part of glass science, since most of the devices are fabricated in semiconductor environments and the glass layers are applied with thin-film techniques like sputtering in clean-room fabrications. However, glassy dynamics, the transition between the two states of material "crystalline" and "amorphous" is essential for the functioning of these rewriteable solid-state memory devices. Current research is ongoing to also apply this method to all-solid-state hard drives.

3.2.2.3 Glass sealings in high-power semiconductors

Glass-to-metal seals in electronics have a long history of more than 100 years. Glasses are excellent insulators and wherever an electric feedthrough in a harsh environment, at elevated temperatures or at high electric field strength is required, a glass seal is a solution. Today, power diodes with reverse voltages in the kilovolt (kV) range, high-power transistors like IGBT's (insulating gate bipolar transistors), thyristors, thermal fuses, electrical feedthroughs e.g. gas tanks or electrical power plants all use glass seals. In these cases, a major topic is to adapt the thermal expansion of the glass to that of the surrounding metals to either achieve a stress-free or a compressive sealing. In all these applications the glass is applied as glass powder, which is, in most cases, pressed with a binder into a preform and later sealed together with the metal parts in a high-temperature process. Depending on the glass composition a large variety of sealing temperatures and thermal expansions can be realized.

3.2.3 Glasses for optical device applications

For thousands of years, mankind has known transparent materials. The German word "Brille", which means eyeglasses, is from its word origin connected to the English word "brilliant", has its origin in the word "Beryll" which is a hexagonal crystal of $Al_2Be_3[Si_6O_{18}]$, a transparent mineral with a weak birefringence. However, understanding the existence of a band gap, which means of an energy range where practically no absorption of electromagnetic radiation occurs, requires quantum mechanics. And quantum mechanics started just in the year 1900 with the quantum hypothesis of Max Planck [22], centuries after the discovery of transparent materials. If such a band gap has an upper energy which is larger than the photon energy of blue photons, the visible photons with the largest energy quanta, the material is transparent in the visible spectral range and can be used for optical applications

$$E_{\text{bandgap}} > h\, v_{\text{blue}} \tag{3.2.1}$$

with $h\, v_{\text{blue}} = h\, c/\lambda_{\text{blue}} = 3.18$ eV, where h is Planck's constant, c the speed of light in vacuum (or air) and v the frequency. The energy is given here in units of electron Volt [eV] (1 eV = 1.602 10^{-19} J). Besides transparency, a second property is crucial for optical materials. This is the (nearly) complete absence of light scattering in the visible spectral range. Contrarily to most ceramic materials, glasses are isotropic and extremely homogeneous on all length scales that are relevant for the interaction with visible light. While glasses have a well-defined structure on atomic length scales in the range of Ångstrom (1 Å = 0.1 nm), they are completely disordered on all longer length scales that are relevant for the interaction with visible light (400 to 800 nm). Therefore, glasses are homogeneous and isotropic for electromagnetic radiation in the visible range and do not show significant light scattering beyond the fundamental Rayleigh scattering limit. This makes glasses the ideal material class for optical applications. Single crystals also have a certain share as optical materials. However, the growth of single crystals with dimensions and shapes large enough for optic lenses requires a large effort. Most crystals are non-cubic and therefore show birefringence [23]. One example of a commercial use of single crystals is CaF_2 [24] for ultraviolet applications, e.g. 193 nm microlithography, or for infrared optics. Also, sapphire, also non-cubic and having a slight birefringence, is used in optics e.g. as a transparent cover material for watches.

3.2.3.1 Optical fibers for light transmission

The use of optical fibers for light transmission is fundamental in modern telecommunications. Figure 3.2.4 shows the functional principle. The general principle of fiber drawing is discussed in Chapter 3.3. The optical fiber is made up of an inner core material carrying the optical signal and an outer cladding layer whose index of refraction is lower than that of the core material. If the inner core of a fiber is so thin, that it is of the order of the wavelength of the light a fiber can be made that transports only one single light mode. In this case the classical picture of total internal reflection breaks down and a wave description of the light modes is needed. These are the single-mode fibers, which are needed for telecommunication applications and which can transport signals over hundreds of kilometers. Electrical signals are converted to light, carried with the speed of light over long distances and converted back to electrical signals at the receiving end. The wavelength of the light chosen to carry the information depends on the transmission window of the core material. At the high-frequency end, the transmission window is limited by electronic transitions of the matrix, and this limit generally lies in the ultraviolet region. The low-frequency end of the transmission window is defined by infrared absorption of vibration modes.

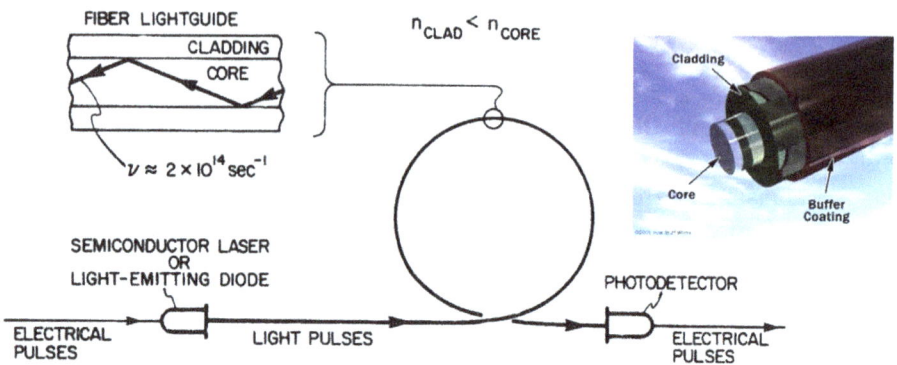

Figure 3.2.4: Working principle of optical fiber signal transmission.

Glassy silica fibers remain the prevalent material, showing signal attenuation performance of less than 0.2 dB km^{-1} at 1.55 μm and 0.3 dB km^{-1} at 1.3 μm. Figure 3.2.5 shows a typical plot of attenuation loss as a function of wavelength [25]. Attenuation losses are caused by a variety of mechanisms: (1) the UV absorption tail at the low-wavelength end, (2) the vibrational absorption tail at the high-wavelength end, (3) the fundamental Rayleigh scattering process and (4) infrared absorption due to vibrations of hydroxyl impurities. The latter effect can be minimized by special synthesis precautions. Since the Rayleigh scattering losses are proportional to the inverse fourth power (λ^{-4}) of the transmission wavelength, attenuation losses can potentially be reduced further by choosing longer transmission signal wavelengths. The latter requires the use of glasses with lower vibration frequencies, such as tellurite, chalcogenide and halide glasses (see Figure 3.2.6). Unfortunately, the low hydrolytic stability of fluoride and chalcogenide glasses requires significant efforts in glass synthesis and fiber preparation and the theoretically estimated limit of 0.08 dB/km is never reached [26]. Typical results reflecting the current state of the art obtained for glasses is 12–50 dB/km at wavelengths of 2–6 μm in As-S and As-Se glasses [26] and about 80 dB km^{-1} at 6.6 μm in the Ge-As-Se system [27].

3.2.3.2 Glasses for optical lenses

All real optical materials show dispersion, which means that the speed of light in a material is changing with changing frequency. An optical prism which splits white light in its spectral components or a rainbow, which arises from the dispersion in water droplets are well-known examples for dispersion in materials. The microscopic origin of dispersion in materials (not only glasses) is discussed in [28].

An optic designer who likes to construct a lens system, which has a minimum of optical aberrations, has to face several challenges:

- chromatic aberrations due to the dispersion of the optical lens materials lead to colored rims at strong contrasts
- spherical aberrations lead to a blurring of the picture
- the total amount of available light is limited and leads to compromises between depth of focus and resolution

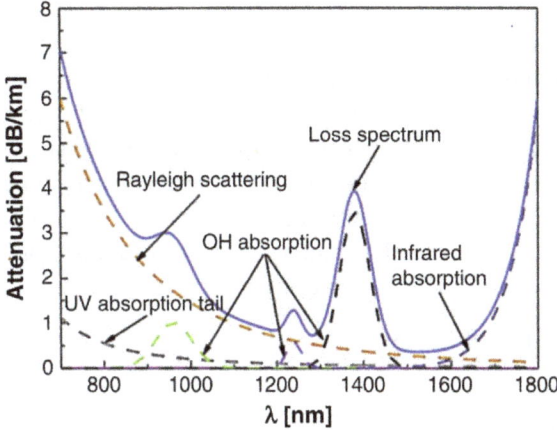

Figure 3.2.5: Typical attenuation loss of glassy silica as a function of transmission wavelength. Sources of the various attenuation losses are indicated. Reproduced with permission from [25].

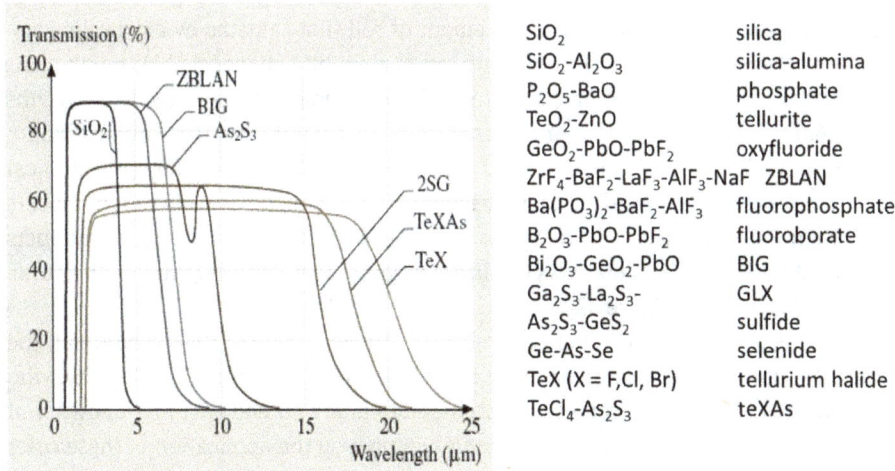

SiO_2	silica
SiO_2-Al_2O_3	silica-alumina
P_2O_5-BaO	phosphate
TeO_2-ZnO	tellurite
GeO_2-PbO-PbF_2	oxyfluoride
ZrF_4-BaF_2-LaF_3-AlF_3-NaF	ZBLAN
$Ba(PO_3)_2$-BaF_2-AlF_3	fluorophosphate
B_2O_3-PbO-PbF_2	fluoroborate
Bi_2O_3-GeO_2-PbO	BIG
Ga_2S_3-La_2S_3-	GLX
As_2S_3-GeS_2	sulfide
Ge-As-Se	selenide
TeX (X = F,Cl, Br)	tellurium halide
$TeCl_4$-As_2S_3	teXAs

Figure 3.2.6: Common fiber glass formulations and their optical transmission windows.

3.2.3.2.1 Strategy to fight chromatic aberrations

In 1733 [29], telescope makers started to learn that one can strongly reduce chromatic aberrations in an optical system, a telescope, by replacing a single lens element by a pair of lenses. Both lenses in the pair, a convex one and a concave one, had to be made from different glasses. The convex one was made from a crown glass, and the concave lens was made from a flint glass. A crown glass means a glass with a relatively low refractive index and a weak dispersion and a flint glass means a glass with a higher refractive index and a strong dispersion. The achromat was invented. In an achromatic lens system the red and the blue light are focused on the same focal point. However, the green light, whose wavelength lies in-between, does not necessarily have to have the same focal point. In order to construct an achromatic lens system, the slope of the refractive index of each of the couple in a lens pair has to be brought in a defined connection to the lens geometry. Historically evolved, optic design still uses the Abbe number of a glass. The Abbe number is defined via the refractive indices $n(\lambda_i) = n_i$ of a material at three different wavelengths, λ_i:

$$v = \frac{n_1 - 1}{n_2 - n_3} = \frac{\beta}{\delta} \approx \frac{n_1 - 1}{\lambda_2 - \lambda_3} \frac{1}{\frac{\partial n}{\partial \lambda}\big|_{n=n_1}} \tag{3.2.2}$$

Here, n_1 is usually the refractive index at a wavelength in the yellow or green spectral range, n_2 is the one in the blue spectral range and n_3 is the one in the red part of the visible spectrum. For a long time and still in most textbooks, the refractive indices at the three wavelengths $\lambda_1 = 587.5618$ nm (the yellow He line, d), $\lambda_2 = 486.1327$ nm (the blue hydrogen line, F) and $\lambda_3 = 656.2725$ nm (the red hydrogen line, C) were used to define the Abbe number n_d. The replacement of roll-film cameras by electronic cameras with their shifted spectral sensitivity led to a slightly changed Abbe number, v_e, being more common. v_e is defined via the refractive indices at the three wavelengths $\lambda_1 = 546.0740$ nm (the green mercury line, e), $\lambda_2 = 479.9914$ nm (the blue cadmium line, F') and $\lambda_3 = 643.8469$ nm (the red cadmium line, C'). For a listing of the most common wavelengths used in optic design, see Table 3.2.1. Even if tunable lasers have been existing for several decades, optic design is still based on these numbers generated based on particular spectral lines of gases. The Abbe diagram n_e as a function of v_e is shown in Figure 3.2.8.

A correction of the green light as well led to the invention of the apochromat, the fully color-corrected lens system. It was Ernst Abbe, professor at Jena University, who formulated the conditions for an apochromat [30] and who had as a partial owner of the companies Carl Zeiss and Schott a strong impact on the application of these color correction schemes in telescopes and microscopes. For an apochromat, not only the first derivative of the wavelength dependent refractive index but also the curvature, related to the second derivative of the wavelength dependent refractive index, $n(\lambda)$, has to be taken into account for the design of an optical system. In practice, this

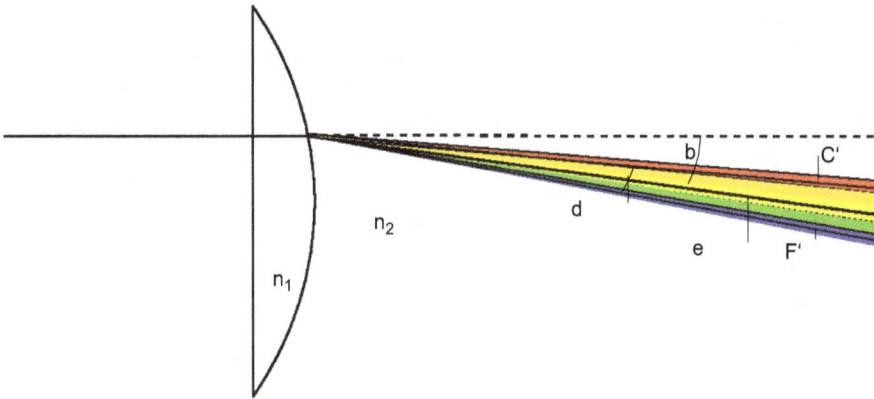

Figure 3.2.7: Illustration of chromatic aberration on a lens element with refractive index n_1 in a medium with refractive index n_2. For $n_2 = 1$ (vacuum or air), the refractive index is related to the angle β and the Abbe number is related to the quotient of β/δ.

curvature is expressed as a partial dispersion, which is, again, defined via discrete absorption lines. They are based on the blue (F) and red (C) hydrogen lines.

$$P_{x,y} = \frac{(n_x - n_y)}{n_F \cdot n_C} \tag{3.2.3}$$

Table 3.2.1: Common spectral lines which are used in optic design andtheir abbreviations.

Wavelength [nm]	Desig- nation	Spectral line used	Element	Wavelength [nm]	Desig- nation	Spectral line used	Element
2325.42		Infrared mercury line	Hg	587.5618	d	Yellow helium line	He
1970.09		Infrared mercury line	Hg	546.0740	e	Green mercury line	Hg
1529.582		Infrared mercury line	Hg	486.1327	F	Blue hydrogen line	H
1060.0		Neodymium glass laser	Nd	479.9914	F′	Blue cadmium line	Cd
1013.98	t	Infrared mercury line	Hg	435.8343	g	Blue mercury line	Hg

Table 3.2.1 (continued)

Wavelength [nm]	Desig-nation	Spectral line used	Element	Wavelength [nm]	Desig-nation	Spectral line used	Element
852.11	s	Infrared cesium line	Cs	404.6561	h	Violet mercury line	Hg
706.5188	r	Red helium line	He	365.0146	i	Ultraviolet mercury line	Hg
656.2725	C	Red hydrogen line	H	334.1478		Ultraviolet mercury line	Hg
643.8469	C´	Red cadmium line	Cd	312.5663		Ultraviolet mercury line	Hg
632.8		Helium-neon gas laser	He-Ne	296.7278		Ultraviolet mercury line	Hg
589.2938	D	Yellow sodium line	Na	280.4		Ultraviolet mercury line	Hg
		(center of the double line)		248.3		Ultraviolet mercury line	Hg

The most common choice for the wavelengths x and y are the blue mercury line, g, and the blue hydrogen line, F. The partial dispersion is called $P_{g,\,F}$. Often, the deviation from a "normal line", $\Delta P_{x,y}$, is also used. It is defined as the deviation from a straight line in the $P_{x,y}$, v_d diagram, which is shown in Figure 3.2.9.

$$\Delta P_{x,y} = P_{x,y} - \left(a_{xy} + b_{xy}n_d\right) \tag{3.2.4}$$

where the coefficients for the linear connection, a_{xy} and b_{xy}, are usually based on the glass types K7 and F2. Glasses which differ significantly from the straight line are of particular interest for color correction.

3.2.3.2.2 Strategy to fight spherical aberrations

Any grinding and polishing process of a glass piece with randomly changing orientations will lead to spherical shapes. Therefore, the most common lens type has a sphere segment as a surface form. This can be either as a convex or a concave surface shape. In typical lens systems, all spatial dimensions are much larger than the wavelength of light. Therefore, classical ray optics can be applied [31].

The situation of refraction in a spherical lens element is sketched in Figure 3.2.10. Snell's law of refraction gives a clear connection between the incoming ray and the diffracted ray:

$$n_1 \sin(\alpha) = n_2 \sin(\alpha + \gamma) \tag{3.2.5}$$

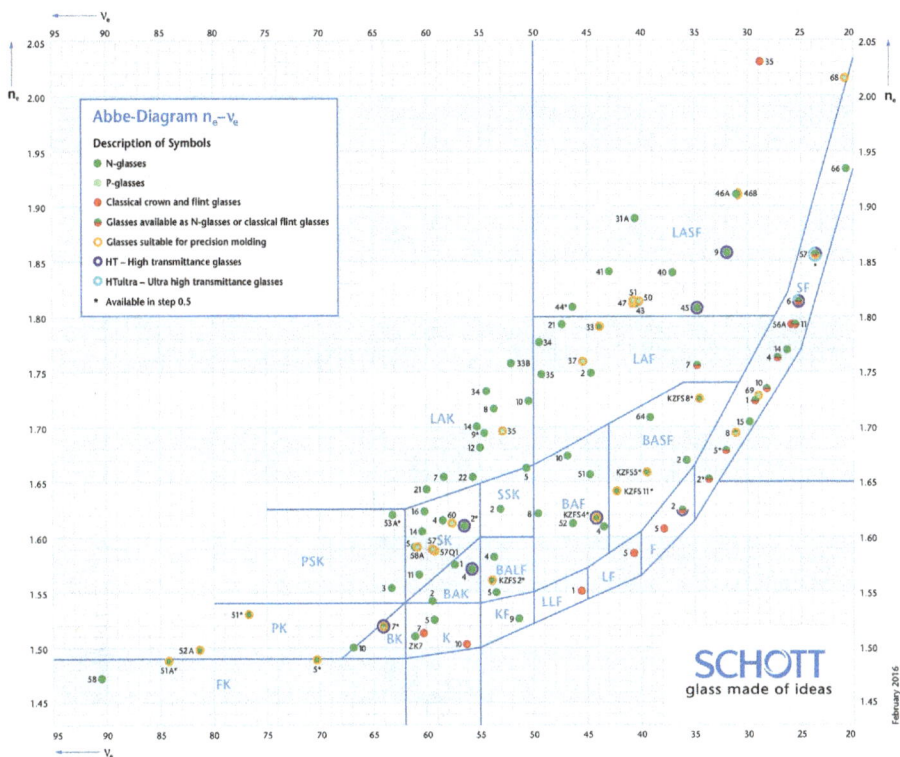

Figure 3.2.8: Abbe diagram based on the three wavelengths n_e, n_F and n_C.

The focal length, f, of the lens can be obtained from a basic trigonometric calculation:

$$R\sin(\alpha) = f\tan(\gamma) \tag{3.2.6}$$

Equations (3.2.5) and (3.2.6) can be easily solved in the case of near central rays where $\sin(\alpha) \sim \alpha$, $\sin(\alpha + \gamma) \sim \alpha + \gamma$ and $\tan(\gamma) \sim \gamma$ hold. In this case, the focal length can be directly expressed as:

$$f = R\frac{n_2}{n_1 - n_2} \tag{3.2.7}$$

which gives for lenses in air or vacuum ($n_2 \sim 1$) the well-known expression $f = R/(n_1{-}1)$. For light rays that have a larger distance to the center of the lens, the approximations for the trigonometric functions are not valid. Here, the expressions get more complicated and as a result such rays see a shorter focal length as shown as a dashed line in Figure 3.2.10.

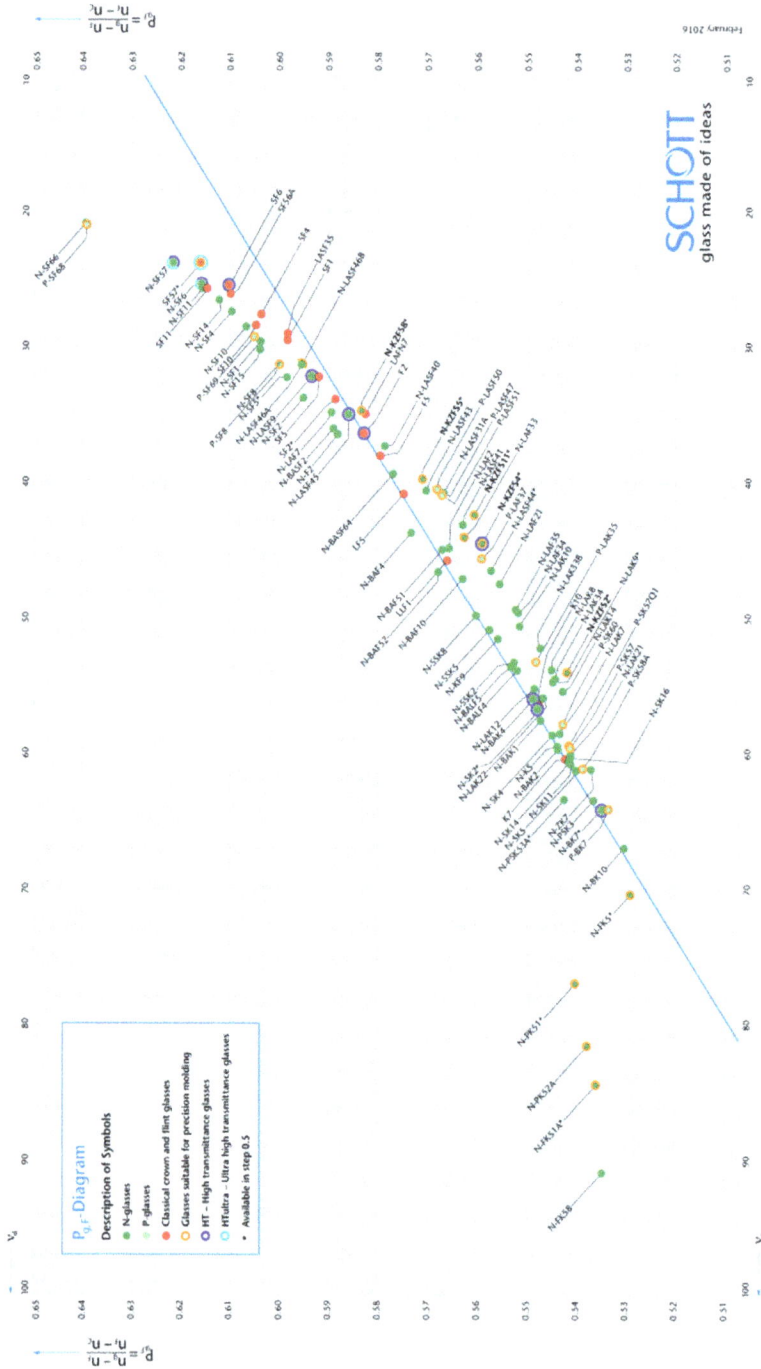

Figure 3.2.9: Diagram for the spatial dispersion $P_{g,F}$ based on the three wavelengths n_g, n_F and n_C.

When using glasses with high refractive indices, one can see from eq. (3.2.7) that the same focal length can be obtained by using smaller radii of curvature, R, of the lens. Therefore a high-index glass material will reduce spherical aberrations.

3.2.3.2.3 The unsatisfiable dream of optic designers

From the previous two subsections, we can see that an ideal lens system is made of an apochromatic pair of lenses with high refractive index (e.g. $n \approx 2$) with an Abbe number sufficiently spaced and with identical curvature in the dispersion of the refractive index. Therefore, a "glass" with a refractive index of $n \approx 2$ and an Abbe number of $v_d \approx 80$ would be an ideal material. As one can see in Figure 3.2.8 there is not a single glass that comes even near to that parameter range. It has clear physics origins that such a glass will never be available. A large refractive index always goes together with a strong dispersion. Among single crystals one can find materials that combine large refractive indices with relatively large Abbe numbers. Diamond has $n_d = 2.41726$ and $v_d = 55.3$ [32]. Also, transparent ceramics and optoceramics can reach record-high values of n_d and v_d. Since it is very difficult to make large optical lens blanks with high homogeneity out of ceramic materials [33] or single crystals, optical materials are still dominated by glasses and the dream of the optic designer for a high n_d high v_d material remains unfulfilled and will be unfulfilled forever.

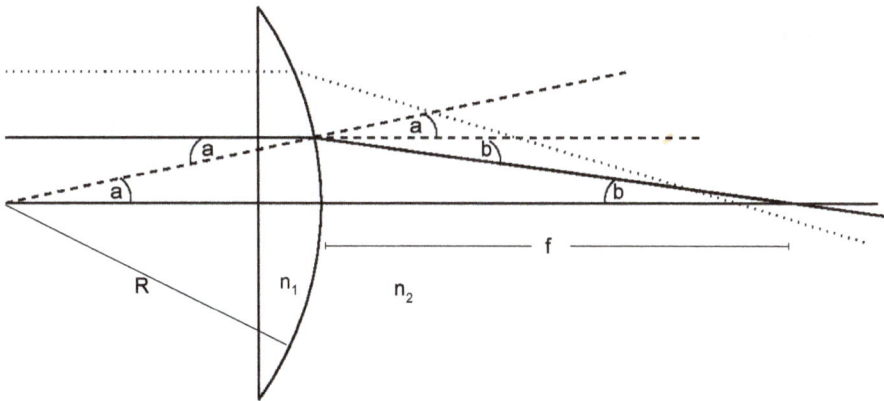

Figure 3.2.10: Illustration of spherical aberration on a spherical lens element with radius R and refractive index n_1 in a medium with refractive index n_2.

3.2.3.2.4 The Sellmeier series for the dispersion of the refractive index

The dispersion of the refractive index in the visible range can be approximated with high accuracy by the Sellmeier formula, which is widely used for characterizing optical materials:

$$n(\lambda)^2 \approx 1 + \sum_j \frac{B_j \lambda^2}{\lambda^2 - \lambda_j^2} \tag{3.2.8}$$

where λ_i are wavelengths at which strong absorptions in the material are assumed and B_j is proportional to the strength of the absorptions at these wavelengths. It describes the dispersion of a material by using infinitely sharp absorption lines (δ-functions), which are far away from the visible range at wavelength λ_j. Normally B_j and λ_j are just fitting parameters, to describe the dispersion of the refractive index over a certain wavelength range. B_j is unit-less and λ_j is usually given in units of μm. The Sellmeier formula (3.2.8) uses three infinitely sharp absorption lines. One in the infrared (IR) wavelength range and two in the ultraviolet (UV) range of the spectrum fit the dispersion of material in the range of optical wavelengths. Even if only used as fitting parameters, the absorptions are connected to the microscopic fundamental absorptions of the material. In some cases $n(\lambda)$ and not $n(\lambda)^2$ is approximated with a Sellmeier formula. Since $n(\lambda)$ as well as $n(\lambda)^2$ are complex differential (analytic) functions both formulae give refractive indices and dispersions with the same accuracy. However, care has to be taken, which quantity is used when using a Sellmeier formula.

3.2.3.3 Filter glasses
Filter glasses are glasses which make use of certain absorption bands to eliminate certain wavelength regions from the spectrum of transmitted light. Depending on the type of absorption and/or by combining different absorptions a wide variety of filter properties can be selected. Most filter glasses are based on optical absorptions of 3d transition metal ions. But there are also filter glasses, where the filter characteristic is based on quantum dots made of chalcogenide systems. Also, interference filters exist where the phase relation of different light waves of different optical paths creates a change in light amplitude. There are low-pass (e.g. UV blocking filters), high-pass (e.g. IR blocking filters) and band-pass filters available. For filter glasses, it is of crucial importance to have a coloration that is just based on absorption while fluorescence should be as low as possible.

Filter glasses played an enormous role in the past, for roll-film camera photography. But also today's modern digital cameras use a blue glass filter as an IR-cut filter in digital cameras. Such an IR-cut filter is of enormous importance for the quality of digital cameras, since a photodiode cannot distinguish between visible and IR photons [34].

3.2.3.4 Photonic and laser glasses
Owing to their excellent optical transparency, their ability to accommodate and disperse rare-earth ions homogeneously and in high concentrations, and their ease of

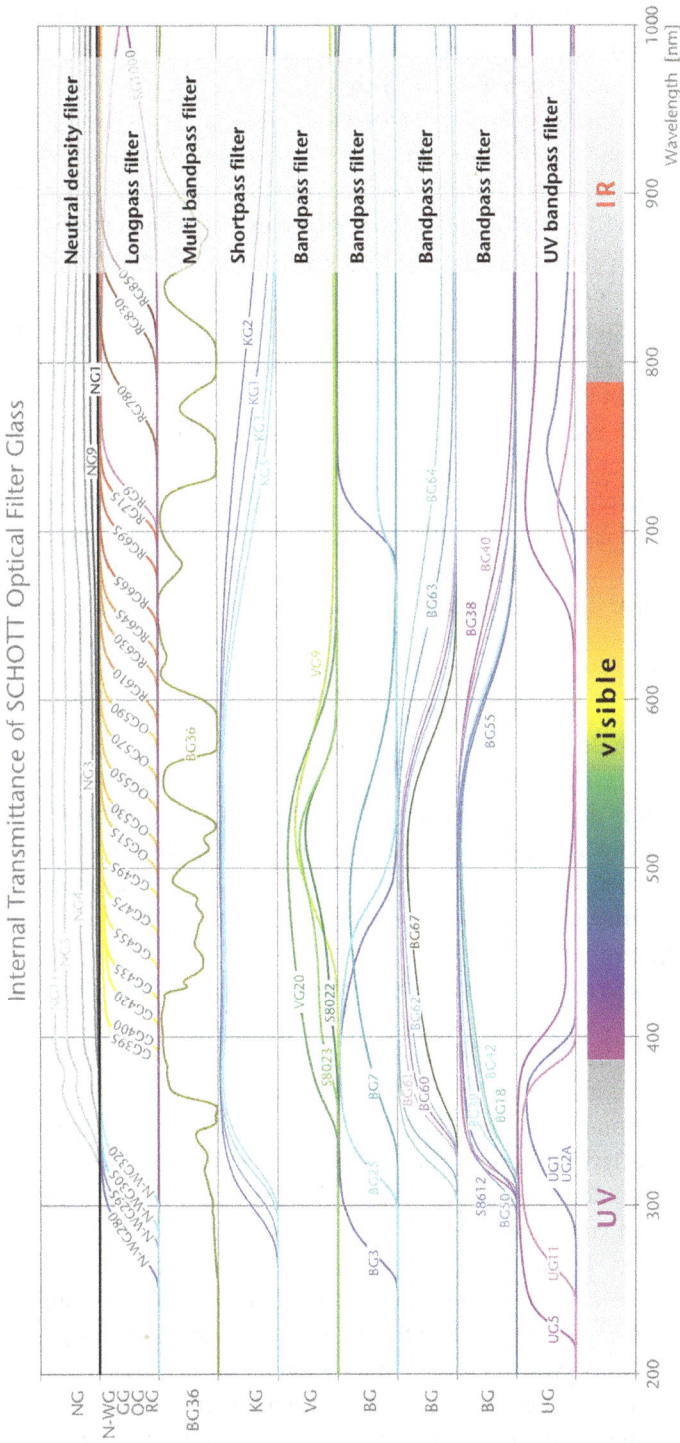

Figure 3.2.11: Spectral sensitivity of different types of filter glasses.

fabricating flexible sizes and shapes, glasses are particularly well-suited host matrices for rare-earth ions, resulting in a plethora of optical and photonic devices. Common matrix formulations (oxides, halides, chalcogenides) and their transmission windows are summarized in Figure 3.2.6. The type of glass matrix to be chosen must be transparent in the wavelength range of the application considered. Rare-earth ions are of great importance in the optical functionalization of glass matrices. The level of doping required is highly application-dependent. Doping at minor levels (0.1 to 5 mol%) is sufficient for photochromic, photothermal refractive and photonic sensing devices (such as chemical or temperature sensors, or scintillators for radiation detection). For example, Ce^{3+} and Eu^{2+} dopants create photochromic properties, darkening the color of glass upon light exposure. The process is reversible and involves light absorption in the ultraviolet and transfer of photo-electrons to nearby traps which absorb in the visible region [35, 36]. The effect is related to reversible photo-oxidation to Ce^{4+} and Eu^{3+}, resulting in intervalence transfer absorption bands in the visible region. The photosensitivity of the cerium valence state is also key to the behavior of photothermal refractive (PTR) glasses, whose optical properties can be modified in a controlled fashion by UV-exposure and subsequent annealing above the glass transition temperature, T_g [37]. The photo-thermo-induced refractive index decrement can be related to the formation of Ag clusters from Ag^+ dopants, via the redox process $Ce^{3+} + Ag^+ \rightarrow Ce^{4+} + Ag$. By applying holographic techniques, optical elements with different spatial profiles of refractive index can be created, including volume Bragg gratings, phase masks and complex holograms, e.g. lenses [38]. Rare-earth doping at higher levels (>10 mol%) is required for high-refractive index glasses, high-power lasers and magneto-optical devices (Faraday rotators) [39]. In general it is desired that the rare-earth ions are homogeneously dispersed, to avoid compromising optical transparency and/or producing spatially inhomogeneous refraction caused by clustering or incipient phase separation. The preparation of such glasses is challenging, posing restrictions on glass compositions, suitable crucible materials and effective melt cooling strategies to avoid phase separation and crystallization [40, 41].

Laser glasses utilize the *RE* ions' emission properties based on the element-specific 4f electronic energy levels. These are populated by optical absorption under resonance conditions (pumping), followed by radiative and non-radiative transfer to intermediate levels and eventual re-emission. Pumping wavelengths range from the near-UV to near-IR region, depending on the active element considered, and emission most commonly occurs in the near- to mid-infrared region. Since the discovery of Nd^{3+} lasing in a glass fiber matrix in 1961 [42], this field has grown to an immense size [43–46]. The optical fiber amplifiers (mainly Er^{3+} doped) constructed from them were vital in building the high-speed backbone of the global telecommunications network, which carries information around the planet. More recently, fiber lasers have become powerful tools in manufacturing, generating multi-kilowatt beams that can cut and weld materials from plastics to metals.

Figure 3.2.12 illustrates some simple lasing schemes supported by glass matrices. The simplest case, involving direct emission from the excited state populated by pumping, is realized in the case of the $^2F_{5/2} \rightarrow {}^2F_{7/2}$ transition of Yb^{3+} ions. In the more common three-level lasers, the system is pumped by a laser from the ground state E_1 to an excited state E_3, from which it relaxes via radiation-less transition to a long-lived electronic state E_2, eventually resulting in population inversion, and intense emission back to the optical ground state. In four-level lasers, the emitting state E_3 is reached via radiation-less transfer from the pumped state E_4, and the stimulated emission from E_3 does not reach the original ground state, but rather another excited state E_2. In this case population inversion between the latter two levels is much easier to accomplish. In the quasi-three-level systems E_3 and E_4 are connected by excited-state absorption, and E_2 is re-populated by significant re-absorption of the emitted laser radiation. In addition, other rare-earth (or transition metal) ion sensitizers can be present, increasing the pumping efficiency through energy transfer mediated by ion-ion interactions. An important example is the enhancement of the $^2I_{11/2}$ excited state emission of Er^{3+} by pumping the $^2F_{7/2} \rightarrow {}^2F_{5/2}$ transition of Yb^{3+} ions in glasses co-doped with Er^{3+} and Yb^{3+}.

Aside from building of lasers based on transitions within bulk glassy materials, the design of optical fiber lasers is important for designing optical amplifiers in telecommunications. As this application normally utilizes the 1.2–1.5 μm transmission window of glassy silica, the design of Nd- and Er-based lasers emitting in this wavelength range is of particular importance. In more recent years, eye-safe lasers operating at wavelengths exceeding 2000 nm have come into focus for many new applications in the medical field, as recently summarized in a comprehensive review [47].

In choosing the most suitable glass matrix for laser design, it is clear that the transmission window (see Figure 3.2.6) must include both the excitation as well as the emission wavelengths. Aside from this obvious fact, the glass matrix influences (1) the exact emission wavelength, (2) the line width of the transition and (3) its intensity. Figure 3.2.13 shows the effect of the glass matrix upon the position and width of the $^4F_{3/2} \rightarrow {}^4I_{11/2}$ transition of Nd^{3+}, illustrating the influence of the ligand field on the energies of the 4f orbitals. While this influence is relatively small due to the shielding from the occupied outer 5s and 5p orbitals, the emission wavelength can sometimes be changed over a range of hundreds of nm by compositional adjustment of the glass matrix.

Regarding the emission intensity, the composition of the glass matrix chosen will have a decisive influence. The relevant figure of merit is the stimulated emission cross-section, which is influenced by multiple parameters relating to the glass composition, structure and vibration characteristics in the *RE* emitter's immediate local environment [47–50]. Important factors are the refractive index, the emission wavelength and the linewidth of the transition. In addition, it is influenced by radiative and non-radiative decay processes, which include contributions of multi-phonon decay, cross-relaxation and energy transfers. The effect of the glass matrix manifests

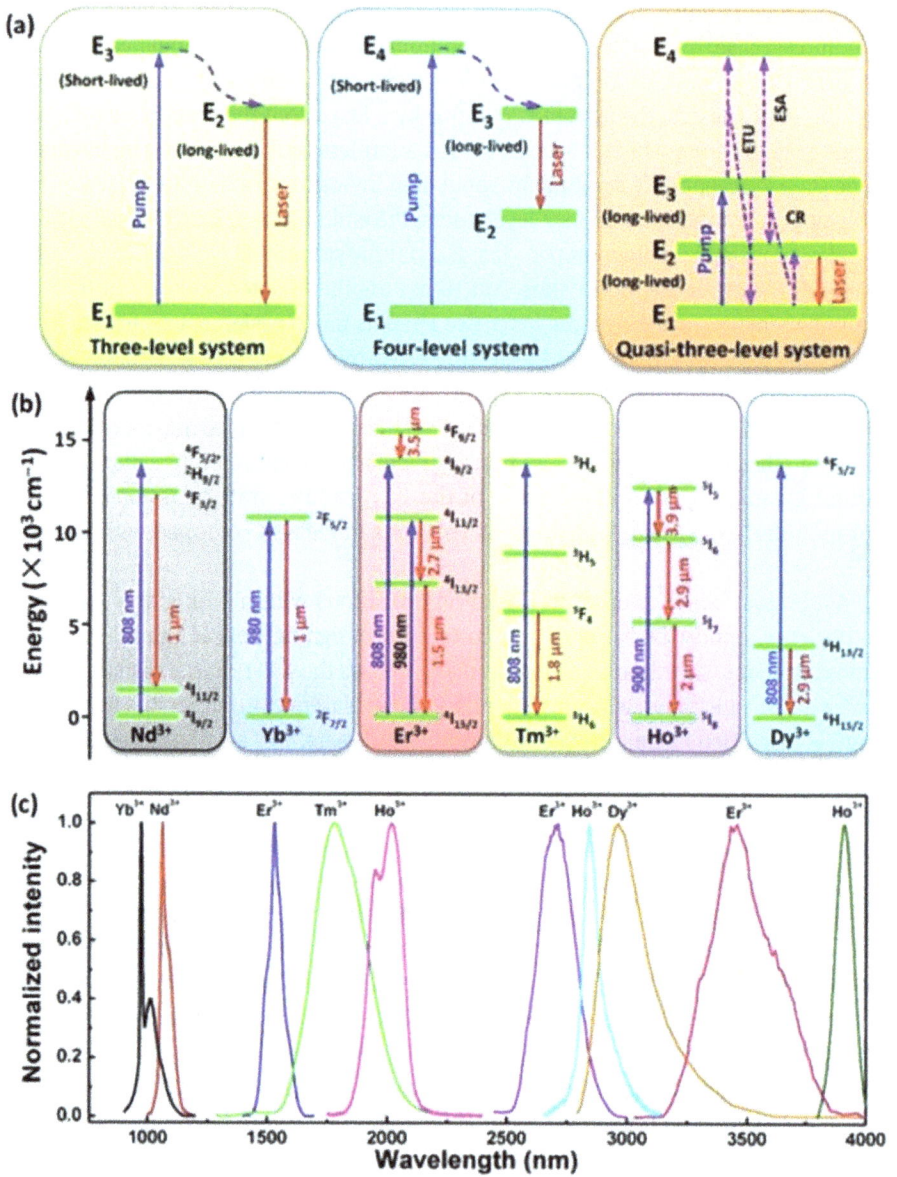

Figure 3.2.12: Common laser schemes involving rare-earth ions in glasses. a) Schemes illustrating three-level, four-level and quasi-three-level systems, b) key energy levels involving some *RE* ion lasers in glasses and c) corresponding laser emission spectra. Reproduced with permission from reference [47].

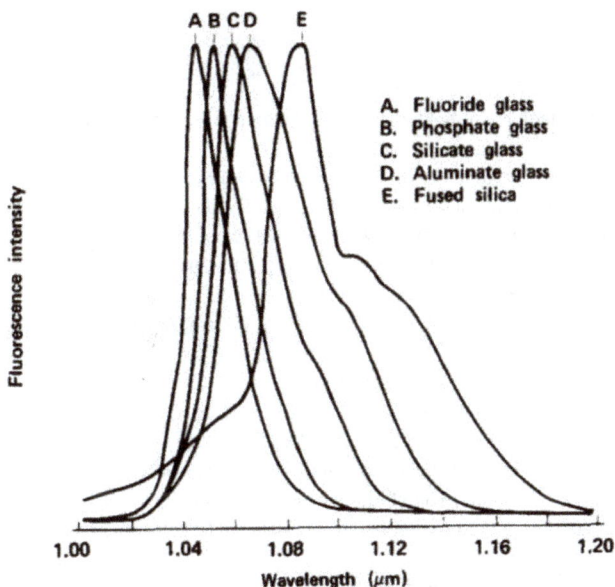

Figure 3.2.13: Influence of the glass matrix on rare-earth emission. Left: emission wavelengths of the $^4F_{3/2} \rightarrow {}^4I_{11/2}$ transition of Nd^{3+}. Reproduced with permission from reference [50].

itself in the way the average excited-state lifetime is limited by radiation-less de-excitation due to (1) energy migration via ion-ion interactions and (2) interactions with phonons associated with the ions' first coordination sphere. With regard to (1), ion clustering in the glassy matrix produces luminescence quenching; thus it is desired that the ions are distributed as homogeneously as possible. The ability of the glass matrices and preparation procedures to disperse the ions homogeneously can be comparatively assessed by measuring the luminescence intensity as a function of ion concentration. Clear deviations from the expected linear relationship will be observed as average ion-ion distances decrease with increasing ion contents, facilitating energy migration. This is called concentration quenching. With regard to (2), depopulation of the emitting state can occur via radiation-less transitions, which are triggered by the vibration dynamics occurring in the local environment of the emitting ion. The effect becomes increasingly severe with increasing phonon energy: emission efficiencies of (high-phonon) oxide glasses tend to be significantly lower than those of (low phonon) fluoride and chalcogenide glasses.

📖 References

[1] Herczog A. J Am Ceram Soc 1964, 47, 107.
[2] Letz M, Hovhannisyan M, Bergmann F, Bai X, Engelmann H, Weidmann G. Appl Phys Lett 2021, 119, 052903.
[3] Lingner J, Funahashi R, Combe E, Letz M, Jakob G. Appl Phys A 2015, 120, 59.
[4] Davis MJ, Vullo P, Kocher M, Hovhannisyan M, Letz M. J Non Cryst Solids 2018, 501, 159–66.
[5] Scrosati B, Hassoun J, Sun YK. Energy & Environ Sci 2011, 4, 3287–95.
[6] Minami T, Hayashi A, Tatsumisago M. Solid State Ion 2006, 177, 2715–20.
[7] Fergus JW. J Power Sources 2010, 195, 4554–69.
[8] Zhang Z, Shao Y, Lotsch B, Hu YS, Li H, Janek J, Nazar LF, Nan CW, Maier J, Armand M, Chen L. Environ Sci 2018, 11, 1945–76.
[9] Feng JK, Yan BG, Liu JC, Lai MO, Li L. Mater Technol 2013, 28, 276–79.
[10] Eckert H, Rodrigues ACM. MRS Bull 2017, 42, 206–12.
[11] Arbi K, Bucheli W, Jimenez R, Sanz J. J Eur Ceram Soc 2015, 35, 1477–84.
[12] Imanishi N, Matsui M, Takeda Y, Yamamoto O. Electrochem 2014, 82, 938–45.
[13] Safanama D, Damiano D, Rao RP, Adams S. Solid State Ion 2014, 262, 211–15.
[14] Zhang P, Wang H, Lee YG, Matsui M, Takeda Y, Yamamoto O, Imanishi N. J Electrochem Soc 2015, 162, A1265–71.
[15] Kundu D, Talaie W, Duffort V, Nazar LF. Angew Chem Int Ed 2015, 54, 3431–48.
[16] Anantharamulu N, Rao KK, Rambabu G, Kumar BV. J Mater Sci 2011, 46, 2821–37.
[17] Guin M, Tietz F. J Power Sources 2015, 273, 1056–64.
[18] Honma T, Okamoto M, Togashi T, Ito N, Shinzaki K, Komatsu T. Solid State Ion 2015, 209, 19.
[19] Okura T, Monma H, Yamashita K. J Electroceram 2010, 24, 83–90.
[20] Okura T. International Conference of Computational Methods in Sciences and Engineering (ICCMSE 2018) AIP Conf. Proc. 2018,2040,020005-1–020005-4.
[21] Wuttig M, Yamada N. Nat Mater 2007, 6, 824.
[22] Planck M. Verh D Physik Ges, Berlin 1900, 2, 202.
[23] Nye JF. Physical Properties of Crystals, soft cover edition, Oxford University Press, Oxford, United Kingdom, 1985.
[24] Letz M, Parthier L, Gottwald A, Richter M. Phys Rev B 2003, 67, 233101.
[25] Ding M, Fan D, Wang W, Luo Y, Peng GD. Basics of Optical Fiber Measurements. In: Peng GD (Ed.) Handbook of Optical Fibers, Springer, Singapore, 2018.
[26] Siryaeve VS, Churbanov MF. J Non Cryst Solids 2017, 475, 1.
[27] Tang Z, Shiryaev VS, Furniss D, Sojka L, Sujecki S, Benson TM, Seddon AB, Churbanov MF. Opt Mater Expr 2015, 5, 1722.
[28] Milonni PW, Eberly JH. Laser Physics, Wiley, Weinheim, 1988.
[29] Hall CM. unpublished, 1733.
[30] Abbe E unpublished, 1886.
[31] Hecht E. Optics, Pearson Education, Incorporated, Hoboken, 2017.
[32] Zaitsev AM. Optical Properties of Diamond: A Data Handbook, Springer, Berlin, 2001.
[33] Ikesue A, Aung YL. Ceramic Laser/Solid-State Laser, Chapter 2, John Wiley & Sons, Ltd, Chichester, 2021, 33–72.
[34] Reichel S, Lentes F-T. Blue glass lens elements used as IR cutfilter in a camera design and the impact of inner quality onto lens performance. In: Mazuray L, Wartmann R, Wood AP, de la Fuente MC, Tissot J-LM, Raynor JM, Smith DG, Wyrowski F, Erdmann A, Kidger TE, David S, Benítez P (Eds.) Optical Systems Design 2012, International Society for Optics and Photonics, SPIE, 2012, Vol. 8550, pp. 211–9.
[35] Locardi B, Guadagnino G. Mater Chem Phys 1992, 31, 45–49.

[36] Cohen AJ, Smith HL. Science 1962, 137, 981.

[37] Cardinal T, Efimov OM, Francois-Saint-Cyr HG, Glebov LB, Glebova LN, Smirnov VI. J Non Cryst Solids 2003, 325, 275–81.

[38] Glebov L. Rev Laser Eng 2013, 41, 684–90.

[39] Borelli NF. J Chem Phys 1964, 41, 3289–93.

[40] Yin H, Gao Y, Guo H, Wang C, Yang C. J Phys Chem C 2018, 122, 16894–900.

[41] Akamatsu H, Fujita K, Nakatsuka Y, Murai S, Tanaka K. Opt Mater 2013, 35, 1997–2000.

[42] Snitzer E. Phys Rev Lett 1961, 7, 444–46.

[43] Gan F. In: Gan F, Xu L (Eds.) Photonic Glasses, World Scientific Publishing Company, Singapore, Chapter 3, Vol. 77, 2006.

[44] Tanabe S. Int J Appl Glass Sci 2015, 6, 305–28.

[45] Tanabe S. J Non Cryst Solids 1999, 259, 1–9.

[46] Rhigini GC, Ferrari M. Rivista del Nuovo Cimento 2005, 28, 1–53.

[47] Wang WC, Zhou B, Xu SH, Yang YM, Zhang QY. Prog Mater Sci 2019, 101, 90–171.

[48] Buddhudu S. Trans Ind Ceram Soc 2014, 64, 69–80.

[49] Jha A, Richards B, Jose G, Teddy-Fernandez T, Joshi P, Jiang X, Lousteau J. Prog Mater Sci 2012, 57, 1426–91.

[50] Weber MJ. J Non Cryst Solids 1990, 123, 208–22.

3.3 Glass fibers

Florian Winter

Glass fibers are known since 1830. They are produced as multifilament man-made yarns and rovings to be the basis of a variety of products. Monofilament glass fibers can be used as optical fiber whereas multifilament fibers are the basis for technical textiles and insulation materials. One factor which makes glass fibers so successful for industrial application is the high-temperature stability and non-combustibility. This chapter will introduce the application of textile glass fibers which are used for various products and industries.

3.3.1 Production, chemistry and properties of glass fibers

The production of glass fibers is made in a glass furnace with electricity and natural gas. The glass composition is defined by mixing several minerals, quartz sand (SiO_2), kaolin (Al_2O_3 main component), limestone ($CaCO_3$) with soda or potassium carbonate as flowing agent. This mixture is heated up to around 1600 °C to get homogenously molten glass. For producing continuous multifilament glass fibers, the molten glass flows through a bushing, which is a metal block with 800 to 2400 tips. The bushing is made up of a Pt/Rh-alloy (90/10, 80/20) to withstand the high temperatures and the quantity of tips depends on the product. The ideal fiberizing temperature of 1200 °C is needed to achieve the right viscosity of the molten glass mixture. When the glass runs through the bushing to form, the filaments it is cooled with much spray water. Directly after filament formation, each of the monofilaments gets in contact with a coating roll to get a homogenous coating which is called size. The size is needed to make the monofilaments more flexible. After sizing application, a defined number of monofilaments are taken as one strand where the filaments are laying parallel to each other and are wound to a cake or a direct roving. Figure 3.3.1 shows the production of glass fibers.

With this procedure, continuous fibers can be produced as zero twist yarns, direct roving, direct chop strands or as a cake. Zero twist yarns and direct roving means a strand of multifilaments laying parallel to each other. After production, the roving goes through a drying process to get off the process water. Chopped strands mean that the continuous multifilament will be chopped to a specific length of several mm and will be packed in a big bag. These chopped strands are sold as wet chopped strands without any drying process. Multifilament cakes are the basis for glass fiber yarns. These yarns are available with filament diameters of about 5 to 16 μm whereas the diameter is defined by the thickness of the single tips in the bushing, the pulling speed and the glass temperature (viscosity). The yarn count of the single yarns can be produced in a range of 11 to ca. 600 tex. Tex is the unit for

https://doi.org/10.1515/9783110733143-021

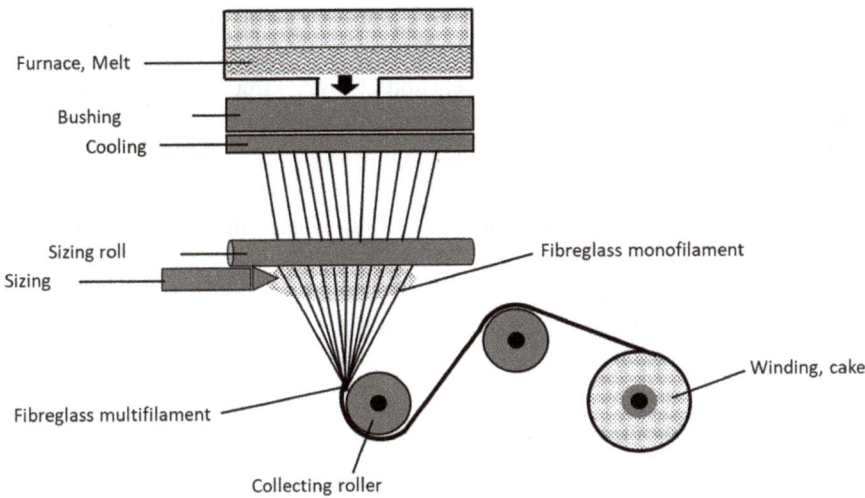

Figure 3.3.1: Production of multifilament glass fibers.

textile yarns and means g km^{-1}. So, a yarn with 134 tex has a weight of 134 g per 1000 m. These pre yarns produced by the glass fiber manufactures are taken by processors to produce plied yarns, texturized yarns or single yarns with specific rotation. Texturizing in this content means the increase of the yarn volume by air pressure to create turbulences.

The sizing of the glass filaments is a very important process during the production of glass fibers. It is applied to the single filaments directly under the bushing. There are different chemical processes available for the sizing. Textile size is mainly composed of specific starch (film former 60–70%) and vegetable oils as lubricants with 20–25%. Further components for textile sizing are cationic softeners (3–5%) silanes (0–5%) and others. For other applications (mainly composites), nonstarch sizing is used which have the following composition within water dispersion:

- Film former: epoxides, polyester, polyvinylacetate, polyurethane or polypropylene 55–65%
- Lubricants 15–20%
- Coupling agents (silanes) 8–13%
- Antistatic, biocides and others 2–6%

The quantity of the dried size applied to the filaments is around 0.1 to 1.2% depending on the product and the final application.

The purpose of the size for a glass fiber is the protection of the fiber in wet conditions. Protection of the fibers during further processing and avoiding of filament brake during the transformation of the yarns while textile processing like weaving, knitting or braiding.

Silane-based yarns are mainly made for matching with specific resins within composites products.

With this process mainly E-glass fibers and ECR-glass fibers (the definitions are listed below Table 3.3.1) are produced with best price performance ratio.

Besides E and ECR glass, there are some more fiber types relevant for the world market. The types of glass fibers are defined via the chemical glass composition. Changes in the glass composition can create better performance in one or more properties.

General properties of glass fibers are:
- Electrical insulation (origin of the name E-glass)
- Low thermal conductivity
- High tensile strength
- Compatibility with organic matrices (depends on the size)
- Dimensional stability
- Fire resistance and high-temperature resistance

Table 3.3.1: Typical chemical composition of different glass fiber types calculated on the oxide content of the elements [1–3].

Oxide/%	E-glass	E-CR-glass	Silica glass	S-glass	AR-glass	C-glass
SiO_2	50–56	59–62	92–95	62–65	60.9–62	62–67
Al_2O_3	12–16	12–15	4–5	20–26	12–16	1–4
CaO	16–25	20–24	1–2	–	16–25	4–6
MgO	≤6	1–4	–	10–15	≤6	4–6
B_2O_3	6–13	≤0.2	–	≤1.2	6–13	3–6
Na_2O	0.3–2.0	<1	<0.5	≤1.1	14.3	15–17
ZrO_2	–	–	–	–	10.2	–
K_2O	0.2–0.5	<1	–	–	2.7	<1
Fe_2O_3	0.3	–	–	–	0.3	0–1
TiO_2	–	–	–	–	0.2	–
fluorine	≤0.7	–	–	–	–	–

Definition of glass types:
E-glass: electric glass fiber made from alumino-borosilicate glass with less than 2% of alkali oxides, most common glass type, used for electrical application and glass reinforced plastic (fiber glass) products.

E-CR-glass: electric glass, chemical resistant, boron-free glass with high resistance to chemicals (especially acids), higher temperature resistance compared to E-glass.

Silica glass: glass fiber with high content of SiO_2, production by leaching process of E- or C-glass fiber to reach the high content of SiO_2, temperature resistance of about 1000 °C but volume shrinkage of about 12%.

S-glass: alumino-silicate glass with magnesium oxide, high mechanical properties.

AR-glass: glass with higher content of Na_2O and additive of zirconium dioxide, high resistance to alkaline solutions, main application in concrete reinforcement.

C-glass: chemical resistant glass fiber, high alkaline content, noncontinuous fiber, production of staple fiber via glass pellets, production of yarns via spinning process to get staple fiber yarns, main applications are woven wallcoverings, thermal insulation and filtration.

The different glass compositions of the fiber types result in different properties and temperature resistances. Figure 3.3.2 shows the viscosity of E-glass in comparison with the temperature in logarithmic scale.

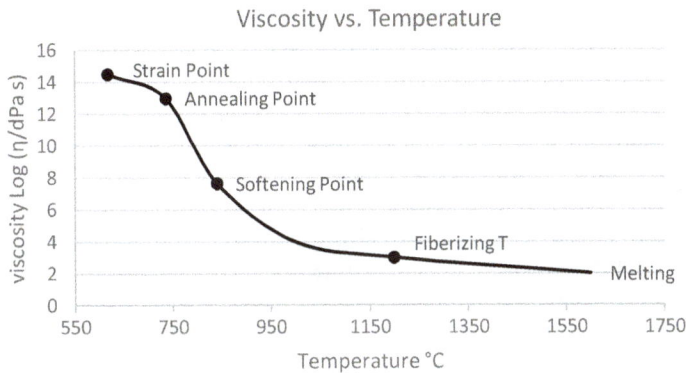

Figure 3.3.2: Viscosity of E-glass fibers.

Important points along the curve are shown in the figure.
- Fiberizing temperature is the ideal temperature for the glass melt to run through the bushing to form the continuous fibers.
- Softening point or Littleton softening point is the temperature where the glass deforms under its own weight and is a standard procedure according to ASTM C338 or ISO 7884-3. For E-glass, it is around 840 °C.
- Annealing point is the maximal temperature where the glass fibers release 95% of their internal stress within 15 min. The calculated value for E-glass is 736 °C.
- Strain point is the maximum application temperature and the glass starts to release its internal stress. For E-glass, this temperature is around 617 °C.

The maximum application temperature for ECR-glass is around 700 °C, for silica ca. 1000–1200 °C due to its high SiO_2 content.

As glass fibers are used for lightweight composite structures, the strength of the fiber is very important. Relevant mechanical values are listed in Table 3.3.2.

Table 3.3.2: Typical mechanical values for E-glass and S-glass fibers [2].

Parameter	E-glass	S-glass
Density [g cm^{-3}]	2.52–2.60	2.45–2.55
E-modulus [GPa]	72–77	75–88
Tensile strength [MPa]	3400–3700	4300–4900
Elongation at break [%]	3.3–4.8	4.2–5.4

Due to its economic advantage, E-glass is the main fiber product for plastic reinforcement. S-glass can be used in products where the high tensile strength is mandatory.

Figures 3.3.3–3.3.6 show some microscopic images of glass fiber filaments and yarns.

Figure 3.3.3: ECR-glass filament with fiber diameter of 13 μm.

3.3.2 Technical textiles and applications

After production of glass fiber cakes or pre yarns, the fibers can be processed to technical yarns and threads. E-glass yarns consist between 200 and 1.600 filaments and are defined by their diameter in microns. The linear mass is expressed in tex

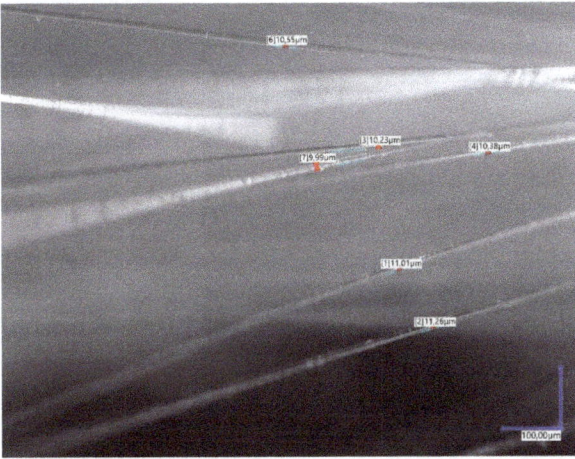

Figure 3.3.4: E-glass-filaments with fiber diameter of 11 µm.

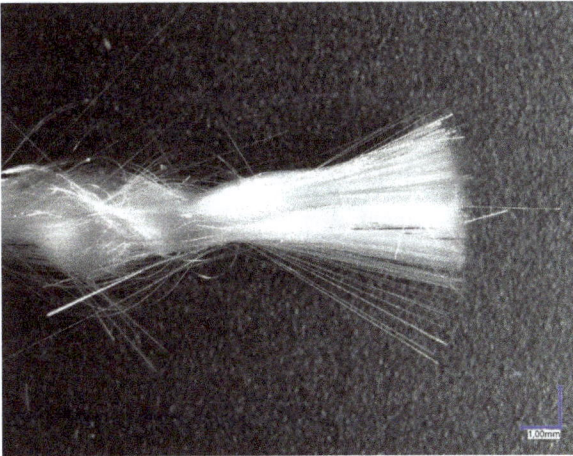

Figure 3.3.5: Glass fiber (multifilament strand).

(g km^{-1}). Single yarns can be processed by direct cabling or ring twisting methods to get 2–10 plied yarns. The yarns are presented on standard bobbins (Figure 3.3.7) which can have different dimensions.

Instead of plied yarns, the glass fiber yarns can be texturized to increase the volume. Due to their volumes, texturized yarns are able to store more air and have a higher insulating and acoustic efficiency than plain yarns. In addition to this, texturized yarns can be used for decorative applications such as wall covering. In principle, there are three different processes available:

Single yarn texturizing
A single yarn is delivered with a certain overfeed to the air jet where the filaments are intermingled.

Parallel texturized yarns
Two or more yarns are delivered parallel with the same overfeed to the air jet where the filaments are intermingled to a new yarn.

Core and effect texturized yarns
One or more core yarns are delivered with a lower overfeed than the effect yarns to an air jet. The result of the intermingling is a very loopy or bulky yarn with excellent insulating properties.

Overspray
For special applications, such as wall covering, texturized yarns with an additional overspray can be produced. One of the advantages is a higher tensile strength and less filamentation in further textile processes [4].

Figure 3.3.6: Texturized E-glass fiber yarn.

Instead of yarns, glass fibers are often used as direct rovings especially in plastic reinforcement applications. Figure 3.3.8 shows some varieties of glass fiber bobbin types or direct rovings.

The textile glass fiber yarns are made for textile processing such as weaving, knitting and braiding. There are manifold fields of technical textiles for glass fibers. For example, knitting and braiding machines are able to produce small tapes or cords of glass fibers to produce gaskets for fire places and industrial ovens. The gaskets are temperature stable and, due to different impregnations more resistant against fraying than pure glass fiber products. Temperature resistance depends on the product and has ranges between 450 and 600 °C. Figure 3.3.9 shows some varieties of gaskets and the application as gasket for oven doors and oven windows.

Figure 3.3.7: Bobbins of ply yarns.

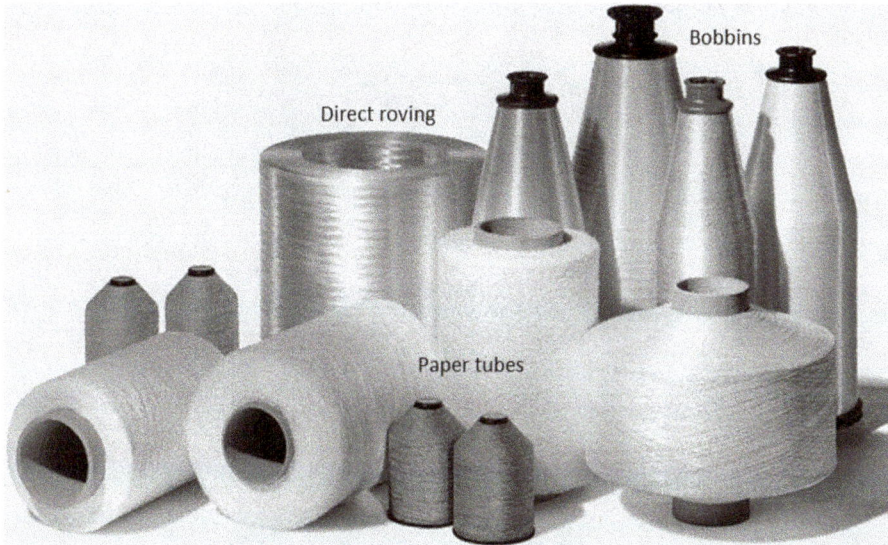

Figure 3.3.8: Glass fibers as direct roving, on bobbins and on paper tubes.

Thousands tons of glass yarns are used in weaving mills. Woven glass fiber fabrics can be very interesting due to the non-combustibility of glass fiber. They are used for textile buildings as roof tops (Figure 3.3.10) or for noncombustible tents. Also fire-protection curtains and blankets are made from woven glass fiber fabrics. When the fabric is PTFE coated, it gets the property of self-cleaning.

Besides the classic textile processes, non-woven glass fiber products can be made. For example glass fiber needle mats which are used for thermal insulations or glass

Figure 3.3.9: Oven gaskets.

Figure 3.3.10: Roof tops made from PTFE-coated glass fiber fabrics at race track in Abu Dhabi.

fiber fleece which are the basis for coated fleeces or decorative non-combustible wall coverings.

3.3.2.1 Thermal and electric insulation

Glass fibers are insulating materials. Both electric insulation and thermal insulation are properties of the glass fibers for the basis of a group of products. Small hoses of glass fiber are produced for cable protection. Braided insulation hoses can be used for the electric insulation in thermal protective switches where the thermal elongation of a bimetal cuts the electric circuit. Figure 3.3.11 shows a protective switch with a glass fiber insulating hose.

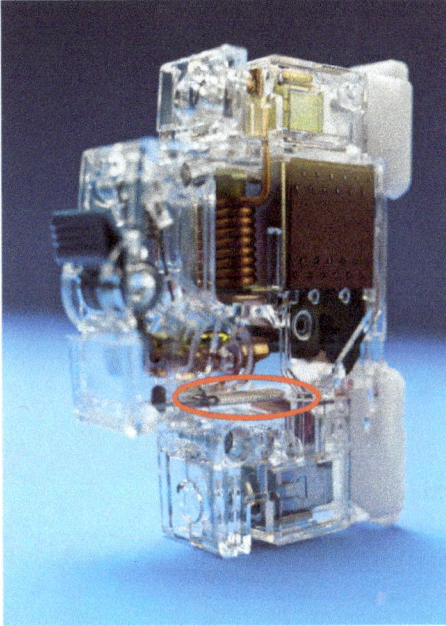

Figure 3.3.11: Thermal protections switch.

High-temperature thermal insulation is used for industrial application as well as for automotive application. Depending on the application, glass fiber fabrics or nonwoven can be used. As mentioned before, nonwoven are typically glass fiber needle mats, veils or felts. To achieve high gradients between the hot surface and the cold surface, insulation thicknesses of 1 mm to more than 20 mm can be used. Generally, glass fiber needle mats can be produced with minimum thickness of 3 mm to maximum 50 mm. They are often produced of cutted roving fibers or production waste of the fiber production.

3.3.2.2 Automotive application

For automotive application, the mentioned glass fiber needle mats are used to insulate the hot end of the exhaust system like diesel/otto particle filters, SCR (selective catalytic reduction) catalysts, exhaust manifolds or just hot pipes. This can be done with insulation half shells made from stainless steel which are 3D formed in the same way like the substrate. These half-shells get an inlay of glass fiber for the insulation (Figures 3.3.12 and 3.3.13). Usually, 2D needle mats are processed to die cut parts which are glued with inorganic water glass based ceramic glue into the shell in 3D form. Normally, this is made manually by workers and due to the die cutting process a huge amount of fiber mat waste is produced. The company Culimeta Textilglas-Technologie GmbH & Co. KG invented a patented [5] process where fibers are processed directly in a water based pulp and 3D formed to get a fiber part which fits perfectly in the half shells. The advantage is that this no-waste process is suitable for a huge amount of parts per year.

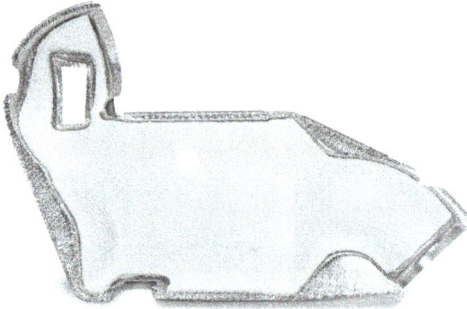

Figure 3.3.12: Formed fiber part in stainless steel half shell, automated process.

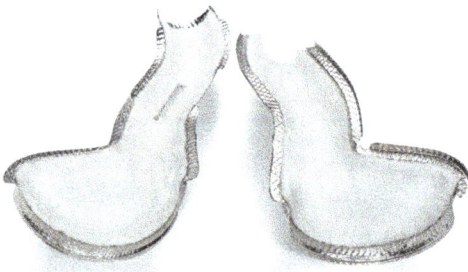

Figure 3.3.13: Glass fiber needle mat in stainless steel half shell, manual process.

To work properly as insulation material for automotive parts, there are some properties which needs to be fulfilled by the material. First of all, it has to be free of organic binder because of the high temperature of the exhaust system. The insulation must be compressable for installation of the shells and to achieve a higher compressed density of the fiber material (matt or formed part), which occurs in better values for the heat conductivity. The fiber material needs to be vibrational consistent because of the motor vibrations. To be sure that the material is suitable, there are some more tests to be done by the OEM (original equipment manufacturer), like hot shaker tests or other lab tests. Therefore, the OEMs use some internal standards like the TL 52725 [6]. The decision which glass fiber chemistry is needed for the insulation depends on the type of motor (diesel or gasoline) and the distance of the insulated part from the motor. Diesel engines produce exhaust temperatures of about 750 to 800 °C while gasoline motors can make higher temperatures of about 950 °C. For truck application, usually E-glass and ECR-Glass fiber materials can be used with temperature resistance of 550–750 °C. Depending on the insulated part, diesel passenger cars can be insulated with ECR-glass fibers as well. Due to the emission standard EURO 6, an efficient catalysis of the exhaust gas is mandatory. The converters need high temperatures for efficient perfor-mance and are mounted within the engine compartment. Therefore, the converter must be insulated. Due to the minimal available space in the engine compartment, plastic parts and gasoline hoses are located next to the converters and must be pro-tected from the heat of the exhaust systems. For theses application in gasoline engines, silica fibers with a temperature resistance of over 1000 °C are used. For diesel applica-tion, AES (alkaline earth silicate) wool can be used. AES is a bio-soluble refractory. The fibers are made of SiO_2, CaO and MgO in a spinning process. Typical densities of fiber mats used for this application are 120–150 kg m^{-3} in the raw material. In the com-pressed state, it is around 180–220 kg m^{-3}. Figure 3.3.14 shows two measurements of the heat conductivity performed with the panel test measurement according to DIN EN 1094-7. The first one is a density of the silica material of around 120 kg m^{-3} and the second with a density of around 200 kg m^{-3}.

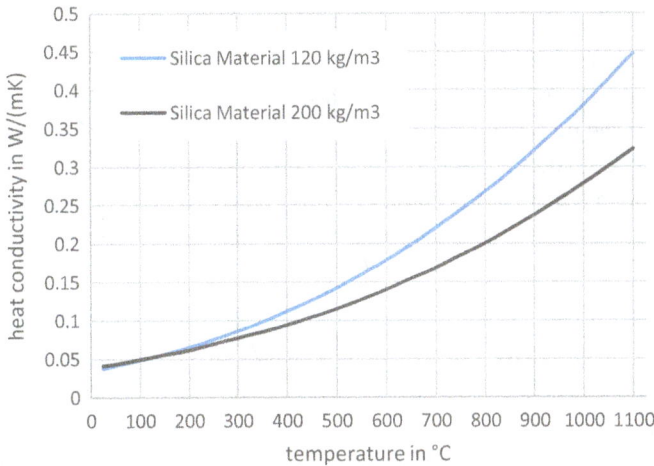

Figure 3.3.14: Heat conductivity measurements of silica insulation materials.

The unit for heat conductivity is W mK^{-1}. The smaller the value, the better is the insu-lation behavior. For the heat insulation, there are three important dimensions. First, the heat transfer by convection is blocked due to the structure with air cavities within the fiber material. The solid-state heat transfer is negligble, due to the insulation be-havior of the glass fiber itself. The third one is the heat transfer by radiation, which is important for high temperatures above 600 °C. This can only be insulated by higher density materials, as can be seen in the graphs.

For acoustic application, the muffler of cars and motorcycles are filled with glass fiber like in Figure 3.3.15, which is another automotive application. For low-emission products, glass fiber bags are knitted and filled with glass fibers to be built in the muffler.

Figure 3.3.15: Knitted glass fiber bag filled with texturized glass fiber strands.

3.3.2.3 Further applications

As mentioned before, glass fiber materials can be used in many applications. A huge number of tons goes into composite materials for lightweight structures in airplanes or boats. The market with the by far highest consumption of E-glass yarns is the electronics market, in which laminated E-glass fabrics are being used for printed circuit boards. A special application regarding to composites is the trenchless sewer rehabilitation (relining), where a glass fiber hose is impregnated with resin and finds application for e.g. underground pipe repairing. This is exemplarily shown in Figure 3.3.16.

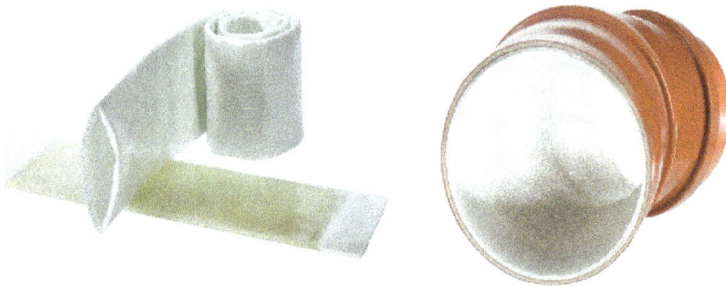

Figure 3.3.16: Glass fiber liner for trenchless sewer rehabilitation.

Other applications are in the fields of construction, infrastructure, industry and transportation:

- Construction means meshes for façade cladding or concrete reinforcement, insect screens, sun protections, wall covering and more.
- Infrastructure means wind energy blades, relining and geo textiles.
- Industrial application means adhesive tapes (reinforcement), filtration papers, grinding wheels, cable reinforcement and more.
- Transportation means timing belts, clutch and brake linings, insulation, cabin interior parts in aircrafts and car body parts.

References

[1]   Teschner R. Glasfasern, Springer-Verlag, Heidelberg, 2013.
[2]   Cherif C. Textile Werkstoffe Für den Leichtbau, Springer, Heidelberg, 2011.
[3]   Owens Corning, system information, OC Advantex 162 A. https://www.owenscorning.com/en-us/composites/products, Accessed February 10 2022.
[4]   Culimeta Facts Yarn processing, www.culimeta.de, Accessed February 10 2022.
[5]   Cuylits Holding, Patent, Verfahren zur Herstellung eines Isolationsformteils: EP 3 215 475 B1.
[6]   Volkswagen TL 52725, Isolationsmaterial zur Außenisolation von Abgasanlagen. https://gwp-kunststofflabor.de/produkt/pruefung-von-isolationsmaterial-nach-tl-52725/. Accessed February 10 2022.

4 Technical gases

Martin V. Dewalsky, Rainer Pöttgen

Many large-scale technical processes use gases as base chemicals [1], but also many niche products are known, where the gas itself is used either with respect to its function or as an intermediate product. The most important applications for technical gases are summarized in the present chapter. Some brief information on a huge number of gases is available in an online encyclopedia (encyclopedia.airliquide.com).

4.1 Hydrogen

Hydrogen became accessible in larger quantities for industrial applications through the chlor-alkali electrolysis process. Currently, it is mainly extracted from fossil fuels. The main processes used are steam reforming, partial oxidation as well as the gasification of coal. Hydrogen is currently being discussed as one of the promising future energy sources. The simple chemical reaction $2H_2 + O_2 \rightarrow 2H_2O$ might be considered as being ecologically intelligent, however, this is only partially true.

The discussion of hydrogen sources is a disputed debate. The sole question for a scientist is the energetic balance, i.e. how much energy do we need to invest into a process and what is the final energy benefit.

The future techniques for high-quantity hydrogen synthesis are (i) catalytic water splitting or (ii) electrolysis. Especially in the latter case, the overpass of energy gradients (most importantly the electric surge) is crucial. Keeping the simple equation $\Delta G = -z \times F \times \Delta E$ in mind, one can at least estimate the loss within the process as a consequence of the electric surge. This does not contemplate overpotentials (i.e., dissipative losses) during electrolysis. An alternative production of hydrogen is possible with electricity produced from regenerative sources (wind, sun, water). Evaluations on the environmental performance for hydrogen have been performed [2] on the eco-balance of the possible regenerative electricity sources [3]. These parameters are decisive with respect to a large-scale future use.

Molecular hydrogen is the source material for fuel cells (probably the most important future application). This is currently driving broad research and test projects in which hydrogen is used as a fuel for city buses, trucks, trains, ships, planes and cars. Thus, the sight of a hydrogen filling station is no longer unusual these days. In any of the mobility devices, hydrogen is mostly stored in gaseous form, at pressures of up to 700 bar. Liquid storage in mobile devices was tested but is currently less in use. The filling processes needed for the various aggregate states of hydrogen are under constant development and strive for automation to make it more convenient, safe and suitable for everyday use. Even though the current ways to produce hydrogen as fuel from fossil sources deliver no environmental benefits, it is nonetheless

https://doi.org/10.1515/9783110733143-022

worthwhile to invest in infrastructure for hydrogen transport. A future fuel management based on hydrogen needs a practicable allocation and logistic.

Similar to helium, hydrogen has an extremely low boiling point and a comparable evaporation enthalpy. As with nitrogen and helium, it is colorless and odorless. As opposed to the other named gases, hydrogen is highly flammable and not inert. The hazard potential of hydrogen is based on its wide flammability range of 4 to 75 vol% in air and its low flammability energy of 0.02 mJ. Methane, in comparison, which is widely used as a fuel in households, has a flammability range of 5.3 to 15 vol% in air and a minimum flammability energy of 0.29 mJ [4].

Due to its flammability, hydrogen is not widely used as freezing agent. In the past *hydrogen mud* was discussed and used as a possible freezing agent in cryogenics. This *mud* is a mixture of liquid hydrogen and hydrogen ice, which is at the triple point in balance with the gaseous phase. It has similar flow properties as the pure liquid phase. The specific cooling capacity is higher, based on the melting heat, as compared to liquid hydrogen [5].

The application field for hydrogen is wide and thus it is used as one of the most important reduction chemicals in chemistry, metallurgy and in the production of semiconductors. It is also used as an alternative fuel source for combustion engines. For example, hydrogen, in liquid form, is a fuel for rockets and on rocket test stands. Hydrogen is used in this application based on its high caloric value to weight (141.8 kJ g^{-1}) ratio.

For the applications, which use high volumes of the gas, hydrogen is normally stored in liquid form. The liquefaction process is hindered by the transformation from ortho-hydrogen (o-H_2) to para-hydrogen (p-H_2) at lower temperatures. This characteristic is caused by the orientation of the nuclear spin during the formation of the hydrogen molecule [5]. The ratio of the modifications changes from 3 to 1 (o-H_2: p-H_2) in the gaseous phase to 0.2:99.8 in the liquid phase. This happens very slowly under energy transfer and the process typically takes more than a hundred days [5]. This transformation in the liquefier needs to be accelerated by a catalyst such as chromium- or iron oxide and the transformation heat dissipated. Otherwise, up to 50% of the liquified hydrogen will be evaporated within the first nine days [5].

In laboratories, hydrogen is used as the carrier gas for gas chromatography. The mixture (H_2 40%/He 60%) is widely used for flame ionization detectors.

During the production of margarine, vegetable oils are treated in an autoclave with hydrogen at temperatures of up to 200 °C and pressures of up to 30 bar. The hydrogen reacts with the oils in the presence of a nickel catalyst and leads to hardening [6].

Deuterium, the heavier isotype, is mainly used in nuclear research and in the synthesis of deuterated chemicals. These can be used for (i) application in nuclear magnetic resonance and (ii) for deuteration experiments for neutron diffraction studies. Radioactive tritium has no application for the civil market.

4.2 Nitrogen and oxygen

The main production of liquid nitrogen and oxygen stems from the air liquefaction process developed by Linde. The gas mixture is separated by fractional distillation. Since the 1960s, liquid nitrogen has been a common operating resource in industry. Liquid nitrogen is easy to store and to apportion, it vaporizes free of residues and its use with sensitive refrigerated cargo is unproblematic. The applications of liquid nitrogen vary from cold shrinkage to quick freeze, from cold milling and up to the conservation of blood and stem cells [7–9].

Even though liquid nitrogen is mostly used as a cryogenic liquified gas, the actual temperature of usage is often not in the so-called cryo-technical area (below minus 120 °C) but in the refrigeration range of minus 120 °C or above. The high effort required for the liquification of nitrogen is balanced by the economical transport costs and the high quality and energy saving applications.

Liquid nitrogen and liquid carbon dioxide are used in quick-freezing of foods, pastry and other prepared bakery and confectionery products. The advantage of quick-freezing of food with liquid nitrogen and carbon dioxide in comparison with conventional freezing is based on the shorter freezing time: cell structures stay intact and the concentration of salts and sugar, which are considered as detrimental to quality, is avoided. During the quick-freezing of foods, different types of freezers are used. In a tunnel freezer, the frozen cargo is precooled in a reverse flow with vaporized nitrogen and afterwards exposed to a liquid nitrogen shower. During quick-freezing, an average of 0.6 to 1 kg of liquid nitrogen is used per kg of frozen goods [8].

Nitrogen is also used to stabilize the ground during construction activities by freezing. In this application, the nitrogen is passed through special freezing tubes which are inserted into the ground in appropriate length and which contain an inner tube with the liquid nitrogen. Through this inner tube, the liquid nitrogen flows down and at the end it is moving up in the outer tube where it vaporizes. As a result, the surrounding ground is cooled and the existing water in the ground is frozen. Other applications with nitrogen in the construction industry are cooling of concrete and asphalt. Gaseous nitrogen is used to protect the steel in prestressed concrete from corrosion damage [6].

Gaseous nitrogen is widely used as an inert gas (for all processes where nitride formation does not occur) for research and industrial applications: metalworking industry (laser fusion cutting and plasma cutting) and electronic industry, in food industry (packaging gas – see section on carbon dioxide), as carrier gas in gas chromatography, for pressure tests of pipelines and tanks.

Nitrogen and argon are used during steel making in the bottom purging process. This process improves the metallurgical results and is beneficial for the cost control of the steel-making process. The gases which flow from bottom to top recirculate the melting and support an average temperature in the furnace. They also remove impurities like oxygen and hydrogen and solid slag particles. This application is likewise used in melting of non-ferrous metals (Chapter 2.2) [6].

We now turn to oxygen. Historically, air was used for combustion processes. With the availability of pure oxygen in bulk form, oxygen-enriched air came into use for many industrial processes. This increases the flame temperature and reduces the cooling effect of the nitrogen present in air and reduced significantly emissions in the exhaust gas.

A typical application for direct oxygen fumigation is the Linz-Donawitz (LD) process (basic oxygen process, BOP) for steel refinement. Other processes are the oxygen bottom Maxhütte (Q-BPO), side-blown argon oxygen decarburization (AOD) converter and vacuum oxygen decarburization converter (VODC), all used for refining highly alloyed steel. The converting processes are used to reduce the carbon in steel by oxidation and to reduce other impurities in the hot metal. Similarly, the oxygen is used in melting processes for non-ferrous metals (Chapter 2.2) and to melt glass [6].

In glass industry, oxygen flame technologies are used to give the final products a perfect surface finish. Other gases used in the glass industry are a nitrogen/hydrogen atmosphere to avoid the oxidation of the tin bath and sulfur dioxide which improves the chemical, physical and mechanical properties of the glass in the float glass process (Chapter 3.1).

Another application area of oxygen is the autogenous technology, including all applications which do use an extremely hot combusting gas-oxygen-flame. As for example, gas welding, autogenous cutting, torch brazing, flame heating, flame spraying and flame annealing [6].

Oxygen is used in the chemical industry in diverse production processes, e.g. ethylene oxide, acetylene, titanium dioxide, propylene oxide and vinyl acetate.

A further application of oxygen is water technology. In lakes and rivers, in periods of oxygen scarcity, the disease susceptibility and mortality rate of the animals is increasing. The introduction of oxygen solves such problems. In fish farming this method is used for economic reasons. An oxygen saturation level between 80% and 120% saturation shows a positive effect on the growth and feed conversion ratio.

In the medical field, oxygen is used as the basis for all anesthetic techniques, restoring of the tissue oxygen tension in many medical assessments, aid resuscitation and supporting of artificially ventilated patients.

4.3 Helium, argon and the heavy noble gases

After its discovery by the French astronomer Pierre Janssen in 1868, helium became a star among the rare gases. The liquefaction of helium was the merit of Heike Kammerling Onnes, who received the Nobel Prize in 1913 for his ground-breaking contributions in lowest temperature investigations [10]. The applications of helium are remarkable. In technology, it is used like never before. In the medical field, as well as in science and technology, the usage of helium is continually increasing. This is readily evident in the research of modern technologies, i.e. in quantum optics and nanotechnology. Possible future applications may be nanolaser, quantum computer or quantum encryption systems as well as electro optical hybrid circuits. Beside this, the focus in research is given to new semiconductor and superconductor systems and solar technology. Helium makes Zeppelins fly, protects deep-sea divers and is on board of the space shuttle on the way to orbit.

In the universe, helium is the second most common element, but on Earth it is an exceedingly rare element. Helium is not produced from air (content of only 0.00052 vol%), since the concentration of a trace gas is extremely energy consuming. Helium is nearly exclusively extracted from natural gas sources with a high helium content (0.2 to 8%) by the help of standard cryogenic technology. These sources are in the mid-west of the USA (Texas, Kansas), North Africa (Algeria), Middle East (Qatar) and Eastern Europe (Poland, Russia).

Liquid helium is, like nitrogen, inert, non-corrosive, odor- and colorless. In many physical properties such as viscosity, density, critical data and heat of vaporization, the values for liquid helium are considerably lower than those for nitrogen [11]. With the same quantity of heat, one can vaporize 1 liter of nitrogen (at standard pressure) but approximately 61 liters of liquid helium! The main application for liquid helium as a refrigerant agent is the cooling of metallic superconductors like niobium-titanium (NbTi) and niobium-tin (Nb_3Sn) with transition temperatures of 10 or 18 K [12]. After the discovery of these materials (at the beginning of the 1960s), several application areas were developed. One can regroup the operation of superconductive inductors into several major research fields like thermonuclear fusion and particle accelerators; energy technology with cryogenerators, superconductive energy storage and superconducting cables; transportation with levitation railways and the magnetohydrodynamic drive; mining with extraction and purification of mineral resources via magnetic separation as well as in analytics and medicine, in structure determination and layer picture production [13]. In the last two application fields, superconductive inductors have already reached the level of commercial use. The nuclear magnetic resonance (NMR) in analytics and the magnetic resonance tomography (MRT) as an imaging method in medicine are commonly used. Approximately half of all liquid helium is used in these technologies [13].

Application fields for gaseous helium in various industries are laser technology, cutting and welding, high vacuum heat treatment, leak testing, filling gas for airbags (also argon is used for this), materials testing, semiconductor production, fiber glass production, balloon filling, diving, coating technology, production of self-cleaning glass and nanotechnology, besides many other niche technologies.

Neon is produced during the air separation process (1.8×10^{-3} vol%) by a special separation of the heat gas (He, Ne, H_2, N_2) from the air separation column. The main application for neon is the filling of fluorescent lamps (also named neon tube); sometimes it is used here as a neon argon mixture. Another application for neon is the (low pressure) filling in plasma displays (mixture neon and xenon (3 to 5 vol%) or neon helium mixtures) [14].

Neon finds application within gas mixtures for gas lasers. In excimer lasers, the chemical excitation of neon (also krypton and xenon can be used) and a reactive halogen (fluorine, chlorine, bromine) in a buffer gas (helium, neon) leads to short-term formation of metastable noble gas halides (e.g. ArF, KrF or XeF). The stored energy is emitted (along with decomposition of the molecule) in the form of electro-magnetic radiation, whereby an ultraviolet laser beam is produced. The mixing ratio of the gases determines the wavelength. It is also a buffer gas in metallic vapor lasers and used in Ne-He lasers.

Argon is produced during the air separation process (approx. 0.93 vol%) and is mainly used as a protective gas. The most common application is the inert gas weld-ing, where the materials to be connected are melted at the interface by an electric arc drawn between the materials and an external electrode. To prevent oxidation of the molten metal, a shielding gas (argon or helium and mixture of both as well as mixtures of argon with hydrogen) covers both the arc and the molten bath. For some materials and special welding techniques, one also uses gases with mild oxi-dation potential. For that application, carbon dioxide is used as shielding and also as a binary mixture with argon or as ternary mixture with argon and oxygen [6].

Another common application for argon is the filling of lamp bodies. For the first light bulbs, the glass plunger was evacuated; nowadays the lamps are filled with an inert gas. A cost-efficient mixture is argon/nitrogen. Krypton and xenon are used in lighting techniques to produce high-intensity, long-life lamps. Both are used to fill halogen sealed-beam headlights. Xenon is used to produce high-intensity light sourced in the UV range. The heavier inert gases allow higher temperatures when the lamp is in operation.

Argon (as well as krypton and xenon and mixtures of these gases – also SF_6, although banned in the EU because of its global warming potential) is used as fill-ing gas in double (triple) glass windows for heat and acoustic noise insulation.

Argon (nitrogen as well) is used in winemaking as an inert gas to protect wine from oxidation. This method is also used for other liquid food products like cooking oil and fruit juice. The best protection does not work if the liquid food is already stored with a high content of dissolved oxygen in the liquid. The dissolved oxygen is cast out from the liquids by nitrogen in form of small bubbles which is brought into the goods with the help of a gas sparger [6]. This method is also used to in-crease the lifetime of fryer oil during frying as the cost of disposal of the used oil is remarkably high.

In heat-treatment operations, argon is used to protect the hot material and in melting processes to shield the liquid metal from reactions with components present in air, mainly oxygen. Argon is used for stirring and inerting purposes in steelmaking and during the argon oxygen decarburization (AOD) operations for stainless steel (see also Chapter 2.2).

Argon is used for analysis and quality control in the industry, R&D and in medical analytics, as plasma gas for plasma emission spectrometry, as blanket gas in graphite furnace atomic absorption spectrometry and as carrier gas in gas chromatography in various detectors. In mixture with methane, argon is used in Geiger counters and in X-ray fluorescence detectors as quenching gas.

Krypton (in air 1.1×10^{-4} vol%) and xenon (in air 9×10^{-6} vol%) are extracted during the air separation process (Linde process).

Xenon is used as an anesthetic agent. These properties have been known for more than 50 years [15]. Even though xenon has almost no side effects, it is rarely used due to its high price. More often, nitrous oxide (N_2O) is used as an inhalation anesthetic (beside some organic gases like isoflurane). The leading theory to explain the mechanism of the anesthetic effect of these gases is their binding to proteins within neuronal membranes and somehow modifying ion fluxes and subsequent synaptic transmission [16, 17]. Even though xenon is an inert gas and does not form covalent bonds with other elements (except under extreme conditions), the large electron shell of xenon is polarizable by nearby molecules and consequently xenon can bind with cell proteins and cell membranes, which is presumably responsible for the anesthetic potency [18]. The nearly non-existing side effects of xenon as an anesthetic gas may result from its monoatomic nature. The other gases used are molecules that consist of multiple atoms, which might explain some side effects (undesired reactivity).

Finally, we briefly summarize some other medical applications. Xenon has been used for decades to study blood flow and gas distribution in the lung. Technical developments have expanded its use in magnetic resonance imaging [19]. Argon plasma coagulation (APC) is an application of gas discharges in argon electrosurgery. It was first introduced in the 1970s [20] and after numerous improvements it is now commonly used in endoscopy. Liquid nitrogen, liquid helium, liquid argon (as well as nitrous oxide and carbon dioxide) are used for cryoablation, which uses the cryo-energy for the treatment of a wide spectrum of diseases [21]. A gas mixture which contains 10% helium and 0.3% carbon monoxide is used in the pulmonary function test to give information about the integrity and size of the alveolar blood membrane [22]. Carbon dioxide is used for insufflation, laparoscopy and as a respiratory stimulant in medical applications.

4.4 Fluorine and chlorine

The main fluorine source is CaF_2 which is treated with sulfuric acid and the resulting hydrogen fluoride is transformed to fluorine electrochemically within a KF/HF melt [1]. Fluorine is used to produce uranium hexafluoride, needed by the nuclear power industry to separate the uranium isotopes, see Chapter 5.8 (nuclear materials). It is also used for the production of sulfur hexafluoride (*vide infra*). Fluorine is basically used for all fluorochemicals, including solvents and high-temperature polymers, such as teflon (poly(tetrafluoroethene) and PTFE).

Nitrogen trifluoride and fluorine are used as chamber cleaning gases in the production of semiconductors. This is an important process to keep chambers in working condition. An excess of chemical reactants and products deposit not only on the substrate, but also on the chamber walls and other equipment inside the process chamber. Because of the sensitive dimensions of electronic devices, even small particles produced from these excess materials can ruin devices under fabrication conditions. In between the process steps, halide gases are plasma-activated to react with and remove the excess materials, like an etching step for the entire inside of the process chamber. Nitrogen trifluoride (NF_3) is synthesized almost exclusively for use in electronics manufacturing. The global production now exceeds 27,000 tons. Fluorine is generated onsite [23].

Another important technique in electronics manufacturing is the etching process. Selective gas-phase etching is a process for removing parts of a single material in a particular shape. Simple gas molecules are used to chemically cut materials into precise shapes at nanoscale dimensions. Almost all etching gases are manufactured on an industrial scale for non-electronic applications. Common etching gases are fluorocarbons (CF_4, C_2F_6, CHF_3), fluorine (F_2), chlorine (Cl_2), hydrogen chloride (HCl), hydrogen bromide (HBr), boron trifluoride (BF_3) and boron trichloride (BCl_3), sulfur hexafluoride (SF_6), sulfur dioxide (SO_2) and nitrogen trifluoride (NF_3) [24].

Chlorine gas is used in water treatment to disinfect drinking water, swimming pools, ornamental ponds and aquaria, sewage and wastewater and other types of water reservoirs. Other gases used in wastewater treatment are oxygen and ozone (see Chapter 13.1.1). Chlorine gas is also used as a disinfectant (see also Chapter 13.1.2), microbiotas/microbicide and algicide in food-processing systems, pulp and paper mill systems and commercial and industrial water-cooling systems. It is used in washing of meat, fresh produces and seeds to control decay-causing microorganisms.

Chlorine gas and chlorine gas mixtures with argon or nitrogen are used in the aluminum industry to remove impurities like Mg, Na, K, Li (salt formation) and in the foundry to remove hydrogen and to improve the removal of the solid particles TiB_2, Al_2O_3, MgO and Al_4C_3 [25].

4.5 Carbon monoxide and carbon dioxide

Incomplete, respectively complete oxidation of carbon and carbon-based sources, leads to the technologically important oxides CO and CO_2. Due to the Boudouard equilibrium, $C + CO_2 \rightleftharpoons 2CO$, carbon monoxide is one of the most important reducing agents in industrial chemistry. The striking applications are the blast furnace process and the water gas equilibrium.

Besides its reducing properties, the huge amount of residual CO of technological processes is used as energy source, since its further oxidation to carbon dioxide delivers a substantial amount of 283 kJ mol^{-1} in thermal power stations.

Furthermore, carbon monoxide is used as synthon for the synthesis of metal carbonyls, isocyanates, methanol, isobutanol, important component in the Fischer-Tropsch process and diverse organic compounds (hydroformylation, Monsanto process, formic acid synthesis). These processes are not subject of the present book.

Carbon dioxide, CO_2, is a trace component in air (ca. 0.04 vol%). Its environmental impact is currently repeatedly discussed. Any tiny reduction of carbon dioxide release needs to be discussed factually. Considering its average molecular velocity of 375 m s^{-1} at 298 K, and presupposing zero intermolecular collision, a carbon dioxide molecule migrates around the equator within ca. 30 h. Thus, any substantial carbon dioxide reduction is a severe worldwide, not local task. This is often completely misunderstood. A small, localized reduction has absolutely no impact [26] at global scale.

Alkaline wastewater (for example, from dairies, tanneries, beverage producers or from the production of detergent or cement) impairs or prevents the clarification processes in biological sewage treatment. This wastewater can be neutralized by injection of carbon dioxide (substituting mineral acids).

In form of dry ice, the solid form of carbon dioxide is used to transport temperature sensitive materials, e.g. chemical reagents, food, vaccines and blood. Another application for dry ice is so-called dry ice blasting, which is becoming one of the most used techniques in cleaning technologies. The dry ice is accelerated in a pressurized air stream at a surface to remove surface contamination. In comparison to other methods, the cleaning with carbon dioxide has the advantage of much lower costs for secondary waste disposal. With this method, one can blast sensitive items like silicon wafers, books, antiques, food and beverage equipment and historical artifacts.

Carbon dioxide can be used in the storage and use of flammable liquids or dusts to prevent fires and explosions that may be caused by sparks, heat or other ignition sources. Carbon dioxide is used to reduce the oxygen concentration below the flammable or explosive limit. As carbon dioxide is denser than air, it can be particularly suitable for the blanketing of hazardous materials. Carbon dioxide is a well-proven fire-fighting agent in sprinkler systems (used as a substitute for halocarbons).

High pressure carbon dioxide (supercritical CO_2) is an excellent solvent which shows high solubility for non-polar compounds and can replace harmful organic solvents. It is used for the extraction of biologically active substances from natural products, such as aromatic essences, e.g. hops and flavorings from spices. Applications can be found in pharmaceutical, cosmetic and food industries. It can also be used to remove unwanted components from products as well, for example to decaffeinate coffee, with the benefit of caffeine recovery.

The growth of plants in greenhouses is promoted by increasing the concentration of carbon dioxide or by irrigation with carbon dioxide enriched water. The ideal carbon dioxide concentration depends on the vegetables or flowers being cultivated but is usually between 0.06 and 0.12% by volume in the atmosphere. Carbon dioxide is used in many greenhouses to allow better growth of the plant, although only around 1% of the carbon dioxide is absorbed by the plants [26].

Carbon dioxide can be added to beverages to create the bubbling/sparkling effect of fizzy drinks. Drawing beer form the barrel, part of the carbon dioxide reacts with the water of the beer and partly forms carbonic acid, slightly increasing the acid content in the beer. The major and thus remaining part of carbon dioxide is only physically dissolved in the beer and delivers a fresh look and taste. For drawing special beer types like Guinness®, one uses a gas mixture of 70 vol% N_2 and 30 vol% CO_2 to protect the taste of the beer. The foam becomes creamier and more durable.

Polymer containers can be produced by blow-molding, where a plastic blank is expanded within a heated mold. The mold is cooled and opened once the polymer has solidified. Cooling the interior with carbon dioxide accelerates this process. Along with the faster cooling, carbon dioxide maintains the quality of the polymer used and reduces internal stresses within the container.

Freezing food, as already covered in Chapter 4.2 along with liquid nitrogen, is just one way of increasing the shelf life of food. For goods that are either at risk of oxidation (e.g. becoming rancid) or where there is a risk of microbial spoilage, the displacement of oxygen is indispensable to improve shelf life. All fatty products and all powdery goods, which because of their large surface area would be eager to absorb oxygen, are particularly at risk. These goods will be covered under inter gas as well when stored and packaged. Depending on the food type, hundreds of different gas types/mixtures are in use: carbon dioxide, nitrogen, argon and mixtures of these [6]. Besides the protection against oxygen, the well know antimicrobial effect on several microorganisms of carbon dioxide is used. A possible mechanism of this antimicrobial activity is the formation of carbonic acid in the cell which decreases the intracellular pH and thus the enzyme activities [27].

Another important application is the use of carbon dioxide as a "matrix" for gas mixtures and as a calibration gas [28].

4.6 Ammonia

The huge amount of industrially used ammonia is produced through the Haber-Bosch process [1, 29, 30] from nitrogen and hydrogen. The most common applications for ammonia are (i) the use as basic material in the chemical industry, (ii) a precursor for explosives (Chapter 5.7), (iii) use in flue gas desulphurization and (iv) manufacturing of fertilizers (Chapter 6.6). Anhydrous ammonia is even used as a fertilizer in pure form. It can deliver a high concentration of nitrogen (82%) and is widely used in the cotton and grain industry. Pure ammonia needs to be handled with care in order to avoid health and safety issues. As it is applied to the soil with special equipment, there is a barrier to use it in other food growing industries. Pure ammonia is initially converted to ammonium salts, which can be better kept in soils and resist leaching. It was found that the application of anhydrous ammonia has a beneficial effect on soil microbes, nitrification bacteria and worms [31].

For more than 150 years, ammonia is used as a refrigerant, mostly in industrial facilities like breweries and in ice sport facilities like ice rinks and bobsled runs (often also in combination with CO_2 as refrigerant).

Ammonia is also in use in geothermic facilities. To generate electricity in the low temperature range between 0 and 200 °C, special generating plants using an operating medium different from water vapor are required. To give an example, the Kalina process uses a mixture of water and ammonia. With a boiling point of −33.7 °C, ammonia evaporates faster than water. By using a different mixture ratio, the medium boils at e.g. 50 °C. This variation allows turbines to run at a broad range of temperatures. The electrical efficiency factor of such a power station is around 12% [32].

Ammonia and hydrogen are used in the manufacturing process of LEDs. These gases are the major portion of the atmosphere in which the crystalline layers are grown. In this process, the purities of the gases are critical, even with a moisture (H_2O) or oxygen trace concentration above a few parts per billion (ppb), oxygen can be incorporated into the crystalline structure of the LED device. This affects the performance and the quality of the LEDs [33].

Ammonia is also discussed as a hydrogen carrier. In comparison with other potential carriers of hydrogen, which is produced using regenerative resources, ammonia does not contain carbon and very importantly does not need to be directly recovered and recycled after the dehydrogenation step [34].

4.7 Nitrogen oxides

Nitrous oxide N_2O is manly used as anesthetic agent (*vide ultra*). Besides this, it is used in atomic absorption spectrometry as a replacement for oxygen in an acetylene flame to reach higher temperatures of up to 2800°C, in food technology as a propellant, based on its sterilizing effect and the good fat solubility (for example with milk products) as the foaming of cream, or in rocket technology as an oxidant. Nitrous oxide is also used in car tuning with N_2O fuel injection increasing the performance of the machine by 20 up to 50% for a short period.

Nitrogen dioxide NO_2 is mainly used for the production of nitric acid. The dimer N_2O_4 is used as an oxidant and in rocket technology.

We have already described the use of gases in medical applications (*vide ultra*). Even the gases used in this field are called *medical gases*, none of them is a drug. The first gas mixture which has reached the authorization to be used as a drug is a mixture of nitrogen monoxide (NO) in nitrogen (400 or 800 ppm NO in N_2). The gas mixture which is launched on the market under the name INOmax® is used to improve the blood oxygen levels of patients like newborn babies with breathing problems associated with pulmonary hypertension [35].

The environmental impact of the emission on NO_x defined as $NO + NO_2$ is widely discussed and will be no subject of the present book. For further reading we refer to a review article [36].

4.8 Sulfur hexafluoride

Sulfur hexafluoride is produced by a direct reaction of the elements (around 10,000 tons a year). Besides the already mentioned application of sulfur hexafluoride in medicine (*vide ultra*), which is more or less a niche application, the main amount is used as an insulating gas (gaseous dielectric medium) for high voltage switchgears due to its excellent properties with respect to electric arc quenching in circuit breakers. SF_6 has much higher density than air and it is thus effectively used as oxidation protection in the magnesium-casting process (as a consequence of the high lattice energy of magnesium oxide, any trace of oxygen would vigorously react with the metal flux). A severe disadvantage of this process is the release of SF_6 to the atmosphere.

SF_6 has a strong negative environmental impact. Its greenhouse potential is around 23,500 times stronger than that of carbon dioxide. For that reason, current research projects focus on the development of suitable substitutes.

4.9 Hydrogen sulfide

Hydrogen sulfide is used in chemical analysis to precipitate heavy metals. The so-called H_2S group is one of the important subjects in analytical chemistry for any kind of chemistry education [37]. The industrially used hydrogen sulfide mostly originates from natural gas sources with a high content of H_2S (so-called *sour gas*); details are given in Chapter 11.2.

Most hydrogen sulfide is directly transformed to elementary sulfur (Chapter 11.2). Gaseous H_2S has only some niche applications. It is used as a calibration gas mixture in petrochemical industry. Hydrogen sulfide acts also as clean sulfur source for the production of II–VI semiconductors. A typical application in thin-film photovoltaics is the generation of copper indium gallium films to adjust the light absorption bandgap.

4.10 Sulfur dioxide

Sulfur dioxide is produced in huge amounts during roasting processes of main group and transition metal sulfides, e.g. SnS_2, PbS, FeS_2, ZnS or CdS (a typical roasting reaction is $2FeS_2 + 11/2O_2 \rightarrow Fe_2O_3 + 4SO_2$) or from other industrial sectors. These SO_2 quantities are directly transferred to industrial plaster ($CaO + SO_2 + \frac{1}{2}O_2 \rightarrow CaSO_4$; see Chapter 1.1) or catalytically to sulfur trioxide for the large-scale production of disulfuric, respectively sulfuric acid. Other sources for SO_2 production are fossil fuels with sulfur content and diverse volcanos (see Chapter 11.2).

Most applications for sulfur dioxide concern gaseous SO_2 (or its aqueous solution, sulfurous acid). However, some interesting usage is also known for liquid SO_2. The latter is a versatile aprotic-polar solvent, which dissolves diverse species. This application mostly concerns research projects in explorative inorganic chemistry [38].

Gaseous SO_2 is widely used as preservative, antioxidant and disinfection species in food industry. Typical treated foods are dry fruits, fruit juices, diverse potato dishes or jam. An important issue is the conservation of wine by direct insertion of gaseous SO_2 or addition of sulfurous acid. This inhibits further fermentation and protects the wine of unwanted bacteria and mold fungus. For home-made wine in small quantities, potassium disulfite (potassium pyrosulfite $K_2S_2O_5$), which releases SO_2 in acidic solution, is used as active SO_2 source.

Another important topic is the disinfection of wine and beer barrels. Formerly, such barrels were treated with SO_2, generated by directly burning so-called sulfur cuts ($1/8S_8 + O_2 \rightarrow SO_2$). This, however, suffers from a bad dosage and risks. Today, sulfurous acid solutions are used for barrel disinfection.

Sulfur dioxide is an important basic chemical for the synthesis of sulfuryl chloride (SO_2Cl_2) and thionyl chloride ($SOCl_2$) and in general for sulfochlorinating processes, e.g. for the synthesis of tensides. Furthermore, SO_2 is used as bleaching agent in paper and textile industry as well as protective gas (oxidation protection) for liquid metal fluxes in foundry technology.

4.11 Carbon tetrafluoride

Carbon tetrafluoride is used as a coolant in low temperature devices; in gaseous form as insulating material in electronics, during dry etching of semiconductor materials and metals e.g. plasma etching of SiO_2, Si_3N_4, as cleaning agent e.g. to remove photoresists and cleaning of drill holes, and as a protective gas for crystal growth processes, e.g. KTb_3F_{10}. These are all typical niche applications.

4.12 Technical gases for semiconductor doping

To modify the conductivity of semiconducting materials, doping with gases, precursor compounds are used. Usually, the gaseous precursor thermally decomposes (e.g. $AsH_3 \rightarrow As + 3/2H_2$) and releases the dopant in the desired amount, since gas pressure can precisely be adjusted. The homogeneous gas volume allows for the exact conditions needed for a well-defined semiconductor doping. Usually, the doping atoms (from the decomposed gas) react on the semiconductor surface and diffuse into the heated substrate. Alternatively, plasma activation by an electric field is used to accelerate diffusion.

Gases used for doping include arsane (AsH_3) and phosphane (PH_3). Precursors for boron doping are boron trifluoride (BF_3), boron trichloride (BCl_3) and diborane (B_2H_6). Diborane is a thermally unstable molecule that will slowly decompose. This can be controlled by storage at low temperatures and mixing it with hydrogen. Germanium from germane GeH_4 is added to silicon thin films to change its conductance by slightly disrupting the order of the silicon crystal structure [24].

Arsane and phosphane are also used to manufacture compound semiconductors for light-emitting diodes (LEDs) by reaction with a metal organic substrate such as trimethyl gallium forming a gallium arsenide/phosphide layer [39]. Phosphane is used in the semiconductor industry for the growth of capping layers [40].

4.13 Silane SiH_4, disilane Si_2H_6 and dichlorosilane SiH_2Cl_2

Silane SiH_4, disilane Si_2H_6 and dichlorosilane SiH_2Cl_2 are used to produce thin-film solar cells. Silane or dichlorosilane in a mixture with hydrogen are deposited on the substrate at high temperatures using chemical vapor deposition or plasma-enhanced chemical vapor deposition. This gaseous phase growth produces thin films of silicon with an amorphous structure. As the global market is asking for more regenerative technologies on low cost, the market share for thin-film solar cells is constantly increasing. Thin-film solar cells promise a price reduction in comparison with solar cells produced from wafer-based monocrystalline or polycrystalline silicon. The lower material consumption at lower temperatures leads to lower production costs. These thin film cells also allow for large area production and can be produced on diverse substrate materials like e.g. plastic foils. Even when thin-film solar cells have a lower efficiency than the cells produced from wafer silicon, they do have a better economy (lower production costs) [41].

References

[1] Bertau M, Müller A, Fröhlich P, Katzberg M. Industrielle Anorganische Chemie, 4. Auflage, Wiley-VCH, Weinheim, Germany, 2013.

[2] Nitsch J. Wasserstoffnutzung: Ökobilanzen, Kosten und Endenergiestrukturen, FVS Themen, 2004, 41–48. ISSN 0939-7582. www.FV-Sonnenenergie.de

[3] Oettli B, Bieler C, Reutimann J, Erdin C. Energiestrategie 2050: Umweltanalyse und Bewertung von Technologien zur Stromerzeugung, https://www.bafu.admin.ch/dam/bafu/de/doku mente/uvp/externe-studien-berichte/energiestrategie2050umweltanalyseundbewertungvon technologienzurs.pdf

[4] Häussinger P, Lohmüller R, Watson A. Hydrogen, Ullmann's Encyclopedia of Industrial Chemistry, VCH Verlag, Weinheim, 1989.

[5] Haefer RA, Frey H. Tieftemperaturtechnologie, VDI Verlag, Düsseldorf, 1981.

[6] Veranneman G. Technische Gase, Verlag moderne Technik, Landsberg am Lech, 2000. ISBN 3_478-93229-7.

[7] Wernicke HJ. Stickstoff, Ullmanns Encyklopädie der technischen Chemie, Verlag Chemie GmbH, Weinheim, 1982.

[8] Veranneman G. Industrielle Anwendung von flüssigem Stickstoff, Linde Sonderdruck Nr. 68. Linde AG, Geschäftsbereich Linde Gas, Seitnerstrasse 70, 82049 Höllriegelskreuth, Germany, www.lindegas.de

[9] Schiffbauer R. Einsatz flüssigen Stickstoffs in der industriellen Fertigungstechnik, Linde Sonderdruck Nr. 159. Linde AG, Geschäftsbereich Linde Gas, Seitnerstrasse 70, 82049 Höllriegelskreuth, Germany, www.lindegas.de

[10] Buckel W, Kleiner R. Supraleitung, 7th edition, Wiley-VCH, Weinheim, 2013.

[11] Hansen H, Linde H. Tieftemperaturtechnik, Springer-Verlag, Berlin, 1985.

[12] Gockel E. Supraleitung, AGF Forschungsthemen 4, Wissenschaftszentrum, Bonn-Bad Godesberg, 1991.

[13] Groll G. Helium: A noble gas with solid applications, International Oxygen Manufacturers Association, 1994. Commemorative on the occasion of the 50-year existence of the International Oxygen Manufacturers Association (IOMA) in 1993. www.iomaweb.org

[14] Kaufmann M. Plasmaphysik und Fusionsforschung. Teubner, Stuttgart, Leipzig, Wiesbaden, 2003. ISBN 3-519-00349-X.

[15] Cullen SC, Gross EG. Science, 1951, 113, 580–82.

[16] Evers AS, Crowde CM, Balser JR. General anesthetics. In: Bruton LL, Lazo JS, Parker K-L. (Eds.) Goodman and Gilman´s, the Pharmacological Basis of Therapeutics, 11th edition, McGraw-Hill, New York, 2006. ISBN-10: 00071468046.

[17] Morgan GE, Mikhail MS, Murray MJ. Clinical Anesthesiology, 4th edition, Lange Medical Books, McGraw Hill, New York, 2006. ISBN-10: 0071423583.

[18] Franks JJ, Horn J-L, Janicki PK, Singh G. Anesthesiology, 1995, 82, 108–17.

[19] Albert MS, Gates GD, Driehys B, Happer W, Saam B, Springer CS Jr, Wishnia A. Nature, 1994, 370, 199–201.

[20] Morrison CF Jr. Valleylab Inc., Electrosurgical method and apparatus for initiating an electrical discharge in an inert gas flow. United Stated patent 4040426, 1977.

[21] Andrade JG, Khairy P, Dubuc M. Circulation: Arrhythmia and Electrophysiology. 2013, 6, 218–27.

[22] Ranu H, Wilde M, Madden B. Ulster Med J, 2011, 80, 84–90.

[23] Stockman P. Etching relies on electronic special gases, https://www.linde-gas.com/en/im ages/Silicon%20Semiconductor%20China%20-%20Enabling%20Electronics%20Manufactur

ing%20Etching%20Relies%20on%20Electronic%20Special%20Gases%20-%20English_
tcm17-482187.pdf

[24] Stockmann P. Creating a semiconductor and the gases that makes it happen, https://www.
linde-gas.com/en/images/Gasworld%20-%20Creating%20a%20Semiconductor%20FEB18_
tcm17-477345.pdf

[25] Aparecido Vieira E, de Oliveira JR, Alves GF, Espinosa DCR, Soares Tenório JA. Mater Trans,
2012, 53, 477–82.

[26] Fangmeier A, Jäger H-J. Wirkungen erhöhter CO_2-Konzentrationen. In: Guderian R. (Hrsg.)
Handbuch der Umweltveränderungen und Ökotoxikologie, Volume 2a: Terrestrische
Ökosysteme, Springer, Berlin, 2001.

[27] Drago E, Campardelli R, Pettinato M, Perego P. Foods, 2020, 9, 1628.

[28] European Industrial Gases Association AISBL, The carbon dioxide industry and the
environment, DOC 101/20, 2020. https://www.eiga.eu/publications/eiga-documents/

[29] Benvenuto MA. Industrial Inorganic Chemistry, De Gruyter, Berlin, 2015.

[30] Holleman AF, Wiberg N. Anorganische Chemie, 103. Auflage, De Gruyter, Berlin, 2016. ISBN
978-3-11-051854-2.

[31] Rogers G. The Effects of Using Anhydrous Ammonia to Supply Nitrogen to Vegetable Crops,
Applied Horticultural Research, Horticulture Innovation Australia Limited, Sydney, 2016. ISBN
0-7341-3888-1.

[32] Ogriseck S. Appl Thermal Eng, 2009, 29, 2843–48.

[33] Tolia A, Travis I. High-purity gases have a key role in LED manufacturing, LED magazine,
April/May 2012, article/16698641.

[34] Jackson C, Fothergill K, Gray P, Haroon F, Makhloufi C, Kezibri N, Davey A, LHote O, Zarea M,
Davenne T, Greenwood S, Huddart A, Makepeace J, Wood T, David B, Wilkinson I. Ammonia to
green Hydrogen project, Feasibility Study, https://assets.publishing.service.gov.uk/govern
ment/uploads/system/uploads/attachment_data/file/880826/HS420_-_Ecuity_-_Ammonia_
to_Green_Hydrogen.pdf

[35] European Medicines Agency, EPAR summary for the Public, INOmax, EMA/121935/2011,
EMEA/H/C/000337, https://www.ema.europa.eu/en/documents/overview/inomax-epar-
summary-public_en.pdf

[36] Air Quality Guidelines. Global Update 2005, World Health Organization, Copenhagen,
Denmark, 2006. ISBN 92-890-2192-6.

[37] Schweda E. Jander/Blasius, Anorganische Chemie I + II, Theoretische Grundlagen und
Qualitative Analyse / Quantitative Analyse und Präparate, 2. Auflage, Hirzel Verlag, Stuttgart,
2016.

[38] Beck J, Steden F, Reich A, Fölsing H. Z Anorg Allg Chem, 2003, 629, 1073–79.

[39] European Industrial Gases Association AISBL, The carbon dioxide industry and the
environment, DOC 163/18, 2018. https://www.eiga.eu/publications/eiga-documents/

[40] European Industrial Gases Association AISBL, The carbon dioxide industry and the
environment, DOC 162/18, 2018. https://www.eiga.eu/publications/eiga-documents/

[41] Roschek T. Microcrystalline Silicon Solar Cells Prepared by 13.56 MHz PECVD Prerequisites
for High Quality Material at High Growth Rates, Berichte des Forschungszentrums Jülich,
No. 4083, ISSN 0944-2952.

Subject index

https://doi.org/10.1515/9783110733143-023

Formula index

https://doi.org/10.1515/9783110733143-024